1/30/92

AN INTRODUCTION TO

MODERN ELECTRONICS

AN INTRODUCTION TO
MODERN ELECTRONICS

William L. Faissler
Northeastern University

JOHN WILEY & SONS, INC.
New York   Chichester   Brisbane   Toronto   Singapore

ACQUISITIONS EDITOR / Cathy Faduska / Clifford Mills
DESIGNER / Kevin Murphy
PRODUCTION SUPERVISOR / Gay Nichols
MANUFACTURING MANAGER / Lorraine Fumoso

Recognizing the importance of preserving what has been written, it is a policy of John Wiley & Sons, Inc. to have books of enduring value published in the United States printed on acid-free paper, and we exert our best efforts to that end.

Library of Congress Cataloging in Publication Data:

Faissler, William L.
  An introduction to modern electronics / William L. Faissler.
    p. cm.
  Includes bibliographical references and index.
    ISBN 0-471-62242-7
    1. Electronics. I. Title.
TK7816.F29 1991
621.381—dc20
                                                    90-20033
                                                    CIP
Printed in the United States of America

10 9 8 7 6 5 4 3 2 1

Preface

I began teaching an introductory electronics course a number of years ago because I believed that the ability of students to do useful electronics in the research laboratory could be greatly enhanced. It was my opinion that the then recent advances in modern solid state electronics had made it possible for students with a reasonably small amount of background knowledge to design, build, and debug useful electronic circuits in the modern research laboratory. What prevented students from doing this was the failure of existing electronic courses and textbooks to respond to the advances in modern electronics.

Since then, over a period of years, I have taught a course in which many people have first learned the basics of modern electronics and have then continued on to design, build, and use various circuits in their own research. In the process of teaching this course, I slowly assembled a collection of my own material to supplement the textbooks I used. Eventually, I realized that I had written my own textbook. This is the result.

This book was written to be used in classes intended for, but not restricted to, physics majors. In fact, in the classes that I teach more than one half of the students enrolled are majoring in a science other than physics. No background in electronics or its related areas is assumed here and, as much as possible, everything needed in an introduction is included. Students who already have some knowledge of a topic can quickly skip over what they have already mastered. To aid in their doing this, the purpose of and material covered in each of the book's sections are made as clear as possible. Anyone who has completed a one-year college science course and who can follow quantitative and mathematical arguments should be able to understand the material presented in this book. Little mathematics beyond the usual college algebra is used. A few short sections of the book use calculus, but these sections can easily be omitted by those who do not have a knowledge of calculus. Moreover, any unusual mathematical techniques are developed in the book.

I strongly believe that anyone involved in electronics must become familiar with the data books supplied by the manufacturers of solid state electronics as well as with the catalogs of the large electronics parts suppliers. Students *must* learn to go to these sources on their own, and the best way for them to learn this procedure is to have to do it from the start. Although there are a few sample data sheets in Appendix B at the end of the book, students will find that they are compelled to consult the data books while they are using this textbook. Appendix A lists many of these books as well as other sources of information.

I also believe that most individuals cannot really learn electronics without hands-on experience. Electronics is more than just lines on paper; after all, the real goal of electronics is working circuits—circuits that do things, not just exist as diagrams. A very large amount of learning takes place while a student is struggling to make a simple circuit work.

In the interest of such learning, many laboratory exercises can be done with only a very limited amount of equipment. A list of this equipment includes the following:

1. A triggerable oscilloscope (the most expensive single item in the list).
2. A digital multimeter and/or an old-fashioned analog multimeter.
3. Two or three small regulated power supplies—plus and minus 15 volts and plus and minus 5 volts are sufficient to do almost everything.
4. Some sort of prototyping system for integrated circuits, available from all of the mail-order hobby electronic and computer outlets. A list of mail-order suppliers is provided in Appendix A. The fancier prototyping systems, costing a few hundred dollars, have the power supplies and several other useful circuits built in.
5. A number of individual components—resistors, capacitors, transistors, integrated circuits, LEDs, and the like. These can be ordered by mail or can be found at local stores like Radio Shack.

In a formal course, the school should make this equipment available in some form. People who are studying on their own may be able to find this equipment at no cost. Many of my students have been part-time students who have done most of the experiments on borrowed equipment where they worked. Radio amateurs, home computer buffs, and others have this equipment already—it is only necessary to look around.

It is not necessary to read this book in the order in which the chapters are presented. In my course, which is about 20 weeks long, we cover the material in the following order:

Chapter 1	
Chapters 2 to 4	Very simple circuit analysis
Chapters 15 to 16	Simple meter circuits as an introduction to the laboratory and to the use of circuit analysis
Chapters 19 to 25	The basics of digital electronics
Chapters 28 to 29, 31, and 34	The basics of operational amplifier circuits
Chapter 38	Power supplies
Chapter 40 to 42	The basics of solid state devices

I have each student work on a small project of his or her own. These projects cause the students to go into one or more areas more deeply and also force them to use the data books to get information. While covering the material listed above, I periodically work in additional topics from circuit analysis, eventually covering Chapters 6 through 14. I combine these topics with the material in Chapter 18 (meter errors), Chapter 30 (analog computer), Chapter 33 (amplifier theory) and Chapters 35 to 36 (D-to-A converters and A-to-D converters) to provide a review and a deeper understanding of the related topics.

William L. Faissler

Contents

Contents

Contents

Chapter 48
Capacitors / 407

Chapter 49
Inductors and Transformers / 416

Chapter 50
Transducers / 419

PART EIGHT
MATHEMATICS / 425

Chapter 51
Complex Numbers / 427

Chapter 1

USING THIS BOOK

INTRODUCTION

Everyone is aware of the electronics revolution that is taking place in the industrialized world. Today, electronics pervades all activities. Not too long ago (well within the memory of many people), home electronics meant the radio; next came television and high fidelity. Then, suddenly, calculators, TV games, home computers, electronic ignitions, and computers built into automobiles all became common. Currently this revolution is taking place so rapidly that the list given above will enable future readers to determine quite accurately when this chapter was written. In fact, if the human race is able to solve the problems caused by destruction of the Earth's environment, there can be no doubt that the most important development in the second half of this century will be the development of solid state electronics and all the technological changes that this development made possible.

The electronic revolution is also taking place in the scientific laboratory and all other workplaces. Today, the preferred method of making almost any kind of measurement is one that permits the use of electronic instruments; for instance, thermocouples are used to measure temperature and electronic balances are used to weigh objects. There are many reasons for this, including the availability of rugged, convenient transducers, the ease of making remote and automatic measurements electrically and, most important, the general ease of making precise, accurate electrical measurements. Moreover the introduction of computer control of many processes is speeding the introduction of all-electronic schemes of measurement into many industrial settings.

This electronic revolution is making it increasingly important for everyone to have at least some understanding of electronics. A few of the reasons why this knowledge is important are listed here.

1. To obtain the intellectual satisfaction of knowing how things work.
2. To help prevent the feeling of complete helplessness when faced with modern electronic devices, especially malfunctioning ones.
3. To be a better, more enlightened consumer.
4. To be able to use instruments correctly, getting more precise information from them while making fewer errors.
5. To be able to build simple electronic circuits in the laboratory, home, or workplace to extend or to modify existing instruments.

The first of these reasons are quite general, whereas the last ones are more specialized. However, they illustrate the usefulness of some knowledge of electronics for everyone.

THE GOALS OF THIS BOOK

This book is an introduction to modern electronics. The main reason for writing this book is to assist you in learning the basic concepts of modern electronics and to do it in such a way that you can continue to learn additional electronics on your own. The field of electronics is changing so rapidly that the details you learn today will soon be obsolete. However, the fundamental laws describing the operation of electronic devices and the methods of analysis used to understand electronic circuits change only slowly, if at all. If you master these fundamentals and the basic vocabulary used in electronics, you will find that you can read the technical literature, understand the new devices that are produced, and use them where necessary.

This book is not intended to be a reference for all time. No effort has been made to be encyclopedic; rather, a very restricted set of devices and applications is considered—all of which are either useful in the normal research laboratory or illustrate important principles. The goal of this book is not to prepare you for every eventuality, but to prepare you to master those eventualities that you will encounter.

To summarize, this book is an introduction. Its purpose is to prepare you to be able to read the current technical literature in the field and to be able to learn more electronics on your own. The goal of teaching you how to learn on your own is, after all, what higher education is all about.

A good working knowledge of electronics has four distinct elements, which are briefly described below. The first element is a knowledge of the basic physical laws that apply to the operation of electronic devices. This includes the basic circuit analysis laws as well as the laws that describe solid state physics. Whereas a complete knowledge of this component would require that a person be a practicing solid state physicist, some familiarity with this material is very useful. This material changes only slowly. At some level the laws do not change; but as new devices are created and as our understanding of solid state physics improves, this type of material grows slowly.

The second element is a knowledge of circuit analysis techniques, which are the mathematical techniques that are used to understand the operation of circuits. This component of electronic knowledge is one that receives very heavy emphasis in the electrical engineering curriculum and that often is almost totally ignored in practical electronic institutes. This is another component that changes only slowly. New analysis techniques are added, but the older ones remain and are sometimes still used. The introduction of the computer and numerical simulation of electronic circuits has made differences in this area.

The third element is a knowledge of the state of the art in electronics. This includes the answer to questions such as, "How do I build an electronic counter that will...?" The collection of the answers to all of these questions defines what is called the **state of the art** at any given time. This type of knowledge is stressed in many practical electronic institutes. Obviously, it is an essential element because it provides the answer to the challenge of "Design a..." or

"Build a...." However, this element of a mastery of electronics is the one that changes most rapidly. The state of the art today is much different than it was 10 years ago or even 5 years ago. This rapid change is what makes it difficult and challenging to learn and to remain current with electronics.

The fourth element is a mastery of the vocabulary. Electronic literature is filled with specialized vocabulary or jargon. The literature cannot be read by someone who does not have a fair mastery of the vocabulary. This vocabulary does not change very rapidly. New terms are occasionally introduced, but mainly old ones are used in new situations.

My approach to teaching electronics is to present a balanced approach to these elements. I want you to be able to do something useful and interesting as soon as possible; and yet, by the end of your study, I also want you to understand electronic terminology and to be able to use simple circuit analysis techniques so that you can continue to educate yourself and keep up with the continuous flow of new electronic devices. This means that a certain amount of state of the art material must be presented, but the basic laws and the methods of analyzing circuits must also be included in a balanced manner. Furthermore, a large portion of this book is devoted to explaining the language of electronics because a mastery of electronic terms is essential for reading the electronic literature on your own.

After completing this book, you should:

1. Be familiar with the basic properties of elementary circuit components and be able to use them in simple circuits. These components include resistors, capacitors, inductors, transformers, and diodes. The associated properties will include voltage–current characteristics and any other information that is commonly provided in the data sheets for the components.
2. Be able to read the data sheets provided by manufacturers and use the information found in them.
3. Be able to carry out elementary circuit analysis and be familiar with how this analysis can be extended to include active devices, such as transistors and op-amps.
4. Be able to design and use some of the basic circuits, including simple amplifiers, oscillators, power supplies, and measuring circuits.
5. Be able to use a small selection of integrated circuits, including operational amplifiers, digital logic, timers, and regulators and have the ability to master and use additional devices.

THE OVERALL DESIGN OF THIS BOOK

There are two problems that must be solved in the design of a good electronics course. First, each of you has a different set of professional goals, personal interests, and level of preparation. Thus no single course design will suit everyone. Second, electronics is a very complex subject; you cannot master it in one pass through the material. For instance, some knowledge about analog circuits is needed to fully master digital circuits. Even more important, efficient learning requires time, motivation, and active participation. Some people find that the motivation needed to master circuit analysis comes from trying to understand analog circuits, whereas others find circuit analysis to be quite interesting by itself. This means there is no unique, correct order to study this material. As a result, it is not necessary or even desirable to read this book in the order in which the chapters are presented.

Several things have been done to make it easier for you to choose a path through this book. The text is divided into eight logical parts. The chapters within each part are all about closely related topics. Each part begins with a short introduction that makes clear what that part of the book is about and what material is covered in each chapter. There are also some comments about the relative nature and importance of each chapter in the part. The introduction to each part also contains a detailed list of prerequisites for each chapter in the section, with references to the chapters that contain the background. Thus for any particular chapter, you can quickly and easily determine what other chapters, if any, you need to study first. Second, as much as possible, each chapter deals with only a single topic. This means that some of the chapters are very short, whereas others are much longer; but in all cases you should be able to study any one chapter in one session. Third, each chapter begins with a list of objectives that should make clear what would be gained from studying the chapter. With this information, you should be able to choose only those parts of the book that are of interest to you. The objectives also can be used as a study guide for the chapter.

There are two types of chapters in this book: those that deal with the fundamental skills and techniques that are used throughout electronics, and those that give background or introduce vocabulary or introduce important circuits that are used in many ways. As much as possible, all of the first type of chapters include problems, exercises, or questions that can be used as study aids. Where possible, answers are provided for the odd-numbered problems. However, the second type of chapter does not lend itself to textbook-type questions. The real test of mastery of these chapters is to read the manufacturers' literature and to use the material in practical situations. In other words, this lack of problems does not mean that the material in these chapters is unimportant.

The second major problem to be resolved in an introductory electronics textbook is that every law or rule described has only a limited range of applicability or is only an approximation to a better and more complex law. These exceptions, restrictions, and refinements to a law have no meaning until the basic law is first mastered. Only after obtaining a basic understanding of a law does it make sense to go back and learn what its limitations are. However, a complete mastery requires an understanding of these limitations and exceptions. There are several possible solutions to this problem: (1) simply dump all the information out at once and leave it to the student to sort it out; (2) repeat the material several times, each time in a more expanded, more correct form; (2) simply ignore the limitations; or (4) do what is done here. The material in a chapter that should be meaningful on the first pass through the material is in the normal format of the book. Refinements, exception, limitations, and other forms of digressions are printed in smaller, darker print, so that they can be easily identified and skipped on the first readings. On later readings, you can worry about what is in these extra sections. This technique has a second advantage—it enables me to satisfy my desire to tell you as much as possible and at the same time makes it possible for you to read the book.

HOW TO STUDY ELECTRONICS

You cannot learn electronics all at once. Gaining a reasonable knowledge of electronics requires going back over the material several times. Only time and repetition will give you a true mastery of the material. Of course, this is not different from anything else you try to learn; for instance, you cannot become an expert skier by taking one series of skiing lessons. The need for repetition applies to the individual chapters, the problems, whole groups of chapters, and the entire body of information about electronics. Repetition or review is crucial in any learning situation.

REFERENCES

I have stressed that this book is an introduction to modern electronics. But what comes next? After studying the text, you should be ready to read most of the normal electronics technical literature, which is large and diverse. A list of references with addresses of sources and other comments is included in Appendix A. The literature includes other introductory electronics text-books, more advanced electronics books, and advanced texts dealing with more specialized topics such as circuit analysis or operational amplifier design and use.

All manufacturers publish catalogs or data books and extensive sets of application notes for their products. These are generally the only source of information for the most recent products and are also a very good source of general information. The larger books frequently have excellent tutorial sections. Some manufacturers publish separate, more extensive tutorials on the techniques and vocabulary used with their products. Also some popular books are available at paperback bookstores or computer stores that are very useful.

There are several popular electronics magazines for home hobbyists. In addition, a number of general-purpose electronics magazines are published for people working in the field. The product announcements and advertisements in these magazines are often as useful or may be more useful than the actual articles. For specific references to all these classes of material, see Appendix A.

PART ONE

CIRCUIT ANALYSIS

The first part of this book deals with circuit analysis. In a very narrow sense, the goal of circuit analysis is to find the current flowing through each element of a circuit. In a more general view, the goal is to completely understand the circuit. For instance, circuit analysis can be used to create a mathematical model of a complex circuit that can then be used to investigate the properties of the circuit in a more complete and efficient manner than would be possible by simply making measurements on the circuit.

Circuit analysis is a very highly developed field. Many electrical engineers spend years studying the various techniques of analysis. More recently, excellent computer programs to analyze circuits are becoming available. However, for most electronics work, only a brief introduction to the subject is needed. Many people find it is most enjoyable and efficient to study the various circuit analysis techniques only when they are needed. The material in this part of the book is organized in a manner to facilitate this.

Chapters 2 through 4 deal with the simplest and most basic concepts of circuit analysis. The concepts and vocabulary introduced in these chapters are used throughout all of electronics. Everyone should study these chapters.

The remaining chapters in Part One cover more advanced and more specialized material. Although some readers may wish to study all of this part at the beginning, others may wish to study additional circuit analysis only when the material is needed. The prerequisites in the other parts of the book make clear what circuit analysis skills are needed for each chapter.

As a brief outline of the material covered in the rest of this section, a short description of what is presented in each chapter follows. Chapters 5 and 6 deal with more advanced methods of circuit analysis. Chapters 7 through 11 deal with various topics concerning alternating current and ac circuit analysis. Chapter 12 introduces step function analysis, and Chapter 13 provides an overall perspective on circuit analysis. Finally, Chapter 14 introduces the topic of dynamic impedance. From a slightly different point of view, Chapters 2 through 9 and 12 introduce formal circuit analysis techniques, whereas Chapters 10, 11, 13, and 14 consist of applications of these techniques to certain special circuits. The first few chapters consider only currents that do not change in time. This means that the circuits involve only resistors and batteries. However, the material is perfectly general; the treatment of currents that vary in time requires only extensions of this material. Nothing has to be unlearned.

As with all other subjects, repetition aids learning. The meaning and utility of some of the topics covered in Part One only become clear after other sections have been covered. Plan on rereading this part several times, first now and then later, after you have covered portions of the rest of the book.

PREREQUISITES

The list below shows the preprequisites for each chapter. Simple algebra is used throughout this part of the book. Any other mathematical skills used are listed in the detailed prerequisites given below. Chapters 2 through 4 are prerequisites for all subsequent chapters.

CHAPTER 2 BASIC PHYSICAL CONCEPTS OF ELECTRONICS
1. Simple algebra.

CHAPTER 3 BASIC CONCEPTS OF CIRCUIT ANALYSIS
1. Basic physical concepts—Chapter 2.

CHAPTER 4 SOME SIMPLE CIRCUITS
1. Basic circuit analysis skills—Chapters 2 and 3.

CHAPTER 5 MESH EQUATIONS
1. Basic circuit analysis skills—Chapters 2 to 4.
2. Ability to solve simultaneous linear equations—Chapter 52.

CHAPTER 6 ADVANCED CIRCUIT ANALYSIS THEOREMS
To understand the chapter:
1. Only the basic circuit analysis skills—Chapters 2 through 4.

To reproduce the examples:
2. Be able to use mesh equations—Chapter 5.
3. Be able to solve simultaneous linear equations—Chapter 52.

CHAPTER 7 ALTERNATING CURRENT: THE SINE WAVE
1. A knowledge of simple trigonometric functions.
2. Complex numbers and simple integrals are mentioned but not used.

CHAPTER 8 ELEMENTS OF AC CIRCUIT ANALYSIS
1. Basic dc circuit analysis including mesh equations—Chapters 2 through 5.
2. A familiarity with the properties of sinusoidal voltages and currents—Chapter 7.
3. Complex numbers are used—Chapter 51.
4. Simple calculus is used in several places in the derivations.

CHAPTER 9 EXAMPLES OF SIMPLE AC CIRCUIT ANALYSIS

1. A knowledge of simple ac circuit analyis—Chapters 7 and 8.
2. The ability to do complex algebra—Chapter 51.
3. A familiarity with decibels—Chapter 53.

CHAPTER 10 RESONANCE

1. A knowledge of simple ac circuit analysis—Chapters 7 through 9.
2. The ability to do complex algebra—Chapter 51.
3. Some of the terminology and results of this chapter closely parallel those of Chapter 12. Reading the two alternately may be useful.

CHAPTER 11 TRANSFORMERS

1. The basics of ac circuit analysis—Chapter 8.
2. Calculus is used in the derivations; no calculus is needed to understand the results.

CHAPTER 12 STEP FUNCTION ANALYSIS

1. The $V\text{--}I$ characteristics for resistors, capacitors, and inductors—Chapter 8.
2. Calculus is used in the details, but the discussions can be understood without calculus.

CHAPTER 13 CIRCUIT ANALYSIS—AN OVERVIEW

1. A knowledge of dc circuit analysis—Chapter 4—ac circuit analysis—Chapter 9— and step function analysis—Chapter 12.

CHAPTER 14 DYNAMIC IMPEDANCE

1. A general knowledge of circuit analysis goals and techniques—Chapter 13.

BASIC PHYSICAL CONCEPTS OF ELECTRONICS

OBJECTIVES

1. Become familiar with the units of charge.
2. Know the definitions of and the units of current, voltage, power, resistance, and conductance.
3. Know Ohm's law.
4. Understand the sign conventions used in circuit analysis.

INTRODUCTION

The basic quantities measured in electronic circuits are defined in this chapter. These quantities are used constantly in the study of electronics. The relationship between some of these quantities and the underlying physical processes occurring in the circuits is also discussed.

CHARGE

Charge and the motion of charge give rise to all electrical and electronic effects. There are both positive and negative charges; the net or total charge of an object is given by the algebraic (signed) sum of all the positive and negative charges in the object. Thus it is possible to take two objects having opposite charges, combine them, and end up with an object having no net charge. In this respect, charge differs from mass. Because mass is always positive, the mass of the combination of two objects is always greater than the mass of either of the objects alone.

Charge can be detected because a charge brought near a second charge experiences a force due to the second charge. This electrostatic force is given by Coulomb's law:

$$F = \frac{kq_1q_2}{r^2}$$

(2-1)

where q_1 is one charge, q_2 is the second charge, r is the distance between the two objects, and k is a constant whose value is given in the following discussion. The force is along the line between the two charges; it is attractive if the charges have different signs and repulsive if the charges have the same sign. This can be remembered as "like charges repel, unlike charges attract."

Charge is measured in units of coulombs (abbreviated as C). The charge of the electron (usually designated as e) is negative and has a magnitude of

$$e = 1.6022 \times 10^{-19} \text{ C}$$

The charge on a proton is positive and has exactly the same magnitude as the charge of the electron. Finally, the value of k is

$$9.0 \times 10^9 \text{ N} - \text{m}^2/\text{C}^2$$

in the mks or SI system of units used in this book and throughout the field of electronics; in these units force is measured in newtons abbreviated as N.

It should be noted that the electrostatic force is a very strong force, especially when it is compared to the other common force, the gravitational force. For instance, two 1-coulomb (C) charges located 1 meter (m) apart experience a force of 9.0×10^9 newtons (N). This force between two 1-C charges is roughly equal to the weight of a battleship!

Charge is a conserved quantity. This means the total or net charge in a volume of space cannot change unless charge is either added or removed from the volume. Charge cannot simply appear or disappear in a circuit. For example, if negative charge is removed from some portion of a circuit, that portion of the circuit is left more positively charged.

Charge is quantized; that is, the charge of any object will always be some positive or negative multiple of the electron's charge. In general, the small value of the electron's charge makes it possible to treat charge as a continuous variable. However, the discrete charge of the electron is an important consideration in some very low-power, low-noise electronic situations.

There are many similarities between charge and mass. Both are fundamental independent properties of matter. The force law for the electrostatic force and the gravitational force has the same form. But there are some important differences: (1) The electrostatic force is much stronger than the gravitational force. (2) There is only one sign of mass, whereas both signs of charge exist. Thus while the gravitational force is always attractive, the electrostatic force can be either attractive or repulsive. (If there were mass of the other sign and "gravitational conductors," then it would be possible to build the equivalent of gravitational "shields" and other items of frequent occurrence in science fiction stories.)

ELECTRIC CURRENTS

An electric current is electric charge in motion. Specifically, the number of coulombs passing through some surface area per unit time is the current flowing through the surface. If ΔQ is the charge passing through a surface during the short time interval Δt, then the current is

$$I = \frac{\Delta Q}{\Delta t} \qquad (2\text{-}2)$$

If the current is changing as a function of time, the Δ in ΔQ and Δt is a reminder that the amount of charge must be measured in a short time interval. The units of current are the **ampere (A)**; 1 coulomb per second is 1 ampere. Commonly used units are

milliampere	(mA)	10^{-3} A
microampere	(μA)	10^{-6} A
picoampere	(pA)	10^{-12} A

In a normal conductor, such as a copper wire, there are many electrons that are free to move about in the wire. Under normal circumstances, these free electrons are randomly moving at quite high velocities because of thermal excitations. When there is an electric current flowing in the wire, there is an overall drift velocity v_d added to the random thermal velocity of the free electrons. Because there are a very large number of free electrons in even a small piece of wire, this drift velocity is generally quite small. As a typical example consider the average random thermal velocity, which is 1.6×10^6 (m/s), whereas the drift velocity due to a current is 10^{-5} m/s, or about 10^{-11} times as large. Figure 2-1 is an attempt to illustrate this random motion of the electrons in a piece of wire.

When a current is being carried by particles having charge q, the current is given by

$$I = Nq \qquad (2\text{-}3)$$

where N is the number of these particles crossing the surface per unit time. For a wire having a cross-sectional area of A (in square meters), a density of free electrons given by n (number per cubic meter), and a drift velocity of v_d (meters per second), the current is given by

$$I = nev_d A \qquad (2\text{-}4)$$

Finally, the common units listed above are an example of a standard problem in electronics. There is no universal, convenient set of electrical units. No matter what system of units is used, actual electrical measurements will result in both very large and very small numbers. To avoid having to deal with numbers having many zeros, for example, 110,000,000 A or 0.000001 A, prefixes are frequently used with all the basic electrical units—as in the examples already given. The common prefixes are as follows:

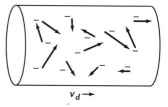

FIGURE 2-1 A schematic representation of the motion of the free electrons in a piece of metal with a small drift velocity v_d to the right, superimposed on the random motion. The lines indicate the velocity of each electron. In a real case, the drift velocity is so much smaller than the normal random thermal velocities of the electrons that the asymmetry in the total velocities are imperceptible.

Symbol	Name	Value
G	giga-	10^9
M	mega-	10^6
k	kilo-	10^3
m	milli-	10^{-3}
μ	micro-	10^{-6}
n	nano-	10^{-9}
p	pico-	10^{-12}
f	femto-	10^{-15}

These prefixes will be used freely throughout the rest of this book.

VOLTAGE

As an electrically charged object is moved about in space, it is necessary to do either positive or negative work on it to overcome the electrostatic forces on it due to the other charges in space. (It may also be necessary to do work to overcome other forces, such as gravity, but we here consider only the work due to the electrostatic forces.) If W_{ab} is the work done moving an object having charge Q from point a to point b, then the electric potential difference between points a and b, V_{ab}, is defined as

$$V_{ab} = \frac{W_{ab}}{Q} \tag{2-5}$$

This quantity has dimensions of work per charge. In the SI system of units used in this book, this is joules per coulomb. This is such a common quantity that it is given the name **volt (V)**. Thus

$$1 \text{ volt} = 1 \text{ joule per coulomb (J/C)}$$

The name "electric potential difference between points a and b" is obviously much too long to be used repeatedly. In practice, the following names will be used interchangeably:

1. Electric potential difference between points a and b.
2. Potential difference between a and b.
3. Voltage between a and b.
4. Voltage drop between a and b.

The nature of the electrostatic force is such that the work done on a charged object in moving around a closed path is zero, no matter what the path. This means that if V_{ac} is the voltage between a and c and V_{ca} is the voltage between c and a, then

$$V_{ac} + V_{ca} = 0 \tag{2-6}$$

because this represents a closed path. Thus

$$V_{ac} = -V_{ca} \tag{2-7}$$

which is just what is expected from the definitions.

Let us consider another example. If V_{ab} is the voltage between a and b, V_{bc} is the voltage between b and c, and V_{ca} is the voltage between c and a, then

$$V_{ab} + V_{bc} + V_{ca} = 0 \tag{2-8}$$

These last two equations can be combined to yield

$$V_{ac} = V_{ab} + V_{bc} \tag{2-9}$$

and

$$V_{ab} = V_{ac} - V_{bc} \tag{2-10}$$

This obscure property has a very practical result. By using one common reference point for all potential measurements, it is possible to forget the reference point and speak about the voltage at some point in the circuit. This becomes the norm in discussing electronic circuits; the common point is called **earth**, **ground**, or **common**. This shorthand notation is fine as long as it does not cause any problems. A voltmeter has two leads; as long as one of these leads is left attached to the same point in a circuit, no problems will occur. But as soon as the lead attached to common is moved to some other point, there is potential for error. If problems exist and the voltages do not add up, you must start thinking in terms of voltage difference between two points and be very careful about which two points.

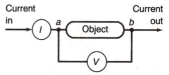

RESISTANCE

Consider the situation shown in the Figure 2-2. The current passing through an object and the voltage across the object are being measured. The points across which the voltage is being measured are labeled a and b. The resistance of the object is defined as

FIGURE 2-2 A simple experiment to measure the relationship between the voltage and the current flowing through an object.

$$R = \frac{V_{ab}}{I} \tag{2-11}$$

The unit of resistance is the **ohm (Ω)**, and

$$1 \text{ ohm} = 1 \text{ volt per ampere} \quad (V/A)$$

Sometimes it is more convenient to deal with the reciprocal of resistance; this quantity is called **conductance**. Conductance has the dimensions of current per voltage; conductance is now measured in units of siemens (S). Until recently these units were called mhos (ohms spelled backward). Mhos are the most common units in the literature today, but siemens are being used in

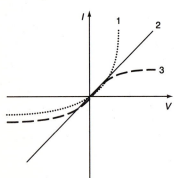

FIGURE 2-3 Possible results of the experiment shown in Figure 2-2 for three different devices. Only device 2 obeys Ohm's law.

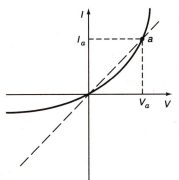

FIGURE 2-4 A graphic method for finding the resistance of a non-ohmic device at point *a* on its *V–I* graph. The slope of the straight line through the origin and point a is I_a/V_a; this is the conductance of the device at point a. The resistance of the device at point a is V_a/I_a, the reciprocal of the slope of the line.

new texts and data sheets. Sometimes the symbol   is used for siemens or mhos. For example, an object that has a resistance of 40 ohms has a conductance of 0.025 siemens.

V–I GRAPHS AND OHM'S LAW

The resistance of an object may depend on many things, including the material and shape of the object, the voltage across the object, the current through it, its temperature, and the illumination of the object. One way to determine how the resistance of an object depends on the current is to use the experimental situation shown in Figure 2-2—you vary the voltage, measure the current, and plot the results on a voltage-versus-current graph, called a *V–I* graph. Several possible results are shown in Figure 2-3.

For any particular voltage value, the resistance of the device at that voltage can be calculated by determining the corresponding current from the graph and by using Equation 2-11. On the graph, this corresponds to drawing a straight line through the point of interest and the origin and then calculating the slope of this line. This slope is the conductance of the device at that particular point; the reciprocal of the conductance is the resistance at that point. Figure 2-4 illustrates this construction. The quantity calculated this way is sometimes called the **static resistance**. The slope of the *V–I* graph at the particular point is related to the dynamic resistance of the device at that particular point. (See Chapter 14 for further discussion of this point.)

The construction makes clear the fact that if the graph is not a straight line passing through the origin, then the resistance is a function of the voltage applied to the device. If the *V–I* graph is a straight line through the origin, the resistance of the object is independent of the voltage. To understand this, recall that the equation for a straight line through the origin on this graph has the form

$$I = bV \qquad (2\text{-}12)$$

where *b* is the slope of the line. If one compares this to Equation 2-11, the comparison shows that *b*, the slope of the line, is simply $1/R$. Thus for this device, *R* does not depend on the voltage applied to the device.

OHM'S LAW

Ohm's law is a statement of the voltage–current relationship for an object whose resistance is constant. It is usually written as

$$V = IR \qquad (2\text{-}13)$$

Although Ohm's law looks the same as the definition of resistance, Equation 2-11, it really is not. The definition of resistance is a general statement. Ohm's law only applies to those situations for which the resistance of the object is independent of the current flowing through it. The *R* in Equation 2-13 is a constant independent of the voltage and current, whereas the R in Equation 2-11 may well vary as a function of current or voltage.

To help remember this distinction, Ohm's law can be restated as follows.

OHM'S SPECIAL CASE

For some materials and for some temperatures, for a restricted range of voltages and currents the resistance of the object is independent of the voltage applied to the object; that is, the voltage–current graph will be a straight line.

Most of the devices used in electronics do not obey Ohm's law. However, resistors, for which Ohm's law is valid, are probably the most common element in electrical circuits, and Ohm's law is used frequently for them. However, if there were only resistors, there would be no electronics. An object that obeys Ohm's law is often said to be **ohmic**. Thus if we refer to Figure 2-3, we see that device 2 is ohmic, whereas devices 1 and 3 are not.

POWER

By rewriting the definition of potential difference (Equation 2-5) to give the work done to move a charge Q through a voltage drop of V volts we obtain

$$\text{Work} = QV \tag{2-14}$$

If a current I flows through a voltage drop of V volts, the power developed is

$$\text{Power} = IV \tag{2-15}$$

The units of this expression can be checked by recalling that volts have units of work per coulomb and current has units of coulombs per second. Thus the product of the two, power, has units of work per second. The SI unit of power is the watt (W). A current of one ampere (1 A) flowing through a potential difference of one volt (1 V) results in one watt (1 W) of power.

In a resistor, energy provided by the source of the current is converted into heat or is **dissipated** as heat. For a resistor having resistance R, a voltage V applied across the resistor, and a current I flowing through it, Ohm's law and Equation 2-15 can be combined to yield

$$\text{Power dissipated in a resistor} = IV = \frac{V^2}{R} = I^2R \tag{2-16}$$

SIGNS AND CONVENTIONS

All of the terms discussed thus far have signs: voltage, current, resistance, and power can all be positive or negative. There are two conventions built into these definitions. They are the following:

1. The direction of motion of positive charges is the positive direction for conventional current. That is, the current is going in the direction the positive charge is going.
2. Positive charge will tend to leave a region having a high (more positive) potential and go to a region of lower (more negative) potential. That is, positive charge will move from a region having a higher voltage to one having a lower voltage.

These conventions give rise to several rules that can be used in circuit analysis. They are as follows:

1. In a circuit, that is, in wires and resistors, conventional current will flow from points having more positive potential to points having less positive potential.
2. In a wire, the current is carried by electrons that have negative charge. The electrons will actually be going in the direction opposite to that of the conventional current. Thus if the electrons are drifting to the right as shown in Figure 2-1, the conventional current is flowing to the left. This confusion is due to a convention established when it was impossible to know which sign of charge was responsible for carrying a current in a metallic wire. This convention, which has been universally adopted by scientists and engineers, was proposed by the early great American scientist Benjamin Franklin.
3. For almost all situations, resistance will be positive.
4. The power developed (dissipated) in a resistor will be positive.

CONDUCTORS, INSULATORS, AND RESISTIVITY

The resistance of a piece of wire depends on the length of the wire, the size of the wire, and the material from which the wire is made. A relationship that is much exercised in elementary physics courses is that the resistance of an object having a uniform cross section made out of an ohmic material is given by

$$R = \frac{L\rho}{A} \qquad (2\text{-}17)$$

TABLE 1-1 Resistivity for Some Common Materials

Material	Resistivity (ohm-meter)
Silver	1.59×10^{-8}
Copper	1.75×10^{-8}
Aluminum	2.84×10^{-8}
Iron	10.0×10^{-8}
Bakelite	2×10^{9}
Glass	2×10^{11}
Rubber	2×10^{13}

where L is the length of the object, A is the cross-sectional area, and ρ is the resistivity of the material. The value of resistivity for common materials ranges over a very large range. A few sample values are given in Table 1-1. In general, materials that have high conductivities, such as copper and silver, are called **conductors**, whereas materials that have very small conductivities, such as glass, are called **insulators**.

PROBLEMS

2-1 A charge of $+6.2$ picocoulombs (pC) is located 18 m from a charge of -3.0 femtocoulombs (fC). Is the resultant electrostatic force attractive or repulsive? What is the magnitude of the force?

2-2 A 5 microcoulomb (μC) charge is located 10 m from a 20-pC charge. Is the force attractive or repulsive? What is the magnitude of the force?

2-3 A wire is carrying 6 A. How many electrons per second are passing through a plane cutting the wire?

2-4 In one type of battery, the current is carried by hydrogen ions (protons) moving through the liquid. If 1 A is flowing through the battery, how long will it take for 1 mole of hydrogen ions to move through the battery? (A mole of protons is 6.0×10^{23} protons.)

2-5 The following voltages are measured in a circuit: $V_{ab} = 5$ V; $V_{bc} = -6$ V; $V_{cd} = 8$ V; $V_{de} = -3$ V. Determine V_{ad} and V_{cb}.

2-6 For the data given in Problem 2-5, determine V_{ac}, V_{eb}, and V_{ea}.

2-7 When 115 V is applied to a heater, a current of 6.5 A flows through it. What is its resistance? What is its conductance? How much power is dissipated in the heater?

2-8 A current of 20 mA flows through a certain diode when 1.6 V is applied to the diode. What are its resistance, its conductance, and the power developed in the diode?

2-9 A toaster dissipates 1000 W at 115 V. How many amperes flow through it? What is its resistance?

The following problems are optional. They illustrate the smallness of the electronic charge and the strength of the electrostatic force. To solve these problems, you will need a knowledge of both elementary physics and chemistry. You can master electronics without being able to do these problems.

2-10 A speck of copper dust has a mass of 1 μg. How many copper atoms does it contain? How many electrons? What is the minimum percentage change in the charge of this piece of dust? Useful data: The atomic number of copper is 29; the atomic weight of copper is 63.5.

2-11 A wire is carrying 1.5 μA. How many electrons per second are passing a plane through the wire? Is the quantization of the electronic charge very important in this case?

2-12 A copper wire has a diameter of 3 mm. Assuming that each copper atom contributes one free electron to help carry the current, what is the drift velocity of these free electrons when the wire is carrying a current of 1 A? The specific gravity of copper is 8.92.

2-13 Two protons are located 1 m apart. What is the gravitational force F_g between them? What is the electrostatic force F_e? How much stronger is the electrostatic force? The mass of a proton is 1.67×10^{-27} kg.

2-14 The earth is bathed in a continuous flux of cosmic rays. These are high-speed particles; some are neutral, some are positive, some are negative, but the overall flux is neutral. Why does this mean that the earth must be neutral?

ANSWERS TO ODD-NUMBERED PROBLEMS

2-1 5.17×10^{-19} N, attractive

2-3 3.75×10^{19} electrons per second

2-5 $V_{ad} = 7$ V; $V_{cb} = 6$ V

2-7 $R = 17.7$ Ω; conductance = 0.057   ; power = 748 W

2-9 $I = 8.7$ A; $R = 13.2$ Ω

2-11 9.38×10^{12} electrons per second

2-13 $F_g = 1.86 \times 10^{-64}$ N; $F_e = 2.3 \times 10^{-28}$ N; 1.2×10^{36} times as strong

Chapter 3

BASIC CONCEPTS OF CIRCUIT ANALYSIS

OBJECTIVES

1. Be able to read simple schematic diagrams.

2. Know Kirchhoff's current and voltage laws.

3. Know the sign convention used with Kirchhoff's two laws.

INTRODUCTION

This chapter covers the basis circuit elements, schematic diagrams, the basic laws of circuit analysis, and the sign conventions used with them. Only the most basic concepts are treated here; more advanced techniques and theorems are covered in later chapters.

CIRCUIT ELEMENTS

The circuits in this chapter are made of combinations of four elements. These are:

1. Wires
2. Resistors
3. Batteries
4. Power supplies

A brief discussion of the properties of each of these elements follows.

Wires

In this chapter, all wires are assumed to be ideal wires with no resistance and no other imperfections. In real circuits, this assumption is not strictly true, but good design and good construction practices tend to make this assumption quite reasonable.

Resistors

Resistors are components that obey Ohm's law; that is, their resistance is independent of the current flowing through them. Resistors are the most frequently used circuit elements in electronics. For now, all resistors are assumed to be perfect; that is, their resistance does not change and they have no other imperfections.

Resistors having resistances from 0.01 to 10^9 ohms (Ω) are available. Common resistors have resistances of 10 to 22×10^6 ohms. To simplify the notation for resistor values, the units of kilohms (k, or kΩ) and megohms (M, Meg, or MΩ) for 10^3 and 10^6 ohms, respectively, will be freely used. A great deal of practical information about types of resistors, their characteristics, and marking schemes is presented in Chapter 47.

Besides fixed resistors, there are also variable resistors—resistors whose resistance can be changed by twisting a knob, pushing a lever, or by some other means. These resistors will be discussed briefly in the next section and in Chapter 47.

The assumptions made here about wires and resistors are generally very good except in very-high-frequency circuits. At very high frequencies, wires and resistors begin to show capacitive and inductive effects, and there may be substantial cross-talk or coupling between neighboring wires. Because of this, analysis of very-high-frequency circuits is a very specialized art and not a topic for an introductory book.

Batteries

Batteries are sources of electromotive force (emf). In less formal terms, they are circuit elements that maintain a more or less constant voltage between their terminals. For now, only ideal batteries will be considered; that is, batteries whose output voltage is constant no matter how much current passes through it.

Inside a battery, a chemical reaction takes place to maintain the potential difference between the terminals. The chemical reaction overcomes the electrostatic tendency for positive charge to move from the positive to the negative terminal through the battery. As current flows through the battery, the chemical reaction moves positive charges from the negative to the positive terminal.

(a)   (b)   (c)

FIGURE 3-1 Schematic symbols for resistors: (a) fixed resistor; (b) rheostat; and (c) a potentiometer.

Power Supplies

A power supply is an electronic equivalent of a battery. Although batteries are used in the examples in this text, power supplies are generally used in the laboratory for reasons of simplicity, capacity, and economy.

SCHEMATIC DIAGRAMS

Schematic diagrams provide a method of representing the electrical properties of a circuit. There are standard symbols for all electronic components in current use. With these standard symbols, most of the information needed to understand and build an electronic circuit can be represented conveniently on

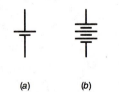

(a)   (b)

FIGURE 3-2 Two schematic symbols for batteries.

paper. The use of a standard set of symbols obviously facilitates communications between people in the field.

Figure 3-1 shows three schematic symbols for a resistor. Part (*a*) is the symbol for a fixed resistor. Part (*b*) is the symbol for a variable resistor having only two terminals; this is called a **rheostat**. Part (*c*) is the symbol for a **potentiometer** (or a **pot**), a very common form of variable resistor having three terminals. In a potentiometer, the resistance between the two outside terminals, the top and the bottom in this drawing, is fixed. The central terminal, called the slider, or wiper, can be moved to various positions, changing the value of the resistance between the slider and both the top and bottom terminals. Drawings involving pots sometimes contain notation to indicate where the wiper will be under some conditions such as: the knob is fully turned clockwise. The value of the resistor is usually written beside the symbol. Sometimes other information about the resistor is also given on the schematic diagram.

Figure 3-2 shows two symbols for batteries. The symbol in (*a*) is most properly used for a single cell, whereas the symbol in (*b*) is for a multiple-cell battery; however, this convention is often ignored. Unless there is some explicit indication otherwise, the longer line represents the positive terminal of the battery. Figure 3-3 shows the nonstandard symbol used for a power supply in this book.

Wires and the connections between components are indicated by lines, straight or otherwise. Unfortunately, this simplest of all symbols gives rise to a common source of confusion. In schematic diagrams, it is often necessary for two wires to cross but not connect; it is also often necessary to indicate the connection between two or more wires. There are two conventions used to tell if two wires simply cross or connect. These are indicated in Figure 3-4.

In the first convention shown in (*a*) of Figure 3-4, nonconnecting wires are indicated by a small loop or bridge, whereas connecting wires simply cross. In the second convention, shown in (*b*), connections are indicated by a dot, whereas nonconnecting wires simply cross. The first convention is somewhat clearer but is harder to draw. The second convention is becoming the more common one because of the increasing prevalence of computer-drawn diagrams. Some people even use a hybrid convention of little bridges for no connections and dots for connections. In any case, when you read a schematic diagram, first determine which convention is being used. In this book, the second convention will be used—connections will be indicated by dots.

Although a schematic diagram may contain enough information to specify completely every component, it does not describe the actual physical layout of the circuit, for example, how the circuit is wired or how long the wires are. There is no necessary correspondence between the orientation of the components in a schematic drawing and the orientation of the components in the circuit. On the drawing, wires can be bent and components moved around at will without changing the circuit that is represented. Often, simply rearranging the drawing can make the circuit look entirely different although, in fact, nothing has been changed but the drawing. Figure 3-5 is an illustration of this fact.

What is important in a schematic drawing is the topological arrangements of the wires and components, that is, the order of the wires, components, and intersections along any path through the circuit. This order must correspond to the actual order in the real circuit, and it cannot be altered when redraw-

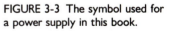

FIGURE 3-3 The symbol used for a power supply in this book.

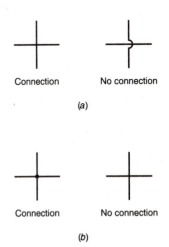

Connection   No connection

(a)

Connection   No connection

(b)

FIGURE 3-4 Two schematic conventions for showing connections.

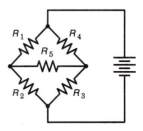

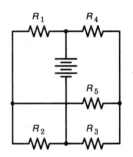

FIGURE 3-5 The same circuit drawn two different ways.

ing the circuit without indicating real changes in the circuit. Study Figure 3-5 carefully from this point of view.

BASIC CIRCUIT LAWS

The basis for all circuit analysis is Kirchhoff's two laws. One of these laws deals with the currents flowing into a node, whereas the other deals with the sum of the voltage changes around a closed circuit. A **node** is defined as the junction of two or more conductors (wires). Currents flowing into a node are positive and currents flowing out of a node are negative. Kirchhoff's current law (**KCL**) states that the algebraic (signed) sum of the currents flowing into a node is zero. Symbolically, this is written

$$\Sigma I = 0 \qquad\qquad (3\text{-}1)$$

where the Greek capital sigma, Σ, means sum of all the currents. The proof of Kirchhoff's current law is based on charge conservation and the fact that charge cannot pile up or disappear at a junction of wires, a simple passive spot in the circuit.

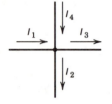

FIGURE 3-6 A four-wire node used as an example of KCL.

For the situation shown in Figure 3-6, Kirchhoff's current law gives

$$I_1 - I_2 - I_3 + I_4 = 0$$

because I_1 and I_4 are flowing into the node and I_2 and I_3 are flowing out of it.

Kirchhoff's current law can be applied to a two-wire node, as shown in Figure 3-7. Applying KCL to this situation gives

FIGURE 3-7 A two-wire node.

$$I_1 - I_2 = 0$$

which yields

$$I_1 = I_2$$

Depending on your point of view, this is either a very profound result or a trivially obvious statement. It says that the current is the same in every element or portion of a circuit having no branches. This fact will be used repeatedly in the discussion that follows with no further comment.

Kirchhoff's voltage law (KVL) states that the algebraic sum of all the potential changes around any closed circuit (or closed loop) is zero. This is sometimes stated as follows: The sum of all the potential rises equals the sum of all the potential drops around a circuit. Symbolically, this is written

$$\sum V = 0 \qquad\qquad (3\text{-}2)$$

This law is really a statement of the conservative nature of the electrostatic field, namely, that the change in electric potential around any closed path must be zero.

KVL APPLICATIONS—SIGNS

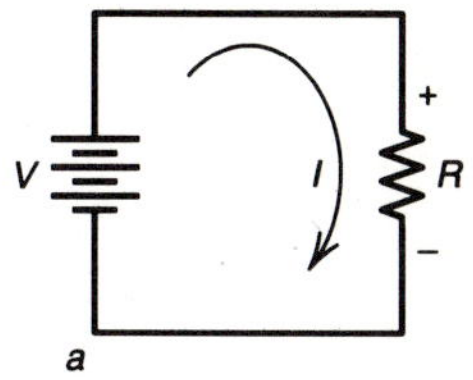

The most difficult part of using KVL is determining the correct sign for each voltage change. The sign conventions can be illustrated by applying KVL to the very simple situation shown in Figure 3-8. The current leaves the positive terminal of the battery and flows through the external circuit to the negative terminal of the battery; thus, the current flows in the direction shown by the arrow in the drawing. The top end of the resistor must be positive with respect to the bottom, since the current is flowing from the top to the bottom. This is indicated by the plus and minus signs on the resistor. Finally, the voltage drop across the resistor is given by Ohm's law:

FIGURE 3-8 A very simple circuit used as an example of KVL.

$$V = IR \qquad (3\text{-}3)$$

Applying Kirchhoff's current law to this situation yields nothing new, since it is already known that the same current must be flowing in each portion of the circuit. Applying KVL requires making an imaginary loop around the circuit to add up the voltage rises and drops. It makes no difference where this loop begins or which direction is taken. To illustrate this situation, the problem will be done twice:

1. Start at point a and go around the circuit in a clockwise direction. The voltage changes are

   i. A voltage rise of V in the battery.
   ii. A voltage decrease (drop) of IR in the resistor.

   So KVL gives

$$V - IR = 0$$

   and solving

$$I = \frac{V}{R}$$

   This, of course, is just the result expected.

2. To show that the direction the loop is traversed makes no difference, start at point a and go around the circuit in the counterclockwise direction. The voltage changes are

   i. A voltage rise of IR in the resistor.
   ii. A voltage drop of V in the battery.

   So KVL gives

$$IR - V = 0$$

and again

$$I = \frac{V}{R}$$

just as before.

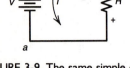
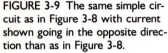

FIGURE 3-9 The same simple circuit as in Figure 3-8 with current shown going in the opposite direction than as in Figure 3-8.

The conclusion is: **It does not matter which way is chosen to go around the closed loops** when applying KVL.

It also does not matter which direction was picked to draw the current. If the current had been drawn going in the other direction, as shown in Figure 3-9, KVL still would have given the correct results. Because the current is going through R in the opposite direction, the voltage drop across R has the polarity shown in Figure 3-9. Starting at point a and going around the loop in the clockwise direction, KVL gives

$$V + IR = 0$$

and

$$I = -\frac{V}{R}$$

The minus sign means the current is actually going in the opposite direction from that shown in the diagram. This will be a fairly frequent situation since, in complex circuits, it is not obvious which way the current will flow until the analysis is done. But, as we have shown, no harm is done if a wrong direction is chosen for a current when starting the analysis—the result for the current will simply be negative.

RULES FOR SIGNS

In the preceding section, the sign conventions needed to apply Kirchhoff's laws to complex circuits have been demonstrated. In simple situations, the results can generally be checked by intuition. However, in complex situations, it will not be possible to check results by using intuition. To get correct results, the sign conventions must be used correctly. To do this, I suggest that you memorize the following set of rules:

KCL

1. Currents flowing into a junction are positive.
2. Currents flowing out of a junction are negative.

KVL

1. If the imaginary loop around the circuit enters a source of electromotive force (battery, power supply, etc.) at the negative terminal and

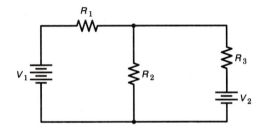

FIGURE 3-10 An example for the use of Kirchhoff's laws.

leaves at the positive terminal, use a positive voltage. If the loop enters at the positive terminal, use a negative voltage.

2. If the loop going through a resistor is in the same direction as the current, use $-IR$. If the loop goes in the opposite direction as the current, use $+IR$.

3. If there are two (or more) currents identified in a resistor, use the sum of the voltage drops with the signs given as in item 2 above.

To give an example of the use of these rules, Kirchhoff's laws will be applied to the situation shown in Figure 3-10. The first step is to pick (guess) the directions of the various currents flowing in each independent branch of the circuit and to give these currents names such as $I_1, I_2,$ and so on. The second step is to indicate which closed loops will be traversed, to give them names, and to pick the directions to be used. All of this has been done in Figure 3-11. Remember, these names and directions are quite arbitrary; but once they have been chosen, the signs of the terms in Kirchhoff's laws are completely determined.

In Figure 3-11, currents $I_1, I_2,$ and I_4 are flowing into node 1, and I_3 is flowing out of it as a result. Applying KCL to node 1 yields

$$I_1 + I_2 - I_3 + I_4 = 0$$

In a similar way, applying KCL to node 2 yields

$$-I_1 - I_2 + I_3 - I_4 = 0$$

Traversing loop 1 in the direction shown in the figure, the voltage rises and

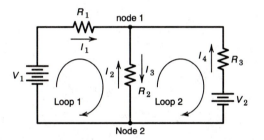

FIGURE 3-11 The example of Figure 3-10 with the currents in each independent portion of the circuit named and the loops named.

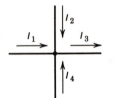

FIGURE 3-12 The circuit for Problem 3-1.

drops encountered are a rise of V_1 in the battery, a voltage drop of $I_1 R_1$ in R_1, a drop of $I_3 R_2$ and a rise of $I_2 R_2$. Thus applying KCL gives

$$V_1 - I_1 R_1 + I_2 R_2 - I_3 R_2 = 0$$

Likewise, applying KVL to loop 2 gives

$$-V_2 - I_2 R_2 + I_3 R_2 + I_4 R_3 = 0$$

PROBLEMS

3-1 If $I_1 = 3.0$ A, $I_2 = 5.0$ A, $I_3 = 1.0$ A in Figure 3-12, find I_4.

3-2 If $I_1 = 4.0$ A, $I_2 = I_3 = 2.0$ A, $I_5 = 3.0$ A in Figure 3-13, find I_4.

3-3 Apply KVL to the circuit in Figure 3-14. (For simplicity in checking the answer, go around each loop clockwise.)

3-4 Apply KVL to the circuit shown in Figure 3-15. (Again, go around the loop clockwise.)

3-5 Going around each loop clockwise, apply KVL to the circuit in Figure 3-16.

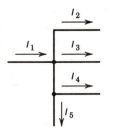

FIGURE 3-13 The circuit for Problem 3-2.

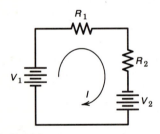

FIGURE 3-14 The circuit for Problem 3-3.

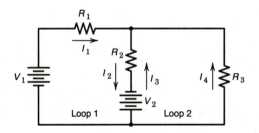

FIGURE 3-16 The circuit for Problem 3-5.

3-6 For Problem 3-3, if $V_1 = 10.0$ V, $V_2 = 5.0$ V, $R_1 = 50$ Ω, and $I = 0.01$ A, what is R_2?

3-7 For Problem 3-4, if $V_1 = 20$ V, $V_2 = 5$ V, $V_3 = 15$ V, $R_1 = 100$ Ω, $R_2 = 25$ Ω, and $R_3 = 250$ Ω, what is I?

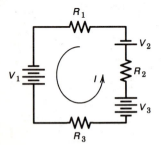

FIGURE 3-15 The circuit for Problem 3-4.

ANSWERS

3-1 -7.0 A

3-3 $V_1 + V_2 - I(R_1 + R_2) = 0$

3-5 loop 1: $V_1 + V_2 - I_1 R_1 - I_2 R_2 + I_3 R_2 = 0$; loop 2: $-V_2 + I_2 R_2 - I_3 R_3 + I_4 R_3 = 0$

3-7 81 mA

Chapter 4

SOME SIMPLE CIRCUITS

OBJECTIVES

1. Be able to tell if circuit elements are in series, in parallel, or neither.

2. Be able to calculate the equivalent resistance value for several resistors in series or in parallel.

3. Be able to recognize a voltage divider and calculate its output voltage.

4. Be able to recognize a current divider and calculate the current in each leg.

INTRODUCTION

In this chapter, the basic laws of circuit analysis are applied to the two simplest arrangements for components—series and parallel. A simple example of each of these arrangements is discussed as a useful, functional circuit.

Again, although only resistors are discussed in this chapter, the results are perfectly general. All the discussions and all the results would be exactly the same if the resistors were replaced by any passive circuit element and the word resistance were replaced by impedance. In Chapter 8, it is suggested that you review these early chapters and make this substitution.

DEFINITION OF SERIES AND PARALLEL CIRCUIT ELEMENTS

By definition, components are in **series** when there are no branching nodes between them. Branching nodes are nodes where three or more conductors join, that is, nodes at which the current divides or has a choice where to go. For components in series, all the current that flows through one component must flow through all the others. Figure 4-1 illustrates this definition.

Components are in **parallel** when they are connected between the same nodes. This means that the current flowing between the nodes need only go

Series

Series

Not series

FIGURE 4-1 Three resistor arrangements, two of which are series arrangements.

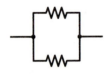

Parallel

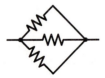

Parallel

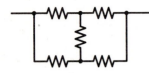

Not parallel

FIGURE 4-2 Three resistor arrangements, two of which are parallel arrangements.

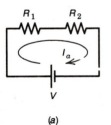

(a)

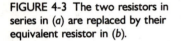

(b)

FIGURE 4-3 The two resistors in series in (a) are replaced by their equivalent resistor in (b).

through one of the components. Figure 4-2 shows several examples of resistors in parallel.

As illustrated in Figures 4-1 and 4-2, there are arrangements of components which are neither series nor parallel. There are many other, more complex arrangements of components; however, series and parallel are relatively common arrangements. Components that are electrically parallel do not have to be geometrically parallel in either the circuit or the schematic diagram—see the second example in Figure 4-2.

THE EQUIVALENT OF RESISTORS IN SERIES

It is possible to substitute one resistor for two or more resistors in series without causing changes in the voltages and currents in the remainder of the circuit; the single resistor is said to be **the equivalent** to the resistors in series. If n resistors are connected in series, the resistance of the equivalent resistor is given by

$$R_e = R_1 + R_2 + \cdots + R_n \tag{4-1}$$

This result will be derived here for the case of two resistors in series. The situation is shown in Figure 4-3; the two resistors in series are to be replaced with one resistor such that the current flowing in the battery does not change. KVL applied to the circuit in part a of the figure yields

$$V - I_a R_1 - I_a R_2 = 0$$

or

$$I_a = \frac{V}{R_1 + R_2} \tag{4-2}$$

For the circuit in (b),

$$I_b = \frac{V}{R_e} \tag{4-3}$$

Since the two currents are to be the same (nothing else in the circuit is to change),

$$I_a = I_b \tag{4-4}$$

Substituting Equations 4-2 and 4-3 into Equation 4-4 gives

$$\frac{V}{R_1 + R_2} = \frac{V}{R_e}$$

or

$$R_e = R_1 + R_2 \tag{4-5}$$

which is just the result given by Equation 4-1 when it is applied to the case of two resistors in series. The generalization of Equation 4-5 to obtain Equation 4-1 is left as a problem.

It is obvious that the equivalent resistance of a series combination of resistors is greater than any of the individual resistors.

THE EQUIVALENT OF RESISTORS IN PARALLEL

It is also possible to replace two or more resistors in parallel by one equivalent resistor without producing changes in the voltages and currents in the circuit. If there are n resistors in parallel, the resistance of the equivalent resistor R_e is given by

$$\frac{1}{R_e} = \frac{1}{R_1} + \frac{1}{R_2} + \cdots + \frac{1}{R_n} \tag{4-6}$$

This result will be derived for the case of two resistors in parallel. The situation is shown in Fig. 4-4. In (a) of the figure, the current through each part of the circuit has been drawn and named. We shall apply KCL and KVL to the circuit.

First, it is clear that I_1 and I_4 are the same current:

$$I_1 = I_4$$

KCL applied to either of the two nodes in (a) gives

$$I_1 - I_2 - I_3 = 0$$

or

$$I_1 = I_2 + I_3 \tag{4-7}$$

KVL applied to the loop including the battery and R_1 gives

$$V - I_2R_1 = 0$$

$$I_2 = \frac{V}{R_1} \tag{4-8}$$

Finally, applying KVL to the loop including the battery and R_2 yields

$$V - I_3R_2 = 0$$

$$I_3 = \frac{V}{R_2} \tag{4-9}$$

Substituting Equations 4-8 and 4-9 into 4-7 gives

$$I_1 = \frac{V}{R_1} + \frac{V}{R_2}$$

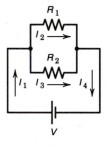

(a)

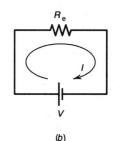

(b)

FIGURE 4-4 The two resistors in parallel in (a) are replaced by their equivalent resistance in (b).

For (*b*) of Figure 4-4 it is clear that

$$I = \frac{V}{R_e} \tag{4-10}$$

The two currents are to be equal:

$$I = I_1 \tag{4-11}$$

Substituting Equations 4-9 and 4-10 into Equation 4-11 gives

$$\frac{V}{R_e} = V\left(\frac{1}{R_1} + \frac{1}{R_2}\right)$$

which yields

$$\frac{1}{R_e} = \frac{1}{R_1} + \frac{1}{R_2} \tag{4-12}$$

which is just Equation 4-6 for the particular case of two resistors in parallel.

For the special case of two resistors in parallel, it is often easier to use the expression

$$R_e = \frac{R_1 R_2}{R_1 + R_2} \tag{4-13}$$

This expression only holds true for two resistors in parallel; for more resistors in parallel, the general expression must be used. Again, deriving the general expression Equation 4-6 is left as a problem.

The equivalent resistor is smaller than any of the individual resistors; in particular, it is smaller than the smallest of the resistors in the parallel combination. This is often a useful check of a numerical result.

MORE COMPLEX NETWORKS

A common type of problem found in elementary courses is to present a great mass of resistors arranged in some complex-looking network and ask the student to reduce it to a single equivalent resistor. In practice, such situations rarely occur; most networks cannot be reduced this way because they are neither series nor parallel. However, such problems do provide lots of practice in recognizing and replacing series and parallel combinations with their equivalents and are a useful step in the learning process.

The procedure for solving such problems is to look for two or more resistors that are in series or in parallel and replace them with their equivalents. Then, redraw the circuit in its reduced form and repeat until no further reductions can be made. Two examples will be given here. To aid in the process, the resistors to be replaced at each step will be enclosed in dashed lines.

The first example to be reduced is shown in Figure 4-5. First, there are two 20-ohm resistors in series; these can be replaced by one 40-ohm resistor. Fur-

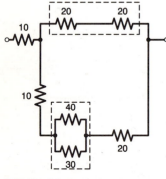

FIGURE 4-5 A sample network to be reduced to a simpler circuit.

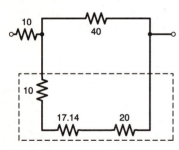

FIGURE 4-6 The circuit of Figure 4-5 after the first two steps of simplification.

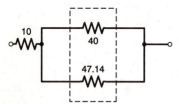

FIGURE 4-7 The circuit of Figure 4-5 after a further simplification.

FIGURE 4-8 The circuit of Figure 4-5 almost completely simplified.

thermore, there is a 40-ohm resistor in parallel with 30 ohms. The equivalent of these two is 17.14 ohms. To prevent errors, the circuit is redrawn with these two simplifications in Figure 4-6.

Now, there are three resistors in series—inside the dotted outline. Replacing these resistors by their equivalent gives the circuit shown in Figure 4-7.

The circuit begins to look much simpler. There are two resistors in parallel. Replacing them gives Figure 4-8.

As a last step, these two resistors are replaced by their equivalent to obtain the result shown in Figure 4-9, a single resistor of 31.64 ohms.

A second example is shown in Figure 4-10. This example is almost exactly the same as the first one, except that one 30-ohm resistor has been added.

This time there are 30-ohm and 40-ohm resistors in parallel; they are now replaced by their equivalent of 17.14 ohms. The 10-ohm resistor is in series with this parallel combination of 30-ohms and 40-ohms—the outer dashed box. To save steps, the 10-ohm resistor is combined with the 17.14-ohm equivalent, and then the circuit is redrawn in Figure 4-11.

The remaining combinations are neither series nor parallel. This circuit can be reduced no further using only the techniques introduced in this chapter. However, the circuit contains only resistors and is equivalent to a single resistor. In the next chapter, an efficient way to find this equivalent resistor will be introduced.

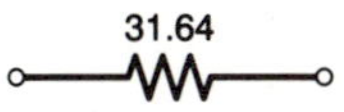

VOLTAGE DIVIDER

This chapter concludes with a discussion of two special circuits, each consisting of only two resistors. Despite the apparently simplicity of these circuits, they are used so frequently they have been given names. They are the voltage divider and the current divider.

The circuit shown in Figure 4-12 is called a **voltage divider**:

$$V_{\text{out}} = V_{\text{in}} \frac{R_2}{R_1 + R_2} \qquad (4\text{-}14)$$

The derivation is quite simple. It is assumed that no current flows from the node in the direction of V_{out}. Ohm's law gives

$$V_{\text{out}} = IR_2$$

and it has already been shown that

$$I = \frac{V_{\text{in}}}{R_e} = \frac{V_{\text{in}}}{R_1 + R_2}$$

Substituting the second of these expressions into the first gives Equation 4-14.

It is easy to see how this circuit gets its name. The input voltage is divided into two parts; one part appears across R_1 and the other across R_2. The part across R_2 is call the output. The output voltage is a fraction of the input voltage, the fraction being determined by the two resistors. R_1 and R_2 could be replaced by the two halves of a potentiometer, with V_{out} being taken from the

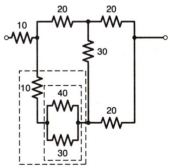

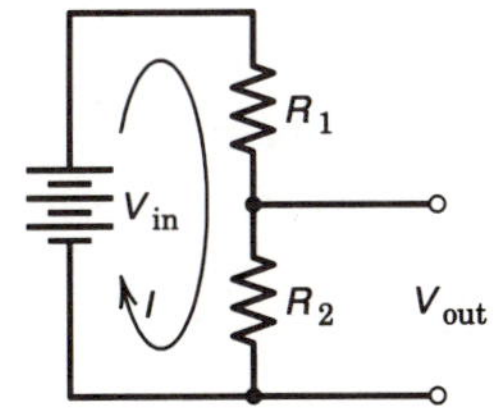

FIGURE 4-9 The completely simplified version of the circuit in Figure 4-5.

FIGURE 4-10 Another network to be reduced to a simpler circuit.

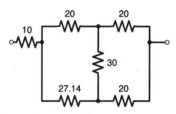

FIGURE 4-11 The simplified version of the circuit shown in Figure 4-10.

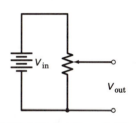

FIGURE 4-12 A voltage divider.

FIGURE 4-13 A voltage divider made with apotentiometer.

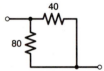

FIGURE 4-14 A current divider.

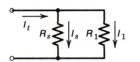

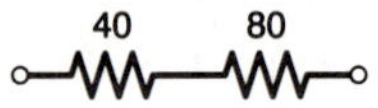

FIGURE 4-15 The circuit for Problem 4-1.

FIGURE 4-16 The circuit for Problem 4-2.

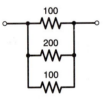

FIGURE 4-17 The circuit for Problem 4-3.

FIGURE 4-18 The circuit for Problem 4-4.

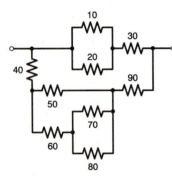

FIGURE 4-19 The circuit for Problem 4-7.

slider as shown in Figure 4-13. Such circuits are very common. Perhaps the most familiar use of this arrangement is the volume control on your radio or tape player.

Finally, as a generalization of this circuit, if there are n resistor in series, the voltage across resistor x is given by

$$V_x = V_{in} \frac{R_x}{R_1 + R_2 + \cdots + R_n} \qquad (4\text{-}15)$$

The analysis of the voltage divider is only correct as long as no current flows into (or out of) the output leads; that is, as long as whatever is attached to the terminals labeled V_{out} presents an infinite resistance (impedance). In practice, infinite resistance does not exist, so there is always some effect due to the loading of the output of the voltage divider. But also in practice, it is generally true that the load resistance is made so large that the error in using Equation 4-14 is acceptably small. This problem of the loading of the output of any circuit is one which will be discussed repeatedly in the remainder of this book.

CURRENT DIVIDER

The two-resistor circuit shown in Figure 4-14 is called a **current divider**. It is easy to show that

$$I_1 = I_t \frac{R_s}{R_1 + R_s} \qquad (4\text{-}16)$$

The proof of this result is left as a problem.

Again, it is easy to see how this circuit gets its name. The current is divided into two parts; one part goes through R_1, the rest goes through R_s, sometimes called the shunt resistor. The fraction going through R_1 is determined by the values of the resistors. As the value of the shunt resistor is decreased, the fraction of the current going through R_1 is decreased. This is also a relatively common circuit. One use of this circuit is described in Chapter 15, on meters.

Hint

The major results in this chapter, Equations 4-1, 4-6, 4-14, and 4-16, are so important and will be used so often that they should be memorized.

PROBLEMS

4-1 Find the equivalent resistance for the circuit shown in Figure 4-15.

4-2 Find the equivalent resistance for the circuit shown in Figure 4-16.

4-3 Find the equivalent resistance for the circuit shown in Figure 4-17.

4-4 Find the resistance for the circuit shown in Figure 4-18.

4-5 Derive Equation 4-1.

4-6 Derive Equation 4-3.

4-7 Reduce the circuit in Figure 4-19 to the simplest possible case using only the material from this chapter.

4-8 Reduce the circuit shown in Figure 4-20 to the simplest possible case.

4-9 Find the output voltage for the circuit in Figure 4-21.

4-10 Find the output voltage for the circuit in Figure 4-22.

4-11 If a 500-kilohm resistor were placed across the output terminals in Figure 4-22, what would the output voltage be?

4-12 Derive Equation 4-16.

4-13 Find the current in each resistor in Figure 4-23.

4-14 Find the current in each resistor in Figure 4-24.

ANSWERS TO ODD-NUMBERED PROBLEMS

4-1 120Ω

4-3 26.7Ω

4-7 29.9Ω

4-9 3.33 V

4-11 22.2 V

4-13 0.667 A in the 50-Ω resistor, 0.833 A in the 40-Ω resistor

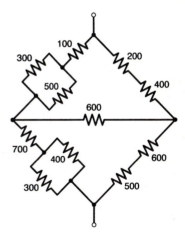

FIGURE 4-20 The circuit for Problem 4-8.

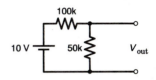

FIGURE 4-21 The circuit for Problem 4-9.

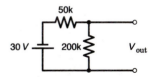

FIGURE 4-22 The circuit for Problem 4-10.

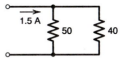

FIGURE 4-23 The circuit for Problem 4-13.

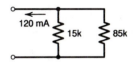

FIGURE 4-24 The circuit for Problem 4-14.

Chapter 5

MESH EQUATIONS

OBJECTIVES

1. Be able to analyze multiloop circuits by using the mesh equation method.

INTRODUCTION

Although the basic laws of circuit analysis introduced in Chapter 3 can be used to analyze essentially any circuit, a direct application of these laws often results in many equations and involves many unknowns. For a circuit of any degree of complexity, a strategy that minimizes the number of variables and equations is needed. One such method, called the **mesh equation method**, or **Maxwell's method**, is introduced in this chapter.

Electrical engineers have devoted a great deal of time and ingenuity to developing such strategies. There are whole books and courses on the methods of circuit analysis. Anyone who wants to know more about the methods of formal circuit analysis than what is presented in this book should go to one of these references; several circuit analysis textbooks are listed in Appendix A. However, the material presented in this book will be adequate for most readers.

Finally, although this chapter is written in terms of constant currents and resistances, everything in this chapter is completely general. Everything can be generalized by replacing the word "resistance" by "impedance".

AN EXAMPLE DONE THE HARD WAY

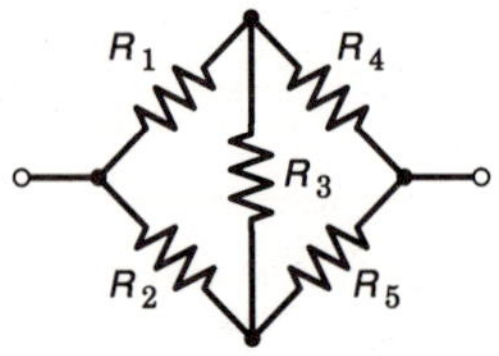

FIGURE 5-1 A simple example of a network of resistors. This circuit is sometimes called a Wheatstone bridge.

To illustrate how circuit problems can easily become very complex, a relatively simple circuit will be analyzed by a straightforward application of KVL and KCL. Later in this chapter, this example is done more directly using the mesh equation method.

Figure 5-1 shows one of the simplest networks of resistors that cannot be reduced to a simpler circuit by combining series and parallel resistors. Yet, since it consists only of resistors, it must be possible to replace the resistors with a single equivalent resistor. To find the equivalent resistance of this circuit, it is necessary to imagine a voltage applied to the circuit as shown in Figure 5-2, calculate the current I, and then calculate the equivalent resistance from

$$R_e = \frac{V}{I} \qquad (5\text{-}1)$$

The procedure is a straightforward but tedious application of KCL and KVL.

1. Draw a unique current through each component or sequence of components in series in the circuit. Give each of these currents a name. Be careful not to use two names for the same current.
2. Apply KCL to get as many equations as possible. The theorem quoted below will show how many independent equations can be obtained from KCL.
3. Apply KVL to get enough additional equations to be able to solve for the desired currents.
4. Solve the resultant system of equations.

The following theorem found in many circuit analysis texts is quoted here without any derivation or further comment.

If there are m currents and n nodes in a circuit, then $n - 1$ independent equations can be obtained with KCL and $m - n - 1$ additional independent equations can be obtained from KVL.

The currents have been drawn and named in Figure 5-3. In this instance there are six currents and four nodes. According to the theorem above, only three independent equations can be obtained by using KCL. Applying KCL to nodes a, b, and d yields

$$I - I_1 - I_2 = 0$$
$$I_1 - I_3 - I_4 = 0$$
$$I_4 + I_5 - I = 0$$

To test the theorem above, we apply KCL to node c, which yields

$$I_2 + I_3 - I_5 = 0$$

This equation is nothing new; it is simply the sum of the three equations above.

Three more equations can be obtained by using KVL. Any three closed loops can be picked. Figure 5-4 shows the three loops used here. (There was no particular reason why these three loops were chosen.) The equations obtained are

Loop I

$$V - R_2 I_2 - I_5 R_5 = 0$$

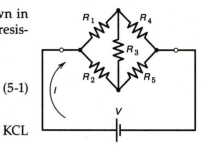

FIGURE 5-2 The circuit used to find the equivalent resistance of the network shown in Figure 5-1.

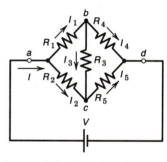

FIGURE 5-3 The circuit of Figure 5-2 with all necessary currents drawn and given names.

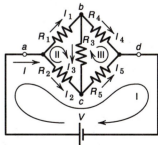

FIGURE 5-4 The three loops used in applying KVL to the circuit in Figure 5-2.

Loop II

$$-R_1I_1 - I_3R_3 + R_2I_2 = 0$$

Loop III

$$-R_4I_4 + R_5I_5 + R_3I_3 = 0$$

Six equations in six unknowns have been obtained by the application of KVL and KCL. Six equations in six unknowns is a relatively difficult problem to solve without a computer. However, these equations are not too complex and, after all, this is an example to illustrate how difficult this method is. There are several tricks that can be used to simplify the equations, but these tricks do not help at all if you cannot think of them. Here, these equations will be solved by a brute force application of Cramer's rule—see Chapter 52.

First, the equations are rewritten in the standard form so that Cramer's rule can be applied:

$$
\begin{aligned}
I \quad -I_1 \quad -I_2 \qquad\qquad\qquad &= 0 \\
I_1 \qquad\qquad -I_3 \quad -I_4 \qquad &= 0 \\
-I \qquad\qquad\qquad +I_4 \quad +I_5 &= 0 \\
R_2I_2 \qquad\qquad + R_5I_5 &= V \\
-R_1I_1 + R_2I_2 - R_3I_3 \qquad\qquad &= 0 \\
R_3I_3 - R_4I_4 - R_5I_5 &= 0
\end{aligned}
$$

Solving for I by Cramer's rule yields

$$
I = \frac{
\begin{vmatrix}
0 & -1 & -1 & 0 & 0 & 0 \\
0 & 1 & 0 & -1 & -1 & 0 \\
0 & 0 & 0 & 0 & 1 & 1 \\
V & 0 & R_2 & 0 & 0 & R_5 \\
0 & -R_1 & R_2 & -R_3 & 0 & 0 \\
0 & 0 & 0 & R_3 & -R_4 & R_5
\end{vmatrix}
}{
\begin{vmatrix}
1 & -1 & -1 & 0 & 0 & 0 \\
0 & 1 & 0 & -1 & -1 & 0 \\
-1 & 0 & 0 & 0 & 1 & 1 \\
0 & 0 & R_2 & 0 & 0 & R_5 \\
0 & -R_1 & R_2 & -R_3 & 0 & 0 \\
0 & 0 & 0 & R_3 & -R_4 & -R_5
\end{vmatrix}
}
$$

Evaluation of these two determinants is a tedious but straightforward process (seeing if you can reproduce this result would be a very good test of your ability to evaluate determinants):

$$
I_e = \frac{V(R_1R_3 + R_1R_4 + R_1R_5 + R_2R_3 + R_2R_4 + R_2R_5 + R_3R_4 + R_3R_5)}{R_1R_2R_3 + R_1R_2R_4 + R_1R_2R_5 + R_1R_3R_5 + R_1R_4R_5 + R_2R_3R_4 + R_2R_4R_5 + R_3R_4R_5}
$$

So, finally, the equivalent resistance is given by

$$R_e = \frac{R_1R_2R_3 + R_1R_2R_4 + R_1R_2R_5 + R_1R_3R_5 + R_1R_4R_5 + R_2R_3R_4 + R_2R_4R_5 + R_3R_4R_5}{R_1R_3 + R_1R_4 + R_1R_5 + R_2R_3 + R_2R_4 + R_2R_5 + R_3R_4 + R_3R_5}$$

MESH EQUATIONS

Obviously, if the methods used above were the only way to solve multiloop circuits, formal circuit analysis would have been the specialty of only the masochistic, at least before the time of computers. Fortunately, there is a somewhat easier way of attacking multiloop circuits than the one used above. This method is called the **mesh equation method**, or **Maxwell's method**. In this section, the procedure used will be outlined and applied to the example. In the following section, there are some general comments about the technique of using this method. This chapter will conclude with another example.

The procedure for using the mesh equation method is as follows:

1. Pick closed current loops called **mesh currents**, or **loop currents**. These may be picked in any way so long as each current forms a closed loop, and so long as each branch of the circuit has at least one current flowing through it and no two different branches have the same current.
2. Apply KVL to each loop. Use the sign conventions given in Chapter 3.
3. Solve for the loop currents. As before, a solution with a minus sign means that the current goes the opposite to the way it is drawn.

To illustrate these rules, the example worked out before will be done using the mesh equation method:

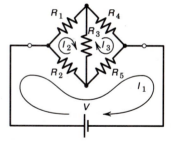

FIGURE 5-5 The mesh currents used to solve the circuit of Figure 5-2.

1. Figure 5-5 shows the circuit with a set of loop currents. Note that there are two currents flowing in R_2, R_3, and R_5.
2. Applying KVL to these loops gives the results below. The loops have been traversed in the same direction as the loop currents and the various terms have been collected.

$$V \quad -I_1(R_2 + R_5) \quad + I_2R_2 \quad\quad + I_3R_5 \quad\quad = 0$$
$$I_1R_2 \quad\quad -I_2(R_1 + R_2 + R_3) + I_3R_3 \quad\quad = 0$$
$$I_1R_5 \quad\quad + I_2R_3 \quad\quad -I_3(R_3 + R_4 + R_5) = 0$$

3. These equations are solved by using Cramer's rule. The first step is to rewrite them in the standard form and changing all the signs on some of the equations to make the diagonal terms positive.

$$I_1(R_2 + R_5) \quad - I_2R_2 \quad\quad -I_3R_5 \quad\quad = V$$
$$- I_1R_2 \quad + I_2(R_1 + R_2 + R_3) \ -I_3R_3 \quad\quad = 0$$
$$- I_1R_5 \quad\quad - I_2R_3 \quad + I_2(R_3 + R_4 + R_5) = 0$$

To find the equivalent resistance, it is only necessary to find I_1. This is given by

$$I_1 = \frac{\begin{vmatrix} V & -R_2 & -R_5 \\ 0 & (R_1 + R_2 + R_3) & -R_3 \\ 0 & -R_3 & (R_3 + R_4 + R_5) \end{vmatrix}}{\begin{vmatrix} (R_2 + R_5) & -R_2 & -R_5 \\ -R_2 & (R_1 + R_2 + R_3) & -R_3 \\ -R_5 & -R_3 & (R_3 + R_4 + R_5) \end{vmatrix}}$$

Although it still is not painless, this expression is much easier to solve. Evaluating these determinants gives

$$I_1 = \frac{V(R_1 R_3 + R_1 R_4 + R_1 R_5 + R_2 R_3 + R_2 R_4 + R_2 R_5 + R_3 R_4 + R_3 R_5)}{R_1 R_2 R_3 + R_1 R_2 R_4 + R_1 R_2 R_5 + R_1 R_3 R_5 + R_1 R_4 R_5 + R_2 R_3 R_4 + R_2 R_4 R_5 + R_3 R_4 R_5}$$

which is just what was obtained above. However, this time it was much less work.

SOME COMMENTS ON THE METHOD

At first, this method of solving circuits often seems quite surprising. It appears as if KCL has been totally ignored and the circuit solved by using only KVL. But this is not true. By drawing all currents as closed loop currents, KCL is automatically satisfied; thus, there is no need to apply KCL to the nodes of the circuit. This eliminates about half the unknowns and equations needed to solve a circuit. It also results in determinants that have fewer null elements (0's). Thus the evaluation of the determinants is often slightly more difficult, but the determinants are of substantially lower order.

There are several ways of determining the number of mesh currents required. All of these methods give the same result when correctly applied to a circuit.

1. For simple circuits, the number is obvious.
2. Pick enough currents so that every element (branch) of the circuit has at least one current through it and no two branches have the same current or the same combination of currents through them.
3. The number of mesh currents is the number of branches minus the number of nodes plus 1 ($B - n + 1$).
4. Imagine cutting the branches of the circuit so that each cut opens a closed path. When there are no closed paths left, the number of cuts is the number of mesh currents required.

The mesh currents may be chosen in any way as long as every branch of the circuit has a current in it and as long as no two branches of the circuit have the same current in them. Outside of this rule, the mesh currents may be chosen in any way to simplify the solution of the problem. Thus if the goal is to find the current through one particular component, then the mesh currents should be picked so that there is only one current through that par-

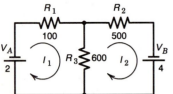

FIGURE 5-6 A second example of use of the mesh equation method.

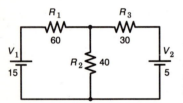

FIGURE 5-7 The circuit for Problem 5-1.

ticular component; this will minimize the number of determinants to be evaluated. For instance, in the example done above, the loop currents were chosen so that only one current went through the battery.

ANOTHER EXAMPLE USING MESH CURRENTS

The problem is to find the current in each resistor in the circuit in Figure 5-6. Here, a set of mesh currents has already been drawn. Applying KVL to this circuit gives (using symbols and not values)

$$\begin{aligned} -V_A - I_1(R_1 + R_3) + I_2 R_3 \qquad\qquad &= 0 \\ V_B \quad + I_1 R_3 \qquad\qquad -I_2(R_2 + R_3) &= 0 \end{aligned}$$

Rewriting these in the standard form gives

$$\begin{aligned} I_1(R_1 + R_3) - I_2 R_3 \qquad\qquad &= -V_A \\ - I_1 R_3 \qquad\qquad I_2(R_2 + R_3) &= \quad V_B \end{aligned}$$

Since numerical values are given and asked for, it will be much easier to use numbers rather than symbols when evaluating the determinants. So

$$\begin{aligned} 700I_1 - 600I_2 &= -2 \\ -600I_1 + 1100I_2 &= 4 \end{aligned}$$

$$I_1 = \frac{\begin{vmatrix} -2 & -600 \\ 4 & 1100 \end{vmatrix}}{\Delta} = \frac{\begin{vmatrix} -2 & -600 \\ 4 & 1100 \end{vmatrix}}{\begin{vmatrix} 700 & -600 \\ -600 & 1100 \end{vmatrix}} = \frac{200}{410{,}000}$$

$$= 0.488 \text{ mA}$$

$$I_2 = \frac{\begin{vmatrix} 700 & -2 \\ -600 & 4 \end{vmatrix}}{\Delta} = \frac{1600}{410{,}000}$$

$$= 3.90 \text{ mA}$$

Thus the current through R_1 is I_1 and that through R_2 is I_2, whereas the current through R_3 is the difference of I_2 and I_1, or 3.42 mA upward.

PROBLEMS

5-1 Find the current in each resistor of Figure 5-7.

5-2 Find the current in each resistor of Figure 5-8.

5-3 Find the current in each resistor of Figure 5-9.

5-4 Find the current in each resistor of Figure 5-10.

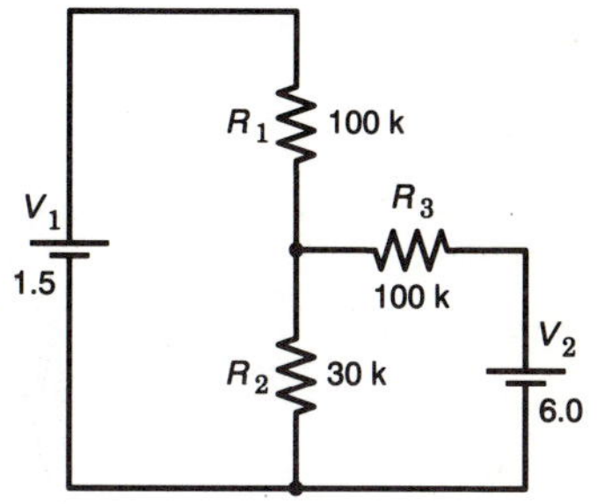

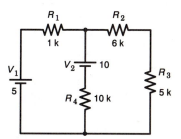

FIGURE 5-8 The circuit for Problem 5-2.

FIGURE 5-9 The circuit for Problem 5-3.

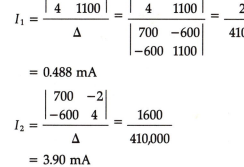

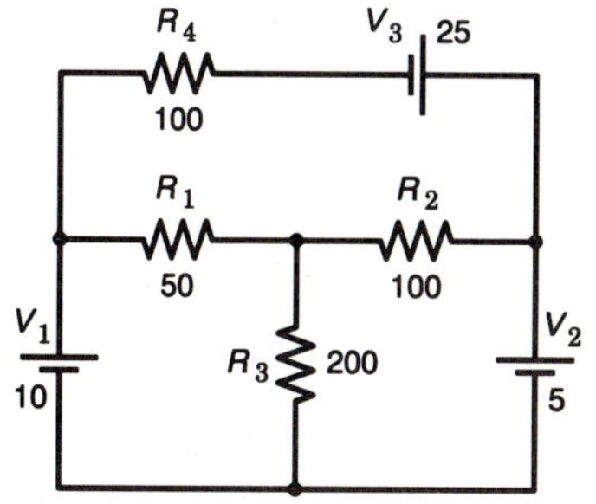

FIGURE 5-10 The circuit for Problem 5-4.

ANSWERS TO ODD-NUMBERED PROBLEMS

5-1 R_1: 0.157 A to the right; R_2: 0.139 A downward; R_3: 0.0185 to the right

5-3 R_1: 0.38 μA to the left; R_2: 0.458 μA to the right; R_3: 0.458 μA downward; R_4: 0.42 μA upward

Chapter 6

ADVANCED CIRCUIT ANALYSIS THEOREMS

OBJECTIVES

1. Become familiar with the concepts of voltage and current sources.

2. Become familiar with Thevenin's and Norton's theorems, the superposition theorem, and the reciprocity theorem.

3. Develop some insight into the purpose of circuit analysis.

INTRODUCTION

The purpose of this chapter is to present a brief introduction to a few of the more advanced topics in circuit analysis. This is not a complete presentation; to learn more about these topics, go to a standard textbook on circuit analysis—see Appendix A for list of references. With the exception of the voltage and current sources and an occasional mention of Thevenin's theorem, the material in this chapter will not be used in the rest of this book.

We begin with the introduction of the concepts of voltage and current sources. Then, Thevenin's and Norton's theorems and two more general theorems are discussed. Following this discussion there is a digression on the purpose of circuit analysis. Finally, the chapter ends with an example that shows how useful Thevenin's theorem can be.

VOLTAGE AND CURRENT SOURCES

Two idealized circuit elements are used in circuit analysis for creating mathematical models of real equipment, such as a transistor or an amplifier. These are the voltage and current sources.

A voltage source is a device that maintains a specific voltage between its terminals, no matter what the current through the device. This may be a dc voltage, a sinusoidal voltage, or some other well-defined voltage. A perfect or ideal battery is an example of a dc voltage source. In general, real batteries are not true voltage sources in this sense.

A current source is a device that maintains a specific current through itself no matter what the resistance of the attached circuit. Again, this may be a dc, a sinusoidal, or some other well-defined current. No true current sources exist, but there are several circuits that are very good approximations to a current source.

A quick survey of the literature reveals that everyone uses a different set of symbols for voltage and current sources. Figure 6-1 presents the most commonly used symbols for these circuit elements; these symbols are used in this book.

Neither a voltage source nor a current source has a fixed voltage–current relationship. That is, the voltage across a voltage source is determined by the nature of the voltage source, whereas the current through the voltage source depends on the rest of the circuit, not by the nature of the voltage source. The voltage across a current source is determined by the rest of the circuit, not by the nature of the current source. Thus when applying KVL to a loop that includes a current source, the voltage across the current source is an unknown

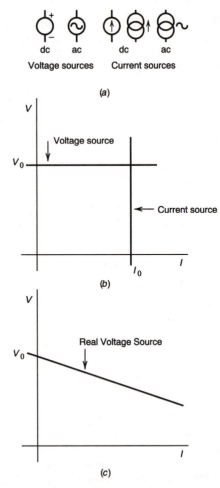

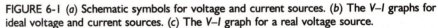

FIGURE 6-1 (a) Schematic symbols for voltage and current sources. (b) The V–I graphs for ideal voltage and current sources. (c) The V–I graph for a real voltage source.

although the current through the source is known. To illustrate this, the *V–I* graphs for a voltage source and a current source are shown in Figure 6-1*b*; for comparison, the *V–I* graph for a real voltage source, for example, a battery, is shown in Figure 6-1*c*.

Circuits involving voltage sources generally are best solved using KVL and the mesh equation technique. Circuits involving current sources are usually solved using KCL. Circuits that include both voltage and current sources can often be very difficult to solve.

THEVENIN'S THEOREM

Thevenin's theorem states:

> Any complex network of linear circuit elements (sources, resistors, and impedances) having two terminals can be replaced by a single equivalent voltage source connected in series with a single resistor (impedance).

There are two ways of describing the situation in which Thevenin's theorem is profitably used. The theorem is useful to find how the current in some resistor in a circuit varies as that particular resistor is varied and the remainder of the circuit remains unchanged. The theorem is also useful to find how the output of some circuit changes as the output is loaded with a resistor. (These two situations are really the same: In the second case, the variable element is considered to be an external load; in the first case, the variable element is considered to be part of the circuit itself.)

The procedure for using Thevenin's theorem is as follows:

1. If the circuit to be replaced by the Thevenin equivalent circuit already has two open terminals, label these terminals V_{Th}. If there is a circuit element at the point where the circuit is to be studied, remove the element from the circuit, replace it by an open circuit, and label the terminals V_{Th}.
2. Compute V_{Th}, the voltage at the open terminals.
3. Replace all the voltage sources in the circuit with short circuits and all current sources in the circuit with open circuits. [Note: This is a conceptual operation, not a laboratory operation.]
4. Compute R_{Th}, the resistance of the circuit looking back into the output terminals after making these changes.
5. The network can then be replaced by the circuit shown in Figure 6-2.

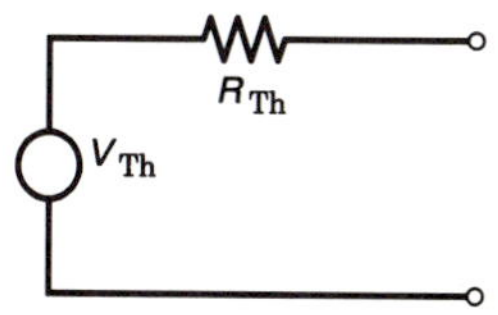

FIGURE 6-2 The Thevenin equivalent circuit.

AN EXAMPLE USING THEVENIN'S THEOREM

A very simple example using Thevenin's theorem will be presented here to illustrate this procedure. A more complex example will be done at the end of the chapter.

The goal is to study the output voltage of a voltage divider as a function of the load resistor R_L, as shown in Figure 6-3. This circuit is analyzed by following the procedures outlined above:

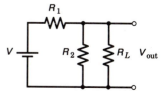

FIGURE 6-3 An example to be analyzed by using Thevenin's theorem.

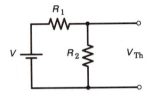

FIGURE 6-4 The circuit of
Figure 6-3 with the load resistor
removed and the output identified.

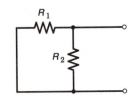

FIGURE 6-5 The circuit of
Figure 6-3 with the voltage
sources replaced by short circuits.

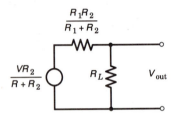

FIGURE 6-6 The Thevenin equivalent of the circuit in Figure 6-3.

1. First, remove R_L and label the resulting terminals V_{Th} as shown in Figure 6-4.
2. Next, compute V_{Th}. In this case, this is trivial, because what is left is a simple voltage divider. So

$$V_{Th} = \frac{VR_2}{R_1 + R_2} \tag{6-1}$$

3. Next, replace the voltage source by a short circuit, obtaining the circuit shown in Figure 6-5.
4. Compute R_{Th}, the resistance looking back into the output terminals; in this case, this is a parallel combination of two resistors. So

$$R_{Th} = \frac{R_1 R_2}{R_1 + R_2} \tag{6-2}$$

5. Finally, the simplified circuit can be drawn as in Figure 6-6.

Now, the original question can be answered. The output of interest is simply another voltage divider, so the answer can be written down directly by combining Equations 6-1 and 6-2 in the voltage divider equation:

$$V_{out} = \frac{VR_2}{R_1 + R_2} \frac{R_L}{R_L + \dfrac{R_1 R_2}{R_1 + R_2}} \tag{6-3}$$

Thus a rather formidable result has been obtained by a series of steps, each of which was simple enough that it could be written down directly without extensive computation. This is not an unusual result. People who creatively use Thevenin's and Norton's theorems (see below) frequently solve the most complex-looking problems with the greatest of ease.

NORTON'S THEOREM

Norton's theorem states:

> Any complex network of sources and resistances (impedances) can be replaced by a current source and a resistor (impedance) in parallel with it.

The procedure for using Norton's theorem is as follows:

1. If the circuit has two terminals, connect these by a short circuit. If the current through some element is to be studied, replace it by a short circuit.
2. Calculate I_{Nor}, the current in this short circuit.
3. Replace all voltage sources by short circuits and all current sources by open circuits. [Again, this is a conceptual operation.]
4. Compute the shunt resistance R_{Nor}, looking back into the circuit after making all these changes. Note that $R_{Nor} = R_{Th}$.
5. Finally, replace the network by the current shown in Figure 6-7.

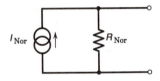

FIGURE 6-7 The Norton equivalent circuit.

AN EXAMPLE USING NORTON'S THEOREM

The example shown in Figure 6-3 will be done again using Norton's theorem. Remember, the goal of this example is to study how V_{out} varies as R_L is varied. Using the procedure for applying Norton's theorem gives the following:

1. First, replace R_L by a short circuit, obtaining the results in Figure 6-8.
2. Compute I_{Nor}; for this case, this is

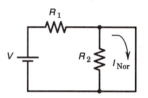

FIGURE 6-8 The first step in analyzing the circuit of Figure 6-3 by using Norton's theorem.

$$I_{Nor} = \frac{V}{R_1} \tag{6-4}$$

3. Replace the voltage sources by short circuits and the current sources by open circuits; for the example the result is shown in Figure 6-9.
4. Looking back into the output of the circuit, calculate R_{Nor}. In this case, this is the parallel combination of two resistors, so

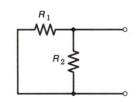

$$R_{Nor} = \frac{R_1 R_2}{R_1 + R_2} \tag{6-5}$$

FIGURE 6-9 The circuit of Figure 6-3 after the sources have been replaced by short or open circuits.

5. Finally, replace the original circuit with the result shown in Figure 6-10.

Once again, this has reduced the original problem to a simple one, that of a current divider, and the answer can be written down reasonably directly. The current through R_L is given by the current divider formula

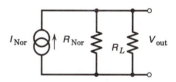

$$I_{R_L} = I_{Nor} \frac{R_{Nor}}{R_{Nor} + R_L} \tag{6-6}$$

and the output voltage is

FIGURE 6-10 The circuit of Figure 6-3 after being simplified by using Norton's theorem.

$$V_{out} = I_{R_L} R_L \tag{6-7}$$

Combining Equations 6-7, 6-6, 6-5, and 6-4 gives

$$V_{out} = \frac{V}{R_1} \frac{R_1 R_2/(R_1 + R_2)}{R_1 R_2/(R_1 + R_2) + R_L} R_L \tag{6-8}$$

which is the same result obtained above by use of Thevenin's theorem—Equation 6-3.

THE RELATIONSHIP BETWEEN THE NORTON AND THEVENIN EQUIVALENTS

Since any network can be replaced with the aid of either Norton's theorem or Thevenin's theorem, there must be a relationship between the two equivalent circuits. This relationship is easy to find by applying Thevenin's theorem to the Norton equivalent circuit. This yields

$$R_{Nor} = R_{Th} \tag{6-9}$$

and

$$V_{\text{Th}} = I_{\text{Nor}} R_{\text{Nor}} \qquad\qquad (6\text{-}10)$$

TWO OTHER THEOREMS

There are two other general circuit analysis theorems that are used extensively in any formal study of circuit analysis. In fact, they are frequently used to prove the existence of the Norton and Thevenin equivalent circuits. These two theorems are simply stated here; they will not be mentioned again. However, they are included here because references to them occur frequently in the technical literature.

THE SUPERPOSITION THEOREM:

The current in any branch of a linear circuit is equal to the sum of the currents produced separately by each source in the remainder of the circuit, with all the other sources set equal to zero.

This theorem simply says that, given a circuit involving n voltage and current sources, the current in some branch of the circuit can be found by solving the n simpler problems, each involving only one source. The solution to the original problem is, then, the sum of the n currents found in these simpler and perhaps easier calculations.

THE RECIPROCITY THEOREM

The partial current in branch x of a linear, dc circuit produced by a voltage source in branch y is the same as the partial current that would be produced in branch y by the same source if it were placed in branch x.

A DIGRESSION ON THE GOALS OF CIRCUIT ANALYSIS

Generally, the goal of practical circuit analysis is to answer "what if" types of questions. For instance, given a big complicated amplifier, how will the operation of the amplifier as a whole be affected if some particular component has a different value, or some transistor has a different gain, and so on. In a sense, the goal of the circuit analysis is to form a mathematical model of the circuit and then do a series of experiments on the model to understand the operation of the circuit.

Generally, however, any circuit which is complex enough to be useful is so complex that the mathematics of the circuit analysis quickly get prohibitively difficult. When this happens, there are two possible approaches—both might well be taken at once. One is to write a program to carry out the analysis and simply let a computer churn away on it. The other possibility is to simplify the situation in ways that will not obscure the effects being studied but will make the analysis easier. In this process, the Thevenin and Norton equivalent circuits are extremely useful. The fact that complex portions of a circuit can be replaced by a single source and a single resistor can often be used to make significant simplifications in circuit analysis.

Of course, Thevenin's theorem can also be used on simple circuits to make their operation more comprehensible. For instance, a real battery can often be modeled quite realistically as an ideal voltage source in series with a resistor—the Thevenin equivalent.

When you analyze circuits given in a textbook, as opposed to analyzing real circuits or circuits that are of interest to you, Thevenin's theorem is most often useful if the circuits involve voltage sources. Norton's theorem is most often useful if the circuits involve current sources. For cases involving both, look for some trick to simplify the circuit. When dealing with real circuits, often just plain hard work is needed to analyze the circuit.

Norton's theorem is most useful when building models of circuit elements that closely represent current sources. Transistors are such elements, and current sources are used in the models used to find equivalent circuits for transistors circuits.

Thevenin's theorem is most useful when building models of circuit elements that closely represent voltage sources. Many transducers are such elements, and voltage sources in series with resistors are often used in models of these items.

A DIGRESSION ON LINEAR CIRCUITS

All four theorems stated in this chapter deal with circuits involving **linear** circuit elements, that is, elements that have voltage–current relationships that depend linearly on the amplitude of the voltage. Although resistors, capacitors, and inductors have such linear V–I relationships, most of the items that make possible modern electronics—diodes, transistors, and the like—do not have linear V–I relationships. Here, however, a mathematical fact comes to the rescue.

Let's look at an operating circuit. If only very small variations in some particular parameter are introduced, then the circuit responds in a linear fashion to these changes. Mathematically, this is the same as saying that any continuous curve can be approximated by short segments of straight lines; to get a better approximation simply use more segments. For instance, the response of an amplifier to small changes in the input signal might be investigated as if the output were a linear function of the input. This is frequently called **small-signal analysis** and will be mentioned in the discussion of dynamic impedance.

A FINAL EXAMPLE

To conclude this chapter, one example will be solved three ways: first, by a direct application of KCL and KVL; then, by using mesh equations; and fi-

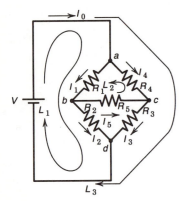

FIGURE 6-11 A Wheatstone bridge.

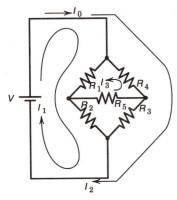

FIGURE 6-12 The mesh currents used in analyzing the Wheatstone bridge circuit.

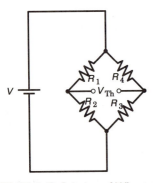

FIGURE 6-13 Solution of Wheatstone bridge circuit using Thevenin's theorem.

nally, by using Thevenin's theorem. As expected, the solution by means of Thevenin's theorem will be by far the easiest.

The circuit shown in Figure 6-11 is the **Wheatstone bridge**. This is a very common circuit used for making precision measurements of resistance values or, when used with inductors and capacitors, for making precision inductance and capacitance measurements. When the circuit is being used, the quantity of interest is the current in R_5 and, in particular, the conditions for which this current is zero. Hence, the problem to be solved here is to find the condition for which the current through R_5 is zero.

First, the problem is solved by using KCL and KVL directly. In Figure 6-11, currents have been drawn through each element, the nodes are labeled, and the loops used here are indicated. The loops have been chosen to simplify the problem to be solved. There are six currents and four nodes; thus, there are three equations from KCL and three from KVL.

Applying KCL to nodes a, b, and c yields

$$I_0 - I_1 - I_4 = 0$$
$$I_1 - I_2 - I_5 = 0$$
$$-I_3 + I_4 + I_5 = 0$$

Applying KVL to loops 1, 2, and 3 yields

$$V - I_1 R_1 - I_2 R_2 = 0$$
$$-I_5 R_5 + I_4 R_4 - I_1 R_1 = 0$$
$$V - I_4 R_4 - I_3 R_3 = 0$$

Rewriting these equations in the standard form gives

I_0	$-I_1$		$-I_4$		$= 0$
	I_1	$-I_2$		$-I_5$	$= 0$
			$-I_3$	$+I_4$	$+I_5$ = 0
	$I_1 R_1$	$+I_2 R_2$			$= V$
	$-I_1 R_1$		$I_4 R_4$	$-I_5 R_5$	$= 0$
			$+I_3 R_3$	$+I_4 R_4$	$= V$

In general, we regard six equations in six unknowns to be a hopeless task if the equations have to be solved by hand. In general, there are two approaches to large sets of linear equations. If the equations have numerical coefficients, that is, if the values of the resistors are given, then many hand calculators can solve the equations. If symbolic values are to be used, as in this case, then one of the many symbolic mathematical programs available for either PCs or mainframe computers can be used. However, this is a very special case, since only the conditions for which I_5 is zero are to be found. This means that only the determinant in the numerator for I_5 need to be evaluated and set equal to zero. Because this determinant will have many zeros in it, this is a feasible undertaking for paper and pencil calculations. The determinant is

$$\begin{vmatrix} 1 & -1 & 0 & 0 & -1 & 0 \\ 0 & 1 & -1 & 0 & 0 & 0 \\ 0 & 0 & 0 & -1 & 1 & 0 \\ 0 & R_1 & R_2 & 0 & 0 & V \\ 0 & -R_1 & 0 & 0 & R_4 & 0 \\ 0 & 0 & 0 & R_3 & R_4 & V \end{vmatrix}$$

After much work, this becomes

$$I_5 = \frac{V(-R_1 R_3 + R_2 R_4)}{\text{denom}} = 0$$

where "denom" is the determinant that is needed to find all values of I_5. But since the goal is to find the condition for which I_5 is zero, the denominator need not be evaluated. The expression can be solved to give

$$-R_1 R_3 + R_2 R_4 = 0 \qquad (6\text{-}11)$$

This is a well-known result. Just in passing, note that if the ratio of R_4 to R_3 is known (not the values, just the ratio), then an unknown resistor, R_1, can be measured in terms of a known resistor, R_2:

$$R_1 = R_2 \left(\frac{R_4}{R_3} \right) \qquad (6\text{-}12)$$

Much of the utility of making measurements with a bridge circuit arises because it is not necessary to measure small currents, but only to detect a null current (or a current minimum). This is always easier than actually measuring a current.

Next, the problem is solved by using the mesh equation method. In the circuit of Figure 6-12, a set of mesh currents has been drawn—these are the same loops as above. Applying KVL to these loops and rewriting the equations in standard form gives

$$I_1(R_1 + R_2) \qquad\qquad +I_3 R_1 \qquad\quad = V$$
$$I_2(R_3 + R_4) \quad -I_3 R_4 \qquad\quad = V$$
$$I_1 R_1 \qquad -I_2 R_4 \qquad -I_3(R_1 + R_4 + R_5) = 0$$

This time, the current wanted is I_3. Again, to find the condition when the current is zero, it is only necessary to evaluate the determinant in the numerator and set this equal to zero. The determinant is

$$\begin{vmatrix} (R_1 + R_2) & 0 & V \\ 0 & R_3 + R_4 & V \\ R_1 & -R_4 & 0 \end{vmatrix}$$

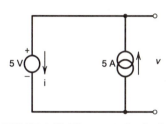

FIGURE 6-14 The circuit for Problem 6-1.

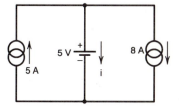

FIGURE 6-15 The circuit for Problem 6-2.

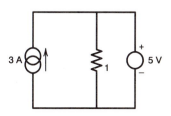

FIGURE 6-16 The circuit for Problem 6-3.

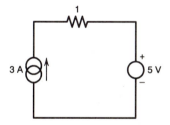

FIGURE 6-17 The circuit for Problem 6-4.

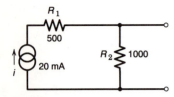

FIGURE 6-18 The circuit for Problem 6-5.

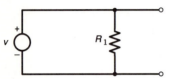

FIGURE 6-19 The circuit for Problem 6-6.

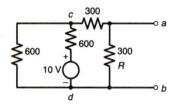

FIGURE 6-20 The circuit for Problem 6-7.

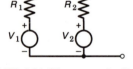

FIGURE 6-21 The circuit for Problem 6-8.

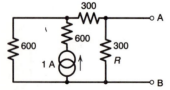

FIGURE 6-22 The circuit for Problem 6-9.

Evaluating this determinant and setting it equal to zero gives

$$V[-R_1(R_3 + R_4) + R_4(R_1 + R_2)] = 0$$

or

$$R_2 R_4 = R_1 R_3 \qquad (6\text{-}13)$$

which is the same result as before, Equation 6-11. This was much easier, but not the easiest way.

Finally, the problem is solved by using Thevenin's theorem. To find the Thevenin equivalent circuit, first replace R_5 with an open circuit and find V_{Th}. The resulting circuit is shown in Figure 6-13.

It can be seen that V_{Th} is simply the difference between two voltage dividers. This can be written down directly:

$$V_{Th} = V\left(\frac{R_2}{R_1 + R_2} - \frac{R_3}{R_3 + R_4}\right)$$

The current will be zero when $V_{Th} = 0$ or when

$$\frac{R_2}{R_1 + R_2} = \frac{R_3}{R_3 + R_4}$$

or

$$R_2 R_4 = R_1 R_3 \qquad (6\text{-}14)$$

This is the same result as Equation 6-13, but this time it took practically no effort.

PROBLEMS

6-1 Find the unknown current i in the circuit of Figure 6-14.

6-2 Find the unknown voltage v and current i in the circuit of Figure 6-15.

6-3 What is the voltage across and the current through each of the circuit elements in the circuit of Figure 6-16?

6-4 What is the voltage across and the current through each of the elements in the circuit of Figure 6-17?

6-5 Find the Thevenin equivalent circuit for the circuit of Figure 6-18.

6-6 Find the Thevenin equivalent circuit for the circuit of Figure 6-19.

6-7 Find the Thevenin equivalent circuit for the circuit of Figure 6-20.

6-8 Find the Norton equivalent circuit for the circuit of Figure 6-21.

6-9 Find the Norton equivalent circuit for the circuit of Figure 6-22.

ANSWERS TO ODD-NUMBERED PROBLEMS

6-1 $i = -3A$

6-3

Element	Voltage	Current
Current source	8 V top positive	3 A upward
R	3 V left positive	3 A to right
Voltage source	5 V as shown	3 A downward.

6-5 $V_{Th} = 20$ V; $R_{Th} = 1000\Omega$.

6-7 $V_{Th} = \frac{5}{3}$ V; $R_{Th} = 200\Omega$

6-9 $I_{Nor} = \frac{2}{3}$ A; $R_{Nor} = 225\ \Omega$

Chapter 7

ALTERNATING CURRENT: THE SINE WAVE

OBJECTIVES

1. Given a plot of a voltage or current versus time, be able to identify it as ac, dc, or other.

2. Be familiar with the terms period, frequency, angular frequency, amplitude, and phase angle and know what they mean with respect to an ac voltage.

3. Given one measure of the amplitude of an ac voltage or current, be able to calculate the other measures of its amplitude.

INTRODUCTION

Thus far, time has not been mentioned in these chapters on circuit analysis. This is equivalent to the assumption that the voltages and currents under consideration do not change; in particular, it implies that the situation has no beginning or end. Clearly, this is only an approximation.

In real circuits, the currents and voltages change as functions of time. In many cases (radios and computers, for instance), the variations are of main interest, whereas in other cases the variations may be so slow that they can be ignored. The nature of the voltages and currents in a circuit can be divided into three classes:

(DC or steady state.)   Cases where the voltages and currents are almost constant with respect to time and time is not considered.

(AC.)   Cases where the voltages and currents are purely sinusoidal in time; this is sometimes called sinusoidal steady state or ac steady state.

(Others.)   Cases where the voltages and currents vary in some more complex and less easily described manner than purely sinusodial.

This is the first of four chapters dealing with circuits involving voltages and currents that vary sinusoidally. This chapter deals with the nomenclature and basic properties of such voltages and currents. The subsequent three chapters deal with their circuit analysis. This present chapter is very close to being purely mathematical, but since it is written so much in terms of circuit analysis, it is included in the circuit analysis section of the book and not in the mathematical section.

Clearly, this division of current variation into three classes is artificial, its main purpose being pedagogical. Obviously, most interesting real circuits will fall into the class called "others." Also obviously, the most difficult class to analyze will be the "others." However, the rules and techniques learned to deal with the two simpler cases can be used to take on the more difficult case. Finally, most real circuits can be usefully approximated by one or the other of the simpler cases.

TERMINOLOGY

A current described by Equation 7-1 is plotted in Figure 7-1:

$$i(t) = I_p \sin(\omega t) \tag{7-1}$$

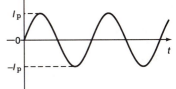

FIGURE 7-1 A sinusoidal current, or an ac current.

As can be seen in the figure, the current increases from 0 to I_p, decreases back to 0, reverses its direction, and continues to decrease to a value of I_p in the opposite direction, and so on. Because of these reversals, or alterations, of direction, this is called an **alternating current**, or **ac**.

Just to make the terminology appear a little foolish, a voltage that varies as

$$v(t) = V_p \sin(\omega t) \tag{7-2}$$

is called an ac voltage (an alternating current voltage!). Similarly, a constant voltage is called a dc voltage.

There is an important convention used in Equations 7-1 and 7-2 that will be used throughout this book in all sections where it is important. Quantities that vary in time, such as the current and voltage given by Equations 7-1 and 7-2, are designated by lowercase (small) letters; fixed quantities, such as the peak voltage V_p, are indicated by uppercase (capital) or Greek letters.

Unfortunately, the terminology in use today is often ambiguous. Here, an ac current is defined as one that is purely sinusoidal in time, whereas a dc current is one that does not change in time. However, some people define ac currents as those that change directions, and dc currents as those that do not change directions. In these terms the current shown in Figure 7-2a would be called dc and the current shown in

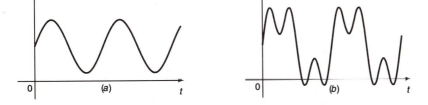

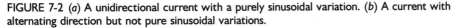

FIGURE 7-2 (a) A unidirectional current with a purely sinusoidal variation. (b) A current with alternating direction but not pure sinusoidal variations.

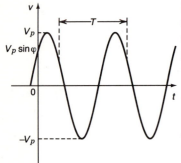

FIGURE 7-3 A plot of a sinusoidal voltage with various important features marked on the plot.

Figure 7-2*b* would be called ac. In the terminology used in this book, both of these cases would be called "other." When reading other material in the electronics field, usually the context will make clear the meaning of the names being used; if it does not, it is probably unimportant.

BASIC PROPERTIES OF THE SINE WAVE

Consider a voltage having a pure sinusoidal shape centered on zero:

$$v(t) = V_p \sin(\omega t + \phi) \tag{7-3}$$

This voltage is plotted in Figure 7-3. (Although all this is written about a voltage, everything said here would equally well apply to a current having a sinusoidal shape.)

The voltage waveform repeats itself after a period of time; this time is marked T in the figure. If the waveform were shifted either to the left or right by T seconds, it would not be possible to tell the difference. This time T, measured in seconds, is the **period** of the sine wave. The **frequency** of the sine wave, measured in **cycles per second**, or **Hertz (Hz)**, is the reciprocal of the period:

$$f = \frac{1}{T} \tag{7-4}$$

The frequency tells how many times per second the voltage goes through a complete cycle of values and returns to the same point in its cycle, whereas, the period tells how long one cycle takes. In the mathematical expression for the sine wave, the **angular frequency** ω, measured in radians, is used. This is given by

$$\omega = 2\pi f = \frac{2\pi}{T} \tag{7-5}$$

Remember neither radians nor cycles have dimensions; they are pure numbers. Thus cycles per second, radians per second, and Hertz are all fancy ways of saying "per second," and all have units of $\sec^{-1}$.

V_p is the **amplitude** of the sine wave described by Equation 7-1. Since the absolute value of the sine is always less than or equal to 1, $|\sin| \leq 1$, the extreme values of the voltage are $+V_p$ and $-V_p$ as shown in Figure 7-3.

The ϕ is the **phase** of the voltage. It is related to the value of the voltage at the time $t = 0$.

$$v(0) = V_p \sin\phi \tag{7-6}$$

In general, in ac circuit analysis, the time $t = 0$ is not well defined, so the phase of one voltage is arbitrary and usually set to zero and the phases of all other voltages and currents in the problem are then determined relative to that first voltage. Although the natural units for phases are radians, they are often expressed in degrees.

Consider two ac voltages both having the same amplitudes and frequencies but different phases, the first being

$$v_1 = V_p \sin(\omega t) \tag{7-7}$$

and the second being

$$v_2 = V_p \sin(\omega t + \phi) \tag{7-8}$$

If ϕ is positive, then the second voltage is said to **lead** the first by ϕ. If ϕ is negative, the second voltage is said to **lag** the first by ϕ. Because a complete cycle is 360° (degrees) (or 2π radians), phase shifts are usually expressed as being between −180 and 180° ($-\pi$ and π radians). Thus a voltage that leads another by 181° also can be said to lag the other by 179°.

To illustrate phase shifts, two voltages are plotted in Figure 7-4. The first is given by

$$v_1 = V_p \sin(\omega t)$$

and is plotted in a solid line. The second is given by

$$v_2 = V_p \sin\left(\omega t + \frac{\pi}{4}\right)$$

and is plotted in a dotted line. Thus the second voltage leads the first by $\frac{\pi}{4}$ radians.

The usual first reaction of a person when seeing this figure is "Wait a minute, you got it backwards! The solid line leads the dotted line because it is to the right of the dotted line."

But remember, this is a plot of the voltage versus time; the fact that the solid line is to the right of the dotted line means the solid line gets to a particular point in its cycle at a later time than does the dotted line. Thus the dotted line leads the solid line in time. There is really no contradiction. Nevertheless, the graph is somewhat confusing. Plots of this nature are frequently seen on the oscilloscope; to determine phase shifts correctly, avoid making this deceptively easy error.

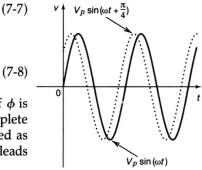

FIGURE 7-4 Two sine waves, the dotted wave leading the solid one.

MEASURES OF AMPLITUDE

When dealing with an algebraic expression for an ac voltage, the amplitude, V_p is the obvious choice as the most convenient parameter to describe how big the voltage is. However, when using meters and when dealing with various components in electronic circuits, there are a number of other measures of the amplitude of an ac voltage (or any voltage waveform) that are often useful. A few of the most common ones are discussed below.

The maximum potential between two successive extreme values is sometimes of interest and is called the **peak-to-peak voltage**. (This is also the easiest value to read from an oscilloscope trace.) For the sine wave, this is

$$V_{pp} = V_p - (-V_p) = 2V_p \tag{7-9}$$

For a sine wave, the **average** value is zero, since the sine function averaged over a full cycle is zero. However, the average of a sine wave over one half of a cycle or the average of the absolute value of a sine wave over a number of cycles is

$$V_{av} = \frac{1}{\pi} \int_0^{\pi} V_p \sin(\omega t)\, dt$$

$$= \frac{2V_p}{\pi} = 0.637 V_p \qquad (7\text{-}10)$$

It is frequently necessary to find the dc voltage that would generate the same amount of heat (power) in a resistor as does the ac voltage when averaged over one full cycle. Power is voltage times current.

$$P = vi = v\frac{v}{R} = \frac{v^2}{R}$$

To find the average power, it is necessary to average the square of the voltage over one cycle and then take the square root to find the equivalent dc voltage. Thus take the Root of the Mean of the Square, or the **rms** value.

$$V_{rms} = \left\{ \frac{1}{2\pi} \int_0^{2\pi} [V_p \sin(\omega t)]^2\, dt \right\}^{1/2}$$

$$= \frac{V_p}{\sqrt{2}} = 0.707 V_p \qquad (7\text{-}11)$$

THE COSINE WAVE

Thus far, all voltages and currents have been written in terms of the sine wave:

$$v = V_p \sin(\omega t + \phi) \qquad (7\text{-}12)$$

But this is also the same as the form

$$v = V_p \cos(\omega t + \theta) \qquad (7\text{-}13)$$

because the sine and cosine differ only by a phase factor. That is,

$$v = V_p \cos(\omega t) = V_p \sin\left(\omega t + \frac{\pi}{2}\right) \qquad (7\text{-}14)$$

TRIGONOMETRIC IDENTITIES

There are a number of trigonometric identities that are often useful when dealing with ac voltages and phase angles. These are

$$(\sin \alpha)^2 + (\cos \alpha)^2 = 1 \qquad\qquad (7\text{-}15)$$

$$\sin(\alpha \pm \beta) = \sin \alpha \cos \beta \pm \cos \alpha \sin \beta \qquad\qquad (7\text{-}16)$$

$$\cos(\alpha \pm \beta) = \cos \alpha \cos \beta \mp \sin \alpha \sin \beta \qquad\qquad (7\text{-}17)$$

$$\sin 2\alpha = 2 \sin \alpha \cos \alpha \qquad\qquad (7\text{-}18)$$

$$\cos 2\alpha = 2 \cos^2\alpha - 1 = 1 - 2 \sin^2\alpha \qquad\qquad (7\text{-}19)$$

COMPLEX NOTATION

In doing ac circuit analysis, ac voltages will sometimes be represented in the complex form

$$v = V_p e^{j(\omega t + \phi)} \qquad\qquad (7\text{-}20)$$

Complex numbers and this expression are considered in detail in Chapter 51.

AC AND THE WALL PLUG

As defined here, ac means pure sine waves having no dc offset. In practice, what comes out of the wall plugs is called ac, but it is generally not a pure sine wave and often has a dc offset (a nonzero average value). However, its direction does reverse. For reference, in the United States the voltage available at a normal wall plug is 60 Hz ac, with an rms amplitude in the range of 110 to 120 volts.

PROBLEMS

7-1 What is the period of a 60-Hz sine wave? Express your answer in milliseconds. (Since electrical power is distributed as 60 Hz ac, this is a number worth remembering.)

7-2 If the angular frequency of a voltage is 6280 radians per second, what are the frequency and the period of the voltage?

7-3 A sine wave has an amplitude of 25 V peak to peak. What is its amplitude? Its average value? The average of its absolute value? Its rms value?

7-4 The nominal 120 V ac that appears at a wall plug has an rms amplitude of 120 V. Assuming it is a pure sine wave, what is its amplitude? What is its peak-to-peak value?

ANSWERS TO ODD-NUMBERED PROBLEMS

7-1 16.7 msec

7-3 12.5 V; average = 0; V_{av} = 7.96 V; V_{rms} = 8.84 V

Chapter 8

ELEMENTS OF AC CIRCUIT ANALYSIS

OBJECTIVES

1. Be familiar with capacitors and inductors.

2. Know the voltage–current relationships for resistors, capacitors, and inductors.

3. Know the definition of impedance.

4. Be able to calculate the impedance of inductors, resistors, capacitors, and combinations of these elements.

5. Have a knowledge of the circuit analysis laws and methods that can be used in ac circuit analysis.

INTRODUCTION

This is the first of two chapters devoted to ac circuit analysis—the analysis of circuits involving pure sinusoidal voltages and currents. There are two common circuit elements that have not been discussed in the previous chapters; these are the capacitor and the inductor. The basic properties and the voltage–current relationships for these two circuit elements are described in the next two sections. Then, the concept of impedance is introduced and all the laws and methods of circuit analysis discussed thus far are generalized to cover ac circuit analysis. In the next chapter, several examples are worked out in detail.

CAPACITORS

In general, any two conductors insulated from each other constitute a **capacitor**. The prototype of all capacitors is two flat parallel conductors separated

60

by a thin dielectric (insulator); a sketch of such a capacitor is shown in Figure 8-1. An older name for a capacitor is **condenser**. This name is rapidly fading out of use, but it still occurs occasionally.

Capacitors remain neutral, as do all other electronic circuit elements. Hence if some positive charge is placed on one plate, a negative charge must also be placed on the other plate, as shown in Figure 8-1. The usual way of thinking is that if an amount of positive charge Q is placed on the upper plate, then an equal amount of positive charge must leave the lower plate, with a net charge of $-Q$ remaining on the lower plate and with no net charge on the whole capacitor. Thus the capacitor remains neutral, although the individual plates are not neutral.

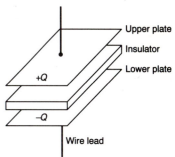

FIGURE 8-1 A parallel-plate capacitor.

In reading electronic literature statements about "capacitors storing charge" are frequently encountered. These statements are referring to the fact that separated charges are stored on each plate; that is, $+Q$ is on one plate while $-Q$ is on the other. Thus when viewed one plate at a time, capacitors store charge, but when taken as a whole, they are neutral.

Because the plates are not neutral, there is an electric field between the plates. Because of this field, there is a potenetial difference between the two plates. It is found that the potential difference between the two plates is proportional to the charge stored on the plates. The constant of proportionality is the **capacitance C** of the capacitor. Commonly, the word **capacity** is used in place of the more proper word **capacitance**. This definition can be written as

$$C = \frac{Q}{V} \quad \text{or} \quad Q = CV \tag{8-1}$$

Capacitance is measured in units of **farads** (F):

$$1 \text{ farad} = 1 \text{ coulomb}/1 \text{ volt}$$

The farad is a very large amount of capacity, so common units are

microfarad	μF	10^{-6} F
nanofarad	nF	10^{-9} F
picofarad	pF	10^{-12} F

It can be shown that the capacity of a parallel-plate capacitor is given by

$$C = \varepsilon \frac{A}{d} \tag{8-2}$$

where A is the area of one of the plates, d is the separation between the plates, and ε is the dielectric constant of the insulator between the plates. This relationship is quite general: the capacity of a capacitor depends on the size of the plates and inversely on their separation. To a good approximation, the essential properties of a capacitor are independent of its shape. Although the actual capacitors used in circuits have many shapes, the flat plate prototype may be used in all examples. Details of practical capacitor construction are found in Chapter 48.

Differentiating the definition of capacity yields

$$\frac{dQ}{dt} = i = \frac{dC}{dt}V + C\frac{dV}{dt}$$

Assuming that the size and shape of the capacitor do not change in time, that is, its capacity is fixed, this becomes

$$i = C\frac{dV}{dt} \tag{8-3}$$

This expression can be integrated to yield

$$v = \frac{1}{C}\int i\,dt \tag{8-4}$$

Equations 8-3 and 8-4 are the voltage–current characteristics for a capacitor; they also show that the voltage across a capacitor cannot change instantaneously.

The schematic symbols for capacitors reflect the flat plate prototype and is given in Figure 8-2.

Fixed value   Variable value

FIGURE 8-2 The schematic symbols for capacitors.

INDUCTORS

A wire carrying a current generates a magnetic field in the space around the wire. As the current varies, the magnetic field varies, and the varying magnetic field induces a voltage in the wire that opposes the original changes in the current. (This follows from Faraday's law.) Under suitable conditions, this voltage is proportional to the change in the current. This defines the self-inductance or, simply, the **inductance L**, of the piece of wire:

$$v(t) = L\frac{di}{dt} \tag{8-5}$$

where v is the voltage caused by the change in the current. This definition of the inductance is also the voltage–current relationship for an inductor. It can be integrated to give

$$i(t) = \frac{1}{L}\int v(t)dt \tag{8-6}$$

Equations 8-5 and 8-6 are the voltage–current relationships for an inductor; they also show that the current through an inductor cannot change instantaneously—note the symmetrical relationships with capacitors:

Inductance is measured in units of **henrys (H)**.

1 henry = 1 volt/(1 ampere)(second)

Common units of inductance are

millihenry	mH	10^{-3} H
microhenry	μH	10^{-6} H
picohenry	pH	10^{-12} H

Commonly used inductances range from a few henries to a few picohenries.

Although any wire used in a circuit will have an inductance, the inductance will be very small and will not be very important except when one is working at very high frequencies. In practice, to get larger values of inductance, it is necessary to arrange the wire in a coil of many turns. Sometimes, the turns are arranged around an iron (or other magnetic material) core to further increase the inductance. Data about practical inductors are found in Chapter 49.

The schematic symbols for inductors are shown in Figure 8-3.

Finally, it is easy to see why capacitors and inductors were not mentioned in the sections on dc or steady state circuit analysis. From the voltage–current relationships for these elements, it can be seen that a capacitor looks like an open circuit (no current) and an inductor looks like a short circuit (a wire) in steady state circuits.

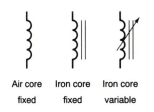

| Air core | Iron core | Iron core |
| fixed | fixed | variable |

FIGURE 8-3 The schematic symbols for inductors.

V–I CHARACTERISTICS: A SUMMARY

The voltage–current relationships for the basic components are summarized in the two tables that follow. In the first table, for each component, the voltage–current relationship given above is restated; then, if a sinusoidal current is passed through the element, the corresponding voltage across the element is given. For instance, if a current of

$$i = I_m \sin(\omega t)$$

is flowing through a capacitor, the voltage across the capacitor is given by

$$v = \frac{1}{C} \int i \, dt$$

$$= \frac{1}{C} \int I_m \sin(\omega t) dt$$

$$= -\frac{I_m}{\omega C} \cos(\omega t)$$

In a similar manner, all of the other elements in Tables 8-1 and 8-2 are calculated.

TABLE 8-1 Calculation of Voltage When the Current is Given

Component	General i	$i = I_m \sin(\omega t)$
Resistor	$v = iR$	$v = RI_m \sin(\omega t)$
Inductor	$v = L\dfrac{di}{dt}$	$v = L\omega I_m \cos(\omega t)$
Capacitor	$v = \dfrac{1}{C} \int i \, dt$	$v = -\dfrac{I_m}{\omega C} \cos(\omega t)$

TABLE 8-2 Calculation of Current When the Voltage Is Given

Component	General v	$v = V_m \sin(\omega t)$
Resistor	$i = \dfrac{v}{R}$	$i = \dfrac{V_m}{R} \sin(\omega t)$
Inductor	$i = \dfrac{1}{L} \displaystyle\int v\,dt$	$i = -\dfrac{V_m}{\omega L} \cos(\omega t)$
Capacitor	$i = C\dfrac{dv}{dt}$	$i = \omega C V_m \cos(\omega t)$

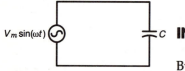

$V_m \sin(\omega t)$

FIGURE 8-4 A simple ac circuit.

IMPEDANCE

By applying some of the results in Table 8-2 to the simple circuit shown in Figure 8-4, it is possible to write down the expression for the current in this simple circuit:

$$i = \omega C V_m \cos(\omega t) = \omega C V_m \sin(\omega t + 90°) \tag{8-7}$$

Thus for this circuit, the magnitude of the current is ωC times the magnitude of the voltage, and the current leads the voltage by 90° ($\pi/2$ radians).

One of the features that made dc circuit analysis relatively easy is the fact that a circuit element could be described by one number, namely, the value of its resistance. Then, given Kirchhoff's laws and various techniques, it is possible to solve most circuit equations in a rather pedestrian manner. It would be nice to be able to deal with ac circuit analysis in a similar manner.

Resistance was defined as $R = V/I$. Proceeding in a naive fashion, why not try to use a similar definition in the case of an ac circuit? Trying this for the circuit shown in Figure 8-4 and using the results in Equation 8-7 gives

$$??? = \frac{v}{i} = \frac{V_m \sin(\omega t)}{\omega C V_m \cos(\omega t)} = \frac{1}{\omega C} \tan \omega \tag{8-8}$$

This quantity varies from minus infinity to plus infinity, taking on all values in between. It doesn't seem like a very good description of anything about the circuit. This should come as no surprise, since this was an attempt to use one number to describe two pieces of information—the relationship between the magnitudes of the voltage and current and the phase difference between these two. Thus Equation 8-8 is an attempt to describe a two-dimensional situation with a one-dimensional number; such attempts are doomed to failure.

Complex numbers provide a natural way of expressing the relationships between currents and voltages in ac circuits by a single expression. Complex numbers are most easily described by a magnitude and a phase angle, just what is needed here. (If you are not familiar with complex numbers, go to Chapter 51 at this point).

The **impedance Z** of a circuit element is defined as the complex ratio

$$Z = \frac{v}{i} \tag{8-9}$$

where the ac voltage across and current through the circuit element are described by complex numbers. Impedance has the same dimensions as resistance, voltage over current, and is measured in the same units, ohms. The reciprocal of an impedance is called the **admittance**.

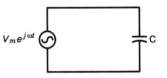

Consider again the simple circuit consisting of an ac generator and a single capacitor shown in Figure 8-5. This time, the voltage generator is described by a complex sinusoidal generator. Using the voltage–current relationship for the capacitor gives

FIGURE 8-5 The simple example of Figure 8-4 with the voltage source shown as a complex voltage generator.

$$i = C\frac{dv}{dt} = CV_m j\omega e^{j\omega t} = j\omega Cv \tag{8-10}$$

Hence, the impedance Z of the capacitor C is

$$Z_c = \frac{v}{i} = \frac{v}{j\omega Cv} = \frac{1}{j\omega C} = -\frac{j}{\omega C} \tag{8-11}$$

where j is the square root of minus 1 (see Chapter 51).

In the same way, it can easily be shown that the impedance Z of an inductor L is

$$Z_L = j\omega L \tag{8-12}$$

and the impedance Z of a resistor R is simply

$$Z_R = R \tag{8-13}$$

Equation 8-10 shows that the current through the capacitor is $j\omega C$ times the applied voltage. Multiplying by j corresponds to rotating by 90° in the complex plane. Thus Equation 8-10 says that the current leads the voltage by 90°, which is just what Equation 8-7 says. From a different point of view, recall that $v = iZ$; the $-j$ in the impedance of the capacitor means that the voltage lags the current by 90° in a capacitor. Likewise, the j in the impedances of an inductor means that the voltage leads the current by 90° in an inductor, and the real impedance for a resistor means that the voltage and current are in phase in a resistor. The phase relationships are illustrated in Figure 8-6.

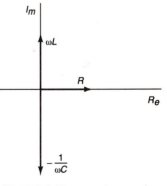

FIGURE 8-6 The impedances of an inductor, a resistor, and a capacitor plotted in the complex plane to show their phase relationships.

GENERALIZATION TIME

It is possible to simply continue on and duplicate everything that has been done in the previous chapters, only this time using complex voltages and currents and using impedances rather than resistances. But this would be a giant waste of time. The previous chapters were written in terms of dc currents and resistances; but they could have equally well have been about ac currents and impedances. Go back and reread the previous chapters, substituting the word impedance for resistance and thinking about both ac and dc. But first, a few general comments are due.

The basic laws of circuit analysis, Kirchhoff's voltage and current laws, apply to all circuits, whether or not the voltages and currents are changing. The variations in the voltages and currents need not be sinusoidal. This is often made more explicit by adding the phrase "at any instant of time" to the laws. Thus, for example, KCL becomes: **At any instant of time, the sum of all the currents flowing into a node must be zero.** Kirchhoff's laws also apply to circuits when the voltages and currents are written in complex form.

Although the basic techniques for applying these laws to circuits do not change, the mathematics gets more complicated. Rather than simple systems of linear equations, systems of complex linear equations or systems of coupled linear differential equations may result.

A hedge is required in the statements above involving the speed of light. As long as the time required for information about any changes in the circuit to reach all parts of the circuit is small compared to the time scale being considered, then the circuit laws apply as stated. This is generally expressed in terms of the wavelengths considered being large compared with the physical dimensions of the circuit. A common situation where such complexities must be taken into account are circuits that operate at very high frequencies such as microwave frequencies. Another interesting example is the power grid of Canada and the U.S.—this is not small compared with the wavelength of 60 Hz.

All the theorems and methods used thus far remain unchanged and equally true and useful when the word resistance replaced by the word impedance. For instance, Thevenin's theorem states that any complex network of sources and impedances having two terminals can be replaced by a single equivalent voltage source connected in series with a single impedance. Of course, in easy cases, the series impedance will be purely resistive.

The concept of impedance implies that the voltages and currents are purely sinusoidal, that is, of a single fixed frequency. If, for instance, a triangular voltage is being applied to a circuit, then impedance is not a useful concept; however, KVL and KCL still apply and can be used. It is because of this kind of complexity that electrical engineers spend years studying circuit analysis.

To connect this material to what was done before, it is only necessary to note that the steady state or dc case is really a special case of ac circuit analysis, the case where the frequency is zero ($f = \omega = 0$). In this instance, the impedance of a capacitor is infinite (∞) and the impedance of an inductor is 0. Furthermore, in this limiting case, complex numbers are not needed; for if there is a capacity in a branch of a circuit, then there is no current in that branch and the branch can be removed. Thus the only impedances left are resistances that are real numbers.

TWO SIMPLE EXAMPLES

As an example of this generalization, consider the voltage divider shown in Figure 8-7. The relationship between the input and output is given by

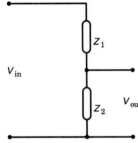

V_{in}

Z_1

Z_2 V_{out}

FIGURE 8-7 A general voltage divider.

$$v_{out} = v_{in}\frac{Z_2}{Z_1 + Z_2} \tag{8-14}$$

The equivalent impedance of several components in series is the sum of the individual impedances. Thus, for the situation shown in Figure 8-8,

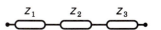

FIGURE 8-8 Three impedances in series.

$$Z_{eq} = Z_1 + Z_2 + Z_3 \qquad (8\text{-}15)$$

The proof of this equation runs as follows:

1. The same current i flows through each element, since they are in series.
2. The total voltage drop across all the components is the sum of the voltage drops across each component. Thus

$$v_T = v_1 + v_2 + v_3$$

3. The voltage drop across each component is given by the impedance Z times the current (really the definition of the impedance):

$$v_1 = iZ_1, \text{ etc.}$$

4. Finally, combining these statements yields

$$v_T = iZ_1 + iZ_2 + iZ_3$$

$$Z_{eq} = \frac{v_T}{i} = Z_1 + Z_2 + Z_3$$

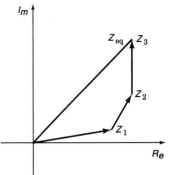

A graphic calculation of the equivalent impedance of three impedances is shown in Figure 8-9. This Figure makes clear that the equivalent impedance differs both in magnitude and in phase from any of the original impedances.

In a similar way, the rule for finding the equivalent impedance of a set of impedances in parallel can be found. Working backward from these rules or directly from the circuit laws, the rules for finding the equivalent capacitance of capacitors in series or parallel can also be found.

FIGURE 8-9 A graphic calculation of the equivalent impedance for three impedances in series. The construction shows both the magnitudes and phases of the various impedances.

POWER

When a sinusoidal voltage is applied to a circuit, on the average, no power is dissipated in the capacitors or inductors. This is not an obvious fact, but it is very important in situations where conservation of power is important. This is due to the 90° phase shift between the voltage and the current. It can be seen for the simple capacitive circuit that has been used throughout this chapter. Remember, the instantaneous power is defined as

$$p = vi$$

The average power is found by averaging this over one full cycle. Thus

$$P_{av} = \int_0^{2\pi} vi \, dt$$

$$= \int_0^{2\pi} [V_m \sin(\omega t)][\omega C V_m \cos(\omega t)] \, dt$$

$$= 0$$

REACTANCE

There is a much less mathematical way of doing ac circuit analysis. This method uses prepared tables, charts, and the concept of reactance. **Reactance** X is defined as the magnitude of the impedance. Thus

$$X_c = \frac{1}{\omega C} = \frac{1}{2\pi f C} \tag{8-16}$$

$$X_L = \omega L = 2\pi f L \tag{8-17}$$

$$X_R = R \tag{8-18}$$

The reactance of a resistor, a capacitor, and an inductor in series is given by

$$X_T = [X_R^2 + (X_C - X_L)^2]^{1/2} \tag{8-19}$$

To find out how much can be done with this nonmathematical formulation of ac circuit analysis, refer to *The Radio Amateur's Handbook* published by the American Radio Relay League.

SUMMARY

In Tables 8-3 and 8-4, essentially all the information needed about the voltage–current relationships and impedances is summarized.

TABLE 8-3 Component *V–I* Relationships

Element	R	L	C
$v(t) =$	iR	$L\dfrac{di}{dt}$	$\dfrac{1}{C}\displaystyle\int i\,dt$
$i(t) =$	$\dfrac{v}{R}$	$\dfrac{1}{L}\displaystyle\int v\,dt$	$C\dfrac{dv}{dt}$
		If $i = I_m \sin(\omega t)$, then	
$v(t) =$	$RI_m \sin(\omega t)$	$\omega L I_m \cos(\omega t)$	$-\dfrac{I_m}{\omega C}\cos(\omega t)$
		If $v = V_m \sin(\omega t)$, then	
$i(t) =$	$\dfrac{V_m}{R}\sin(\omega t)$	$-\dfrac{V_m}{\omega L}\cos(\omega t)$	$\omega C V_m \sin(\omega t)$
X (Reactance)	R	ωL	$\dfrac{1}{\omega C}$
Phase shift	0	Current lags by 90°	Current leads by 90°
Z	R	$j\omega L$	$\dfrac{-j}{\omega C}$

TABLE 8-4 Equivalent Components

Element	Parallel	Series
Z	$\dfrac{1}{Z_{eq}} = \dfrac{1}{Z_1} + \dfrac{1}{Z_2} + \cdots$	$Z_{eq} = Z_1 + Z_2 + \cdots$
R	$\dfrac{1}{R_{eq}} = \dfrac{1}{R_1} + \dfrac{1}{R_2} + \cdots$	$R_{eq} = R_1 + R_2 + \cdots$
L	$\dfrac{1}{L_{eq}} = \dfrac{1}{L_1} + \dfrac{1}{L_2} + \cdots$	$L_{eq} = L_1 + L_2 + \cdots$
C	$C_{eq} = C_1 + C_2 + \cdots$	$\dfrac{1}{C_{eq}} = \dfrac{1}{C_1} + \dfrac{1}{C_2} + \cdots$

FIGURE 8-10 The circuit for Problem 8-3.

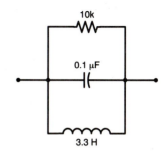

FIGURE 8-11 The circuit for Problem 8-4.

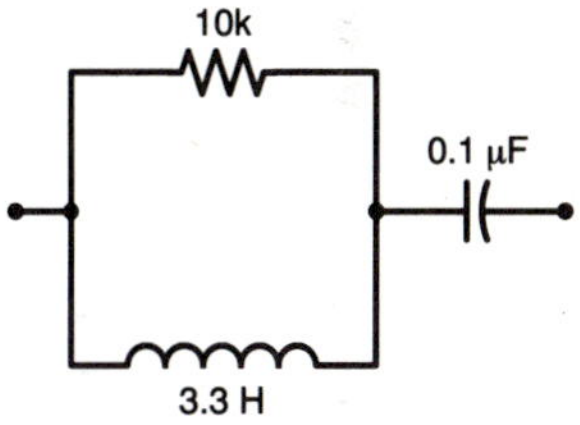

PROBLEMS

8-1 Calculate the impedance of (a) 0.1 μF at 100 and 1000 Hz; (b) 4.0 μF at 100 and 1000 Hz; (c) 1 mH at 100 and 1000 Hz.

8-2 Calculate the impedance of (a) 33 nF at 25k and 50k Hz; (b) 6 mH at 100 and 1000 Hz.

8-3 Find the impedance of the circuit shown in Figure 8-10 for a frequency of 440 Hz.

8-4 Find the impedance of the circuit shown in Figure 8-11 for a frequency of 440 Hz.

8-5 Find the impedances of the circuit shown in Figure 8-12 at a frequency of 440 Hz.

FIGURE 8-12 The circuit for Problem 8-5.

ANSWERS TO ODD-NUMBERED PROBLEMS

8-1 (a) $-15.9j$ kiloohms at 100 Hz, $-1592j$ ohms at 1000 Hz; (b) $-392j$ ohms at 100 Hz, $-39.8j$ ohms at 1000 Hz; (c) $0.628j$ ohms at 100 Hz, $6.28j$ ohms at 1000 Hz

8-3 $10,000 + 5,506j$ ohms

8-5 $1157 - 419j$ ohms

Chapter 9

EXAMPLES OF SIMPLE AC CIRCUIT ANALYSIS

OBJECTIVES

1. Be familiar with some of the methods that can be used in ac circuit analysis.

2. Be familiar with the performance of the high-pass and low-pass filters—the simplest ac voltage dividers.

3. Be able to analyze simple ac circuits.

INTRODUCTION

This chapter contains several examples of ac circuit analysis together with a discussion of the results of the analysis. First, a simple voltage divider circuit will be analyzed three different ways, and the nature of the solution will be discussed in some detail. This discussion should be of interest even to those readers who do not want to follow the details of the circuit analysis. Next, a simple parallel combination of impedances will be analyzed and used in a simple voltage divider circuit. Finally, a somewhat more complex ac circuit will be partially analyzed by means of mesh equations.

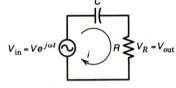

FIGURE 9-1 A high-pass filter circuit.

A SIMPLE VOLTAGE DIVIDER

The circuit shown in Figure 9-1 is a simple ac voltage divider. This circuit is also called a **high-pass filter**. The goal of this section is to find the expressions for the voltages drops and currents in this circuit. In the next section, the nature of these expressions will be discussed.

Without analyzing the circuit formally, it is possible to learn a great deal by simply inspecting the circuit. At low frequencies, the capacitor will have a very high impedance (tending to infinity as the frequency tends to zero); thus, at very low frequencies, the output voltage will be a small fraction of the input voltage. At high frequencies, the impedance of the capacitor will be small

70

(tending to zero as the frequency becomes very large); thus at high frequencies, the output voltage will be essentially equal to the input voltage.

From these two observations, it is possible to draw a very rough sketch of the magnitude of the ratio V_{out}/V_{in}. This is shown in Figure 9-2. The purpose of actually doing the circuit analysis is to be able to generate the exact shape of this curve.

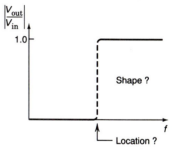

FIGURE 9-2 A rough sketch of the relative output of the high-pass filter obtained by qualitative analysis.

The circuit in Figure 9-1 is an ac voltage divider. In the previous chapter, the expression for the output voltage of this circuit was written down as an example of the generalization of dc circuit analysis to the ac case. The result was Equation 8-14; using it, the output voltage is

$$V_{out} = V_{in}\frac{Z_R}{Z_R + Z_C}$$

$$= V_{in}\frac{R}{R - (j/\omega C)} \tag{9-1}$$

If the circuit is not recognized as a voltage divider, the easiest way to proceed is to use mesh equations. Since this is a one-loop circuit, the process is relatively easy. The loop current is already drawn in Figure 9-1; going around the loop in the same direction as the current gives the following results:

$$V_{in} - iZ_C - iZ_R = 0$$

$$i = \frac{V_{in}}{Z_R + Z_C} = \frac{V_{in}}{R - (j/\omega C)} \tag{9-2}$$

The output voltage is given by

$$V_{out} = iZ_R$$

$$= V_{in}\frac{R}{R - (j/\omega C)} \tag{9-3}$$

which is the same as the previous result, Equation 9-1.

As a final step in the analysis of this circuit, with Equation 9-2 and the definition of the impedance, the equivalent impedance of this circuit can be found as follows:

$$Z_{eq} = \frac{v}{i} = \frac{V_{in}}{\dfrac{V_{in}}{R - (j/\omega C)}} = R - \frac{j}{\omega C} \tag{9-4}$$

which is exactly what would have been obtained by a direct application of Equation 8-15.

A DISCUSSION OF THESE RESULTS

The **gain** of any circuit is the ratio of the output voltage to the input voltage. Thus for the voltage divider, the gain G is

$$G = \frac{V_{out}}{V_{in}} = \frac{R}{R - (j/\omega C)}$$

This should be written in rationalized form, so

$$G = \frac{R}{R - (j/\omega C)} \frac{R + (j/\omega C)}{R + (j/\omega C)}$$

$$= \frac{R^2 + (jR/\omega C)}{R^2 + [1/(\omega^2 C^2)]} \qquad (9\text{-}5)$$

It is traditional to consider the gain in polar form, that is, to consider its magnitude and its phase angle. The magnitude of the gain can be worked out to be

$$|G| = \frac{R}{\{R^2 + [1/(\omega^2 C^2)]\}^{1/2}} \qquad (9\text{-}6)$$

and the phase angle is

$$\phi = \tan^{-1} \frac{1}{R\omega C} \qquad (9\text{-}7)$$

These results look rather formidable. However, by considering some limiting cases, it is possible to gain some insight into their behavior.

At very low frequencies,

$$\omega C \approx 0 \qquad \frac{1}{\omega C} \gg R \qquad \frac{1}{R\omega C} \gg 1$$

so the gain is $|G| \approx 0$, and the phase shift is

$$\phi = \tan^{-1} \infty = 90° = \frac{\pi}{2} \text{ radians}$$

At very high frequencies,

$$\frac{1}{\omega C} \approx 0 \qquad \frac{1}{R\omega C} \approx 0$$

so the gain is $|G| \approx 1$, and the phase shift is

$$\phi = \tan^{-1} 0 = 0$$

Note that these results are just what was obtained by inspection at the very beginning of this chapter.

The frequency where $\omega C = 1/R$ is the frequency at which the gain changes most rapidly with frequency. This is the frequency at which one half the power input to the circuit is available at the output. It is called the **half-power frequency**, or the **half-power point**, or the **3 db frequency** (see Chapter 53 for details about decibels). At the half-power point, the gain G is

$$|G| = \frac{R}{(R^2 + R^2)^{1/2}} = \frac{1}{\sqrt{2}} = -3\ dB$$

and the phase shift is

$$\phi = \tan^{-1}\frac{R}{R} = 45°$$

From these results, it is possible to sketch more accurate graphs of the magnitude of the gain and the phase shift versus frequency. Rather than do this, however, we present careful plots of the gain and phase shift versus frequency in Figure 9-3. The gain is plotted versus frequency on both a linear and a loga-

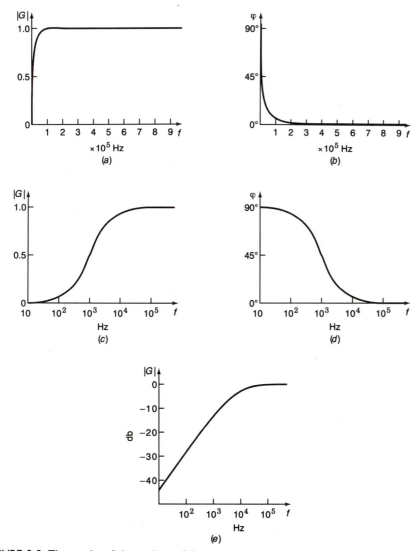

FIGURE 9-3 The results of the analysis of the circuit in Figure 9-1 for the case where $R = 1000\ \Omega$ and $C = 0.01\ \mu$F. (a) The gain plotted versus a linear frequency scale. (b) The phase shift versus a linear frequency scale. (c) The gain plotted versus a logarithmic frequency scale. (d) The phase angle plotted versus a logarithmic frequency scale. (e) The gain (in db) plotted versus a logarithmic frequency scale (Bode plot).

rithmic scale. The gain, expressed in decibels, is also plotted versus a logarithmic scale. Likewise, the phase shift is plotted versus a linear and logarithmic scale. It is easy to see that the most useful graphs have the phase shift and the gain measured in decibels plotted versus the log of the frequency. Only in this way can the full details of all the changes be presented on one graph. The plot of the log of the gain (the gain in db) versus the log of the frequency is often called a **Bode plot**. (These plots were generated for R = 1000 ohms and C = 0.1 μF.) Finally, it should be clear why this circuit is called a high-pass filter and what the plots for a low-pass filter will look like.

A PARALLEL COMBINATION OF IMPEDANCES

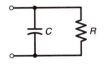

FIGURE 9-4 a simple parallel circuit.

The simple ac voltage divider—two impedances in series, the circuit discussed in the previous two sections—is certainly one of the most common simple ac circuits. A second very common configuration is two impedances in parallel, as shown in Figure 9-4. For such a simple circuit, little can be done except calculate the impedance of the parallel combination. This can be done by a direct application of the rules given in the tables at the end of Chapter 8. Thus

$$\frac{1}{Z_{eq}} = \frac{1}{Z_R} + \frac{1}{Z_C}$$

$$= \frac{1}{R} + \frac{1}{-j/\omega C} \tag{9-8}$$

Solving this to get Z_{eq} in rationalized form gives

$$Z_{eq} = \frac{R}{1 + j\omega CR} = \frac{R - j\omega CR^2}{1 + \omega^2 C^2 R^2} \tag{9-9}$$

Although Equation 9-9 may not look very simple, consideration of two extreme cases will make its nature clear. At low frequencies, that is, as ω tends to 0, the terms involving ω in both the numerator and denominator can be ignored. In this case, Equation 9-9 simplifies to

$$Z_{eq} = R \quad \text{(for low frequencies)} \tag{9-10}$$

At high frequencies, that is, as ω becomes very large, the term involving only R in the numerator and 1 in the denominator can be ignored, and in this case Equation 9-9 simplifies to

$$Z_{eq} = \frac{-j\omega C^2}{\omega^2 C^2 R^2}$$

$$= \frac{-j}{\omega C} \quad \text{(for high frequencies)} \tag{9-11}$$

Thus, at low frequencies, this combination acts like a pure resistor; at high frequencies, it acts like a pure capacitor. Only at intermediate frequencies does this circuit look like some combination of both a resistor and a capacitor.

(It should be noted that these qualitative comments could have been obtained by directly considering the results of putting one large and one small imped- ance in parallel—the smaller element always dominates the combination.) Ob- viously, similar expressions can be obtained for different parallel combinations of impedances.

A MORE COMPLEX VOLTAGE DIVIDER

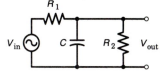

FIGURE 9-5 The parallel circuit of Figure 9-4 included in a voltage divider.

The next obvious step in complexity is to combine the two circuits discussed thus far. Figure 9-5 shows one possible combination. This is a voltage divider consisting of a resistor in series with a parallel combination of a resistor and a capacitor. The expression for the output can be obtained by combining the voltage divider, Equation 8-14, with Equation 9-9. After some algebra the re- sult for the output voltage is

$$V_{out} = V_{in} \frac{R_2/(1 + j\omega CR_2)}{R_1 + [R_2/(1 + j\omega CR_2)]}$$
$$= V_{in} \frac{R_1 R_2 + R_2^2 - j\omega C R_1 R_2^2}{R_1^2 + 2R_1 R_2 + R_2^2 + \omega^2 C^2 R_1^2 R_2^2} \quad (9\text{-}12)$$

Obviously, this result is sufficiently complicated that very little can be said without extensive numerical evaluations of the expression for specific values. However, by combining the quantitative discussions of the previous sections, the overall nature of the response of this circuit can be determined. At very low frequencies, the parallel combination of the resistor and the capacitor acts like a resistor. Thus, for low frequencies, this circuit reduces to the circuit shown in Figure 9-6, a resistive voltage divider. The phase shift is 0°, and the output is given by the voltage divider equation for resistors,

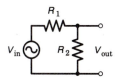

FIGURE 9-6 The equivalent to the circuit of Figure 9-5 at low frequencies.

$$V_{out} = V_{in} \frac{R_2}{R_1 + R_2} \quad (9\text{-}13)$$

At very high frequencies, the parallel combination of the resistor and the capacitor acts like a capacitor and the circuit reduces to that shown in Fig- ure 9-7. This is clearly related to the circuit discussed in the section on the Simple Voltage Divider above and is a low-pass filter—see Problem 9-1 for the details.

Thus simple qualitative arguments can be used to determine the overall nature of the frequency response of this circuit. To obtain detailed plots of the response would require picking specific values for the components and doing extensive numerical calculations. Even so, not much further informa- tion would be gained for the effort.

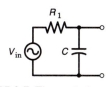

FIGURE 9-7 The equivalent to the circuit of Figure 9-5 at high frequencies.

A MORE DIFFICULT EXAMPLE

The following example is solved by using mesh equations. The purpose is simply to demonstrate that the procedures used in ac circuit analysis really

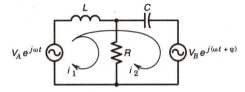

FIGURE 9-8 Another circuit to be analyzed.

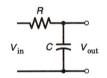

FIGURE 9-9 The circuit for Problem 9-1.

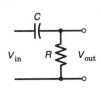

FIGURE 9-10 The circuit for Problem 9-2.

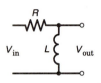

FIGURE 9-11 The circuit for Problem 9-3.

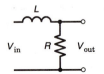

FIGURE 9-12 The circuit for Problem 9-4.

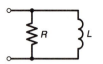

FIGURE 9-13 The circuit for Problem 9-5.

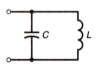

FIGURE 9-14 The circuit for Problem 9-6.

are no different than those used in dc circuit analysis—only the results are more complex. The problem is to find the current in the resistor in Figure 9-8.

The mesh currents have been drawn to minimize the work required. The mesh equations are

$$V_A e^{j\omega t} - i_1(Z_L + Z_R) - i_2 Z_1 = 0$$

$$V_A e^{j\omega t} - V_B e^{j(\omega t + \phi)} - i_1 Z_L - i_2(Z_1 + Z_C) = 0$$

Rearranging these yields

$$i_1(Z_L + Z_R) + i_2 Z_L = V_A e^{j\omega t}$$

$$i_1 Z_L + i_2(Z_L + Z_C) = V_a e^{j\omega t} - V_B e^{j(\omega t + \phi)}$$

$$i_1 = \frac{\begin{vmatrix} V_A e^{j\omega t} & Z_L \\ V_A e^{j\omega t} - V_B e^{j(\omega t + \phi)} & Z_L + Z_C \end{vmatrix}}{\begin{vmatrix} Z_L + Z_R & Z_L \\ Z_L & Z_L + Z_C \end{vmatrix}}$$

$$= \frac{V_A(Z_L + Z_C) - [V_A e^{j\omega t} - V_B e^{j(\omega t + \phi)}]}{Z_L^2 + Z_R Z_C + Z_L Z_C + Z_R Z_C - Z_C^2} \qquad (9\text{-}14)$$

Equation 9-14 is the desired solution. Clearly, it is very complex. Since this was merely a practice circuit, there is no point in continuing the analysis.

PROBLEMS

Calculate the magnitude of the gain and the phase shift as a function of frequency, and sketch the graphs of these for the following four circuits:

9-1 The circuit in Figure 9-9.

9-2 The circuit in Figure 9-10.

9-3 The circuit in Figure 9-11.

9-4 The circuit in Figure 9-12.

9-5 Find the impedance of the parallel combination of elements given in Figure 9-13. Discuss your results.

9-6 Find the impedance of the parallel combination of elements given in Figure 9-14. Is the result suprising?

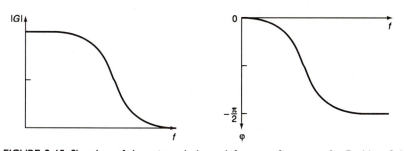

FIGURE 9-15 Sketches of the gain and phase shift versus frequency for Problem 9-1.

ANSWERS TO ODD-NUMBERED PROBLEMS

9-1

$$|G| = \frac{1/\omega C}{\{R^2 + [1/(\omega^2 C^2)]\}^{1/2}}$$

$$\phi = \tan^{-1}(-R\omega C)$$

This is a low-pass filter. Its 3 db point is at $R\omega C = 1$. The graph is shown in Figure 9-15.

9-3

$$|G| = \frac{\omega L}{(R^2 + \omega^2 L^2)^{1/2}}$$

$$\phi = \tan^{-1} \frac{R}{\omega L}$$

This is a high-pass filter. The 3 db point is at $R = \omega L$, and the plots are the same as those in Figure 9-3.

9-5

$$Z_{eq} = \frac{\omega^2 L^2 R + j\omega L R^2}{R^2 + \omega^2 L^2}$$

Chapter 10

RESONANCE

OBJECTIVES

1. Be familiar with the phenomenon of resonance.

2. Be able to calculate the resonant frequency of a simple series resonant circuit.

3. Know some of the conditions under which resonance occurs.

INTRODUCTION

One dramatic feature of ac circuits is resonance. A simple example of a resonant circuit will be analyzed and discussed. The chapter ends with a discussion of several situations where resonance occurs and how it can be used.

The primary purpose of this chapter is to present information about the phenomenon of resonance and the terminology used to describe it, without the need for new skills in ac circuit analysis. As a result, the details of the analysis are omitted, and no problems are given at the end of the chapter.

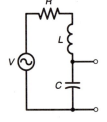

FIGURE 10-1 An *RCL* circuit as example of a series resonant circuit. The amplitude of the sinusoidal generator is *V*.

THE *RCL* CIRCUIT

The *RCL* circuit shown in Figure 10-1 is an example of a **series resonant** circuit. By using the methods developed in Chapter 9, it is possible to work out the expressions for the total impedance Z, the current i, the output voltage V_{out}, and the gain G of this circuit. These are

$$Z = R + j\omega L - \frac{j}{\omega C} \tag{10-1}$$

$$|Z| = \left[R^2 + \left(\omega^2 L^2 + \frac{1}{\omega^2 C^2} - 2\frac{L}{C} \right) \right]^{1/2} \tag{10-2}$$

$$i = \frac{V}{Z} = \frac{V}{R + j\omega L - (j/\omega C)} \tag{10-3}$$

$$V_{\text{out}} = iZ_c = \frac{-jV/\omega C}{R + j\omega L - (j/\omega C)} \tag{10-4}$$

$$G = \left|\frac{V_{\text{out}}}{V}\right| = \frac{1}{\omega C\{R^2 + [\omega L - (1/\omega C)]^2\}^{1/2}} = \frac{1}{\omega C|Z|} \tag{10-5}$$

In general, these expressions are sufficiently complex that not very much becomes evident by a quick study. However, at resonance, things are much simpler. The **resonant frequency f_0** is defined as that frequency at which

$$Z_L = Z_C$$

This is given by

$$j\omega_0 L = \frac{j}{\omega_0 C}$$

or

$$\omega_0^2 = \frac{1}{LC} \tag{10-6}$$

or

$$\omega_0 = \sqrt{\frac{1}{LC}} \tag{10-7}$$

At resonance, the sum of the impedances of the capacitor and the inductor is zero. As a result, all the general equations become much simpler; Equations 10-1 to 10-5 become

$$Z = R \tag{10-8}$$

$$i = \frac{V}{R} \tag{10-9}$$

$$V_{\text{out}} = -\frac{jV}{R\omega_0 C} \tag{10-10}$$

$$G = \frac{1}{R\omega_0 C} \tag{10-11}$$

Plots of the general solution for the magnitude of the impedance and the voltage gain versus frequency are given in Figures 10-2a and 10-2b, respectively. These plots show the dramatic changes that occur in the response of the circuit near resonance. There is a dip in the impedance, which becomes purely resistive at resonance. There is also a peak in the voltage gain at resonance. These are discussed in more detail in the next section.

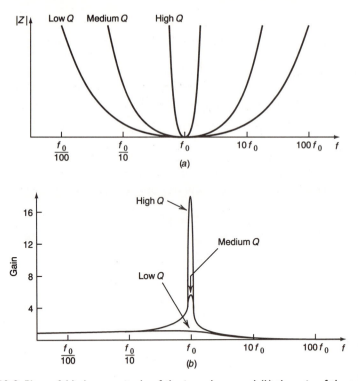

FIGURE 10-2 Plots of (a) the magnitude of the impedance and (b) the gain of the series resonant circuit of Figure 10-1. Here, f_0 is the resonant frequency and Q is defined in the next section of the chapter.

GENERAL DISCUSSION OF RESONANCE

Perhaps the most striking feature of these plots is that the gain is greater than 1; that is, the output voltage is greater than the input voltage (and this is a passive circuit). This is caused by the phase differences in the voltages across the inductor and capacitor. Although the same current flows through the resistor, the inductor, and the capacitor, the voltage across the inductor leads to current by 90°, whereas the voltage across the capacitor lags the current by 90°. At resonance, the two voltages have the same magnitude and add to zero; mathematically,

$$\sin(\omega_o t + 90°) + \sin(\omega_o t - 90°) = 0$$

This means that the current is limited only by R, which can be made small, allowing the current to be quite large. If V_{out} were taken across the resistor, the gain would peak at 1 at resonance. To get a voltage gain out of a series resonant circuit, the output must be taken across either the inductor or the capacitor. Since $V_C = iZ_C$ and Z_C may be large, the magnitude of V_C may be quite large. Since $V_{out} = V_C$ for this example, V_{out} may be quite large, in fact, larger than V_{in}.

At resonance, the current flowing in the circuit for a fixed input voltage V is determined only by R, whereas the resonant frequency is determined only by L and C. Thus the magnitude of the current at resonance and the resonant frequency can be independently changed. The **quality factor Q**, the ratio of the inductive or capacitive reactance to the resistance, determines the details of the shape of the voltage gain and impedance plots. The greater the Q of the circuit, the sharper is the peak in the voltage gain and the narrower is the dip in the impedance graph. Several cases are shown in Figure 10-2. A circuit with a high Q can be used to discriminate between two relatively close frequencies, having a high voltage gain for one and a low gain for the other:

$$Q = \frac{X_L}{R} = \frac{X_C}{R} = \frac{\omega L}{R} \qquad (10\text{-}12)$$

The Q of a circuit is also a measure of how long an oscillation will continue once it has been excited. The higher the Q, the longer the oscillation will continue. The plots in Figure 10-3 show the response of two resonant circuits differing only in Q.

Intuitively, the L and C determine the resonant frequency, whereas R and L determine the decay time constant for the circuit. The response of the circuit is then made up of a sine wave multiplied by a decaying exponential with the appropriate decay time (see Chapter 12).

On this same intuitive level, it is evident that if the decay time becomes short enough, the resonance will decay away before even a single cycle has been completed. In this case, the resonant effects have been damped out of the circuit. (Note that this is essentially the same discussion that occurs at the end of Chapter 12 when discussing the response of the RCL circuit to a step function voltage.)

OCCURRENCES OF RESONANCE

The rapid variation in the gain of a resonant circuit with frequency is the feature that makes resonant circuits both useful and harmful. Any time there is a need to pick out a signal at one frequency and suppress nearby frequencies, resonant circuits are used. The most familiar examples are in radio and TV receivers, where the tuning circuits consist of resonant circuits whose resonant frequencies can be changed, or **tuned**, to the frequency of the station desired. In a tuned circuit, the resistance R may be made just as small as possible; the limiting case is simply the resistance of the wire used to make the inductor and the rest of the circuit. This makes the frequency selection as sharp as possible—in a radio it improves the ability to reject adjacent stations.

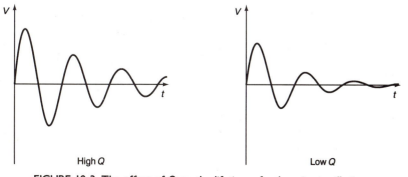

FIGURE 10-3 The effect of Q on the lifetime of a decaying oscillation.

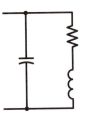

FIGURE 10-4 A representation of a real resistor.

Unwanted resonances can be a problem in circuits. Any circuit that has an R, C, and L may have an unwanted resonance. In practice, unfortunately, all circuits have resistance, inductance, and capacitance. Pure or ideal resistors, inductors, and capacitors do not exist. Any wire has an inductance, and any two points on the wire constitute a capacitor. Thus a resistor such as the common ones used in the laboratory consisting of a chunk of carbon with two wire leads stuck onto its ends really is a combination of resistance and of a small, unwanted **parasitic** inductance and a capacitance. This is shown by the equivalent circuit for a resistor is given in Figure 10-4. In this figure, the various components are assumed to be ideal or pure.

A similar diagram could be given for a real capacitor or inductor, the only difference being the relative values of the components. The ability to manufacture and use components called resistors is simply a reflection of the fact that it is possible to manufacture components for which the resistive element dominates the capacitive and inductive effects for most situations.

As a result, all circuits will have parasitic resonances. If the circuit is used at frequencies near the parasitic resonant frequency, the resonance may distort the frequency response of the circuit. Fortunately, the Q values of such parasitic resonant circuits are generally so low that they do not cause any trouble. However, when working at very high frequencies, or very high power, this may not be true.

Any time a circuit includes a real inductor, the resonant effects are likely to be serious. Of course, if the resonance causes problems, it can always be damped out by adding a resistor to the circuit to decrease the Q, to the point where the resonant effects are tolerable—but, of course, this resistor may have other undesired side effects.

A high-fidelity system also illustrates the problems of resonance. The ultimate goal of the amplifier and speaker combination is to cause equal sound levels in the room for equal input voltages at all frequencies. The inductance of any output transformer introduces a resonant frequency that can easily distort this frequency response. Worse yet, the speaker itself has a mechanical resonance which transforms into an electrical resonance which in turn is coupled to the amplifier. Thus the amplifier must actively contend with the effects of its own and the speaker's resonances, hopefully eliminating all traces of them. For good amplifiers, you will find damping factor specifications that specify how well these resonances are suppressed. In addition, you will find speaker specifications referring to how the speaker's resonances are controlled—acoustical damping, and so on.

Chapter 11

TRANSFORMERS

OBJECTIVES

1. Be familiar with the properties and uses of tranformers.
2. Be able to calculate the voltage ratio, current ratio, and impedance ratio for a transformer given the turns ratio, and vice versa.

INTRODUCTION

The transformer is another common passive circuit element. The fundamental properties of transformers and the most rudimentary features of circuit analysis with transformers are discussed in the first two sections of this chapter. The last two sections contain an optional derivation of a general impedance matching theorem and a sketch of the derivation of the properties of transformers for those who have the interest and mathematical skills to follow the arguments.

BASIC PROPERTIES OF TRANSFORMERS

A transformer consists of two or more coils of wire placed near each other so that most of the magnetic field generated by one coil passes through the other coil. The transformer may have an air core or a metallic core of some nature; the coils may be tightly coupled together or loosely coupled. The details of the design of the transformer depend on the use that is to be made of it. The schematic symbols for transformers are given in Figure 11-1.

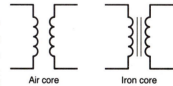
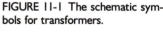
Air core Iron core

FIGURE 11-1 The schematic symbols for transformers.

The most important parameter describing a transformer is the ratio of the number of turns in each coil. The basic relationship for a transformer having two coils is

$$\frac{di_1}{dt} = \frac{N_2}{N_1}\frac{di_2}{dt} \tag{11-1}$$

where N_2 is the number of turns in coil 2, N_1 is the number of turns in coil 1, i_1 is the current in coil 1, and i_2 is the current in coil 2. This relationship says

83

that the rate of change of the current in one coil is proportional to the rate of change of the current in the other coil, with the turns ratio as the scale factor.

It is much easier to think about this relationship for ac currents, because the magnitude of di/dt is simply a constant times the magnitude of i. For the case of ac currents where I_1 is the amplitude of the ac current in coil 1, Equation 11-1 becomes

$$I_1 = \frac{N_2}{N_1} I_2 \tag{11-2}$$

As a result, for ac currents the ratio of the current flowing in the two coils varies inversely as the turns ratio of the two coils. There is more current flowing in the coil that has fewer turns. Equation 11-1 also makes clear that there is no relationship between the dc currents flowing in each of the two coils.

The transformer is a passive device; that is, it does no work and absorbs no energy (to the extent that the coil has no resistance). Thus the power available on each side of the transformer must be equal:

$$v_1 i_1 = v_2 i_2 \tag{11-3}$$

where v_1 is the voltage across the terminals of coil 1 and v_2 is its coil 2 counterpart.

Solving this equation for the voltage ratio,

$$\frac{v_2}{v_1} = \frac{i_1}{i_2} \quad \text{(instantaneous)} \tag{11-4}$$

then taking the derivative of both sides, and using Equation 11-1 gives

$$\frac{dv_2/dt}{dv_1/dt} = \frac{N_2}{N_1} \tag{11-5}$$

Again, for ac voltages this becomes particularly simple. If V_1 is the voltage appearing across coil 1, then Equation 11-5 becomes

$$V_2 = \frac{N_2}{N_1} V_1 \tag{11-6}$$

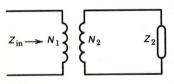

FIGURE 11-2 A transformer used to change or transform an impedance value.

This equation says the voltages appearing across the terminals of the two coils vary directly as the turn ratio of the coils. Thus there is more voltage appearing across the terminals of the coil having more turns, but there is less current flowing in this coil.

Consider the situation shown in Figure 11-2. The impedance Z_{in} looking into the circuit is given by

$$Z_{in} = Z_1 = \frac{v_1}{i_1} \tag{11-7}$$

Since the voltage at the terminals of the second side of the transformer is applied to the impedance Z_2, the voltage–current relationship for the second side of the transformer is known:

$$Z_2 = \frac{v_2}{i_2} \qquad (11\text{-}8)$$

Using Equations 11-1 and 11-5 to replace v_1 in terms of v_2 and i_1 in terms of i_2 and the turns ratio gives

$$
\begin{aligned}
Z_1 &= \frac{v_2 N_1/N_2}{i_2 N_2/N_1} \\
&= \frac{v_2}{i_2}\left(\frac{N_1}{N_2}\right)^2 \\
&= Z_2\left(\frac{N_1}{N_2}\right)^2 \qquad (11\text{-}9)
\end{aligned}
$$

Thus when viewed from the input of the transformer, the impedance attached to the output of the transformer is multiplied by the square of the turns ratio.

THE USE OF TRANSFORMERS

There are two main uses of transformers: changing ac voltage levels and impedance matching. The most familiar and common use of transformers is to change the amplitude of the ac voltages—Equation 11-6. The power company delivers 2.3 kV on the wires outside my house but I want 220 V inside the house. A transformer on the pole makes the necessary transformation. Equation 11-6 tells what the turns ratio must be; many additional engineering details tell how many turns are actually needed, what size of wire to use, and the like. In a similar manner, transformers are used to change normal 117 V to whatever voltages are required in power supplies.

Other than transformers in power supplies, the most common use of transformers in electronic circuits is to provide impedance matching. Any source producing an ac voltage has some internal source impedance Z_s, (the Thevenin equivalent impedance). The load circuit to which this source is attached has some input impedance Z_L. It is a general theorem (see the next section for a proof) that the maximum power will be transferred from the source to the load when the two impedances are the same. In general, both impedances are not equal; but by the use of a coupling transformer, the load impedance can be transformed to appear the same as the source impedance. Thus in Figure 11-3, by selecting the turns ratio of the transformer correctly using Equation 11-9, the load impedance Z_L can be made to appear the same as the source impedance Z_S. Note that the magnitudes can be made the same. If both Z_L and Z_S are resistive, there is no problem but if one or the other is reactive, then there may never be a perfect match between the source and the load.

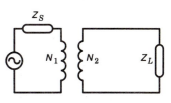

FIGURE 11-3 A transformer used to transform an impedance to allow for maximum power transfer.

In any application of transformers, there is a definite direction of signal flow (signal in to signal out) or power flow. The input side of the transformer is generally called the **primary side**, or **primary winding**, and the output side is called the **secondary**. For power transformers, there may be more than one secondary; sometimes one of the windings is called the **tertiary winding**.

But these names are only nomenclature; transformers are really symmetrical. If the turns ratio and other engineering details are suitable, the secondary can be used as the input.

Transformers have inductance. As a result, a transformer can be part of a resonant circuit. Furthermore, the primary and secondary can be parts of resonant circuits tuned to different frequencies, yet these are coupled together. This is a relatively common situation in radio receivers and is one reason that circuit analysis involving transformers quickly acquires a formidable complexity.

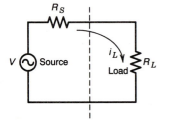

FIGURE 11-4 The circuit used to prove the power transfer theorem. transformers.

PROOF OF POWER TRANSFER THEOREM

The next two sections are optional. The results have already been discussed in this chapter. However, those who can follow these derivations will obtain a deeper understanding of the results.

Consider the circuit shown in Figure 11-4. A voltage source produces a voltage V and has a Thevenin equivalent resistance of R_S. The goal is to find the condition on R_L that will result in the maximum power being delivered to the load from the source. Since the reactive components of the impedance dissipate no power, only the resistance need be considered.

The power P delivered to the load is

$$P = V_L I_L = (I_L R_L) I_L \tag{11-10}$$

and the current flowing in the load is given by

$$I_L = \frac{V}{R_S + R_L} \tag{11-11}$$

So the power is

$$P = \frac{V^2}{(R_S + R_L)^2} R_L \tag{11-12}$$

To find the maximum, take the derivative of this expression with respect to R_L and set it equal to 0:

$$\frac{dP}{dR_L} = \frac{V^2}{(R_S + R_L)^2} - \frac{2V^2 R_L}{(R_S + R_L)^3} = 0$$

$$R_S + R_L = 2R_L$$

$$R_L = R_S \tag{11-13}$$

Therefore, to obtain the maximum power delivered into the load, the load resistance must be equal to, or **matched** to, the source resistance. At this point, one-half the power is delivered to the load and one-half the power is dissipated in the source. This is the best that can be done.

DERIVATION OF TRANSFORMER EQUATION

What follows is an outline of a derivation of the basic equation describing a transformer, Equation 11-1. The first part of this derivation is a review of the basic theory of inductance. When the current in a circuit changes, the magnetic field in the region around the circuit changes. Thus the magnetic flux through the circuit changes, and an electromotive force (emf) is induced in the circuit. If the magnetic permeability of the circuit is a constant, the induced emf is proportional to the change in the current; thus

$$V_L = L\frac{di}{dt} \tag{11-14}$$

where L is the **self-inductance** of the circuit. In a coil of N turns in which the same flux passes through all the turns, this becomes

$$V_L = N\frac{d\phi}{dt} \tag{11-15}$$

where $d\phi/dt$ is the rate of change of the magnetic flux through one turn of the coil. Combining these two expressions gives

$$L = N\frac{d\phi}{di} \tag{11-16}$$

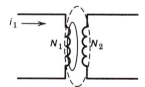

FIGURE 11-5 The situation used to derive the basic relationships for transformers.

Consider the two coils shown in Figure 11-5. When a current flows in coil 1, a magnetic field is set up in the surrounding space. Some of this magnetic field passes through coil 2 and some of it does not. In the figure, two lines of magnetic flux are shown; one of the lines of flux links coil 2 and one does not. The part of the flux that does not link the two coils is called the **leakage flux**; the part of the flux that does link the two coils is called the **linkage flux** ϕ_{12}. The linkage flux is proportional to the current i_1:

$$\phi_{12} = \alpha i_1 \tag{11-17}$$

Faraday's law says that the voltage induced in coil 2 is

$$v_2 = N_2\frac{d\phi_{12}}{dt} \tag{11-18}$$

Using the relationship between ϕ_{12} and i_1 in this expression yields

$$v_2 = N_2\frac{d\phi_{12}}{dt} = N_{21}\alpha\frac{di_1}{dt} = m\frac{di_1}{dt} \tag{11-19}$$

where m is the **mutual inductance**. Solving this expression for the mutual inductance yields

$$m = N_2 \frac{d\phi_{12}}{di_1} \tag{11-20}$$

But the entire problem is symmetrical—the mutual inductance does not depend on which coil is the first one considered. This argument could have started by considering the current in coil 2; in this case, the result obtained would have been

$$m = N_1 \frac{d\phi_{21}}{di_2} \tag{11-21}$$

Combining these two results and using the symmetry of the situation ($\phi_{12} = \phi_{21}$) gives

$$N_2 \frac{d\phi_{12}}{di_1} = N_1 \frac{d\phi_{21}}{di_2} \tag{11-22}$$

$$\frac{N_2}{N_1} = \frac{di_1}{di_2} \tag{11-23}$$

which directly yields Equation 11-1.

PROBLEMS

11-1 What is the turns ratio of a transformer that reduces 13.6 kV to 220 V?

11-2 The primary winding of a transformer has 100 turns; the secondary has 2000 turns. If the primary is connected to 110 V, what is the secondary voltage? If 20 mA flows in the secondary, what is the primary current?

11-3 A voltage generator has an internal impedance of 1000 Ω. What turns ratio in a transformer is needed to connect this generator to a load of 100 Ω?

11-4 The output stage of a high-fidelity amplifier has an internal impedance of 162 Ω. If this is to be connected to an 8-Ω speaker, what must be the turns ratio of the output transformer for maximum power transfer?

ANSWERS TO ODD-NUMBERED PROBLEMS

11-1 61.8
11-3 3.162

Chapter 12

STEP FUNCTION ANALYSIS

OBJECTIVES

1. Be able to calculate the time constant for an RC or RL circuit.

2. Be able to sketch graphs of the currents and voltages in RC and RL circuits when subjected to step function voltages.

3. For more advanced students: Be able to write down and solve the circuit equations for simple circuits subjected to step function voltages.

INTRODUCTION

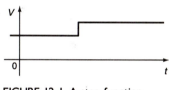

FIGURE 12-1 A step function voltage.

The response of simple circuits to step function voltages is discussed in this chapter. **Step function voltages** are voltages that change suddenly from one steady value to another. A typical step function voltage is illustrated in Figure 12-1. Although step function voltages are neither ac nor dc, some of the techniques used to analyze both ac and dc circuits can be used in this case. The material discussed in this chapter is one particular form of **transient analysis**.

Obviously, step function voltages are idealizations of real situations, just as the ac and dc examples presented in preceeding chapters are idealizations. However, the response of circuits to step function voltages emphasizes certain ways of looking at these circuits. (An overview of all types of circuit analysis is contained in the next chapter.)

The application of the basic circuit analysis laws to situations involving step functions gives rise to differential equations. To analyze these circuits completely, these equations must be solved. Fortunately, in introductory electronics only the most simple cases of circuits subjected to step functions are encountered. The three most important circuits are worked out and discussed in considerable detail in this chapter. If you do not know calculus and cannot follow the derivations, don't worry; go on to the solutions and the discussion. The discussions are the important part of this chapter, not the derivations.

The problems at the end of this chapter illustrate only the most important aspects of the solutions for these three circuits, not the details of solving the circuit equations.

THE *RC* CIRCUIT—PART I

A circuit very similar to the first example given for ac circuit analysis will be analyzed in this section; the circuit is shown in Figure 12-2. This circuit consists of a battery, a resistor, and a capacitor in series with a two-position switch that can be used to connect or disconnect the battery from the rest of the circuit.

The initial condition of the circuit is shown in Figure 12-2*a*. It is assumed that the switch has been in the down position for a very long time; this means there is no voltage across the capacitor, no charge on the capacitor, and no current flowing in the circuit.

At some instant of time, the switch is moved to its upper position; by definition, this is time $t = 0$. The situation with the switch closed is shown in Figure 12-2*b*. The goal of the analysis is to find the current and voltages in the circuit as a function of time from the moment the switch was closed.

Without writing down and solving any equations, it is possible to learn a great deal about the solution of this problem. At the instant the switch is closed, there is no voltage drop across the capacitor because there is no charge on the capacitor; this is an example of the fact that the voltage across a capacitor cannot change instantaneously. KVL shows that the entire voltage must appear across the resistor. Thus the instant the switch is closed, the current changes very rapidly (instantaneously) from zero to an initial value of V/R. As the current flows, charge builds up on the capacitor as shown in Figure 12-2*b*. This means that there is an increasing voltage drop across C, and hence the voltage drop across R must be decreasing (according to KVL). Thus as time goes on, the current must decrease. Eventually, the voltage across the capacitor will build up to V; at this point, there will be no current flow and nothing will change thereafter. So, without writing down or solving any equations, three facts have been learned about the behavior of this circuit to this step function:

1. Just after $t = 0$, the current has the value V/R.
2. After a very long time, the current will have dropped to zero.
3. Because solutions to physical problems are always smooth, there will be some sort of gradual transition between the initial and final state. Several potential transitions are shown in Figure 12-3. The goal of the com-

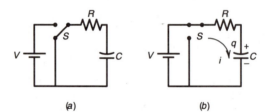

(a)   (b)

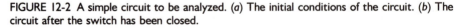

FIGURE 12-2 A simple circuit to be analyzed. (*a*) The initial conditions of the circuit. (*b*) The circuit after the switch has been closed.

plete circuit analysis will be to work out the exact shape of this curve. (The initial change in current from its original value of 0 to the value of V/R also was smooth but very rapid.)

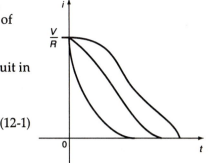

To get a complete solution to this problem, KVL is applied to the circuit in Figure 12-2b. This yields

$$V - iR - \frac{q}{C} = 0 \tag{12-1}$$

But recalling and using the definition of current,

$$i = \frac{dq}{dt} \tag{12-2}$$

FIGURE 12-3 A sketch of the current versus time for Figure 12-1. Three conceivable results are shown.

this becomes the following differential equation:

$$V - R\frac{dq}{dt} - \frac{q}{C} = 0 \tag{12-3}$$

An alternative way to approach the problem is to use the voltage–current relationship for the capacitor,

$$q = \int i \, dt \tag{12-4}$$

Using this expression in Equation 12-1 gives the integral equation

$$V - iR - \frac{1}{C} \int i \, dt = 0 \tag{12-5}$$

In general, when analyzing the response of circuits to step functions or any transient situations, integral–differential equations arise. To find the functional form of the current, these equations must be solved.

An optional brief outline of one method of solving the differential Equation 12-3 will be presented here for those who can follow it. The important material in this chapter begins again with the discussion of Equation 12-15. This differential equation can be solved by guessing a solution and verifying that it is a solution. Assume a solution of the form

$$q = B + K^{\alpha t} \tag{12-6}$$

Substitute this into the differential Equation 12-3 and obtain

$$V - RK\alpha e^{\alpha t} - \frac{B}{C} - \frac{K}{C}e^{\alpha t} = 0 \tag{12-7}$$

This equation can only be satisfied for all times if the terms involving and not involving the exponential add up to zero separately. Thus

$$V - \frac{B}{C} = 0 \tag{12-8}$$

and

$$-RK\alpha - \frac{K}{C} = 0 \tag{12-9}$$

The first of these gives

$$B = VC \tag{12-10}$$

whereas the second gives

$$\alpha = \frac{-1}{RC} = \frac{-1}{\tau} \tag{12-11}$$

Substituting these two results back into the assumed solution yields

$$q = CV + Ke^{-t/RC} \tag{12-12}$$

where K is determined from the initial conditions of the problem, that is, from the fact that at the time $t = 0$ the charge on the capacitor was zero. Using this fact in the equation above gives

$$q(0) = 0 = CV + Ke^{-0} \tag{12-13}$$

$$K = -CV \tag{12-14}$$

The solution of Equation 12-3 is

$$q = CV(1 - e^{-t/RC}) = CV(1 - e^{-t/\tau}) \tag{12-15}$$

where

$$\tau = RC \tag{12-16}$$

The quantity τ is called the time constant for the circuit and is the characteristic time unit for dealing with this problem.

The current is given by the derivative of the charge on the capacitor

$$i = \frac{dq}{dt} = \frac{V}{R}e^{-t/RC} \tag{12-17}$$

These solutions are plotted in Figure 12-4. These plots should be compared with those of Figure 12-3.

The initial differential equation was obtained by using the fact that KVL must be satisfied at all times. This means that the voltage drops across the resistor and capacitor should add up to V at all times. Since it may not be obvious that this is true, it will be verified here as a check.

The voltage drop across the capacitor is

$$V_C = \frac{q}{C} = V(1 - e^{-t/\tau}) \tag{12-18}$$

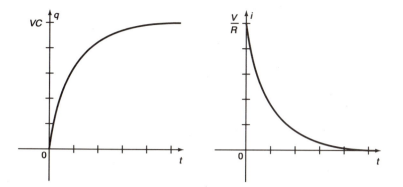

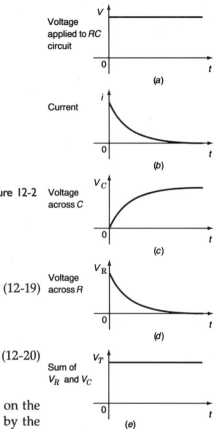

FIGURE 12-4 Plots of the charge on the capacitor and the current in the circuit of Figure 12-2 as a function of time after the switch is closed.

The voltage drop across the resistor is

$$V_R = iR = Ve^{-t/\tau} \qquad (12\text{-}19)$$

The sum of these two voltage drops is

$$V_T = V_C + V_R = V \qquad (12\text{-}20)$$

as it should be.

To further illustrate the solution, all relevant quantities are plotted on the same time scale in Figure 12-5. The time scale for this problem is set by the **time constant** $\tau = $ **RC**. In more detail, all RC circuits will respond to a step function voltage in the same way, the only difference being that some circuits will respond faster, some slower than others. The rate of the response is determined by the time constant. If time is measured in units of RC, then the response of all RC circuits is the same—that is, one graph can be used for all values of R and C.

At time $t = \tau = RC$, the current has fallen to a value of $e^{-1} = 0.368$ of its initial value. The following table shows how the current falls as a function of time.

FIGURE 12-5 Plots of some interesting quantities for the circuit of Figure 12-2 as a function of time after the closing of the switch. (a) The total voltage applied to the RC combination. (b) The current flowing through the resistor. (c) The voltage across the capacitor. (d) The voltage across the resistor. (e) The sum of the voltage across the resistor and capacitor.

t (in units of τ)	$e^{-t/\tau}$
1	0.368
2	0.135
3	0.0498
4	0.0183
5	0.00674
10	4.54×10^{-5}
$\vdots$	$\vdots$

The answer to the question: How long do you have to wait for the current to drop to zero? depends on the accuracy you demand. For most situations, after 5 time constants everything has come close enough to its final value; very precise work might require waiting for 10 or more time constants.

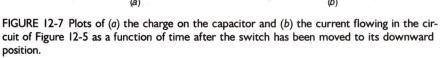

FIGURE 12-6 The same circuit as in Figure 12-5 after the switch has been moved to its downward position.

THE *RC* CIRCUIT—PART 2

Consider the situation shown in Figure 12-2. After the switch has been in the upper position for a very long time (measured in terms of RC times constants), the switch is moved back to the lower position. The situation now is shown in Figure 12-6. For this problem, the time $t = 0$ is taken as the time at which the switch was moved to the down position. The initial condition is that of $q(0) = CV$. From the drawing it can be seen that the current is discharging the capacitor, so that

$$i = \frac{dq}{dt} \tag{12-21}$$

Applying KVL to the circuit gives

$$iR - \frac{q}{C} = 0 \tag{12-22}$$

or

$$R\frac{dq}{dt} = \frac{-q}{C} \tag{12-23}$$

By using the same methods as before, it can be shown that the solution to this equation is

$$q = CVe^{-t/RC} \tag{12-24}$$

and

$$i = \frac{-dq}{dt} = \frac{V}{R}e^{-t/RC} \tag{12-25}$$

These results are plotted in Figure 12-7. Again, the same constant, $\tau = RC$, sets the scale for the problem. If the switch were to flip back and forth between

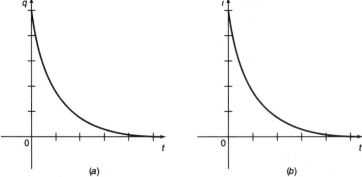

(a)   (b)

FIGURE 12-7 Plots of (a) the charge on the capacitor and (b) the current flowing in the circuit of Figure 12-5 as a function of time after the switch has been moved to its downward position.

the two settings at some regular rate that is slow compared to the time constant RC, and if the voltage across the capacitor were displayed on an oscilloscope, the display would be similar to that shown in Figure 12-8.

FIGURE 12-8 The voltage across
the capacitor in Figure 12-2 if the
switch were to switch slowly back
and forth from one position to
another.

TWO POINTS OF VIEW OF THE *RC* CIRCUIT

Consider the two arrangements of the RC circuit shown in Figure 12-9. For
the configuration used in Figure 12-9a, the output voltage is given by

$$V_{out} = \frac{q}{C} = \frac{1}{C} \int i \, dt \qquad (12\text{-}26)$$

Since the current is related to the voltage applied to the circuit, that is, since
i is proportional to V_{in}, this circuit can be regarded as integrating the input
voltage.

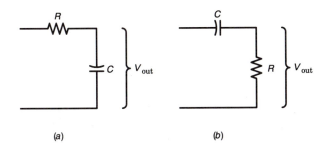

(a)   (b)

FIGURE 12-9 Two variations on the RC circuit. In version (a), the output voltage is taken
across the capacitor; in version (b), the output is taken across the resistor.

For the configuration in Figure 12-9b, the output voltage is given by

$$V_{out} = iR = R\frac{dq}{dt} \qquad (12\text{-}27)$$

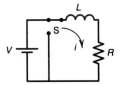

FIGURE 12-10 A simple RL circuit
analyzed in the text.

Since the charge on the capacitor is related to the voltage applied to the circuit, this circuit can be regarded as differentiating the input voltage.

These two ways of looking at RC circuits are often useful in understanding what an RC combination will do at some point in a circuit. This will be
discussed further in Chapter 13.

THE *RL* CIRCUIT

In Figure 12-10, the capacitor in the circuit of Figure 12-2 has been replaced
by an inductor. The switch S has been in the down position for a very long
time so that there is no initial current. As before, at time $t = 0$ (by definition),
the switch moves to its upper position. At that instant, the current is zero;
and because there is an inductor in the circuit, the value of the current cannot change instantaneously Applying KVL to the circuit yields

$$V - L\frac{di}{dt} - iR = 0 \qquad (12\text{-}28)$$

This is essentially the same differential equation as previously, only this time it is for i, not q,

Solving in a similar manner as before, the solution is

$$i = \frac{V}{R}(1 - e^{-t/\tau}) \tag{12-29}$$

where the time constant is given by

$$\tau = \frac{R}{L} \tag{12-30}$$

After a very long time, the switch is moved back to its lower position. (In the laboratory, it must be made certain that the circuit is never open at any time. This means using a "shorting" type switch.) The solution for the current this time is

$$i = i_0 e^{-t/\tau} = \frac{V}{R} e^{-t/\tau} \tag{12-31}$$

These solutions have the same form as those for the RC circuit, except that the currents and voltages have been interchanged. Again, the time scale is set by a single time constant, $\tau = R/L$. Thus with suitable relabeling, the graphs for the voltages and currents for the RC circuit can be used for the RL circuit.

THE *RCL* CIRCUIT

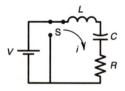

FIGURE 12-11 The *RCL* circuit discussed in the text.

The situation considered is shown in Figure 12-11. Again, after being in the down position for a very long time, the switch S is moved to its upper position. Applying KVL to this situation yields

$$V - iR - \frac{q}{C} - L\frac{di}{dt} = 0 \tag{12-32}$$

or

$$V - iR = \frac{1}{C} \int i\, dt - L\frac{di}{dt} = 0 \tag{12-33}$$

Differentiating yields

$$R\frac{di}{dt} + \frac{i}{C} + L\frac{di^2}{dt^2} = 0 \tag{12-34}$$

This is the equation of the damped harmonic oscillator. Working out its solutions in detail is a favorite problem in any differential equations course or intermediate mechanics course. The solutions will not be given in detail here, but their general nature will be discussed.

It turns out that there are three different classes of possible solutions, depending on the relative values of R, C, and L. These depend on whether the quantity

$$\left[\frac{1}{LC} - \left(\frac{R}{2L}\right)^2\right]^{1/2} \tag{12-35}$$

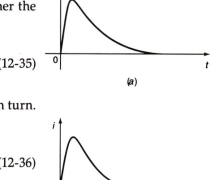

(a)

is zero, real, or imaginary. Each of these will be discussed and plotted in turn.

If

$$\left(\frac{R}{2L}\right)^2 > \frac{1}{LC} \tag{12-36}$$

then the oscillator is said to be **overdamped**. If it is given an initial push, or, in the electrical version, when the switch is changed, it returns to its initial state only slowly. Such a response is shown in Figure 12-12a.

If

$$\left(\frac{R}{2L}\right)^2 = \frac{1}{LC} \tag{12-37}$$

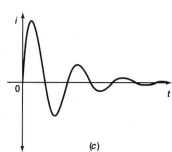

(b)

then the oscillator is said to be **critically damped**. If it is given an initial push, the oscillator returns to its initial position as fast as possible, with no overshoot (ringing or oscillating). This response is shown in Figure 12-12b.

If

$$\left(\frac{R}{2L}\right)^2 < \frac{1}{LC} \tag{12-38}$$

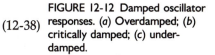

(c)

FIGURE 12-12 Damped oscillator responses. (a) Overdamped; (b) critically damped; (c) underdamped.

then the oscillator is said to be **underdamped**. If it is given an initial push, it returns to its initial position via a series of decreasing oscillations. Such a response is shown in Figure 12-12c.

When designing a damped mechanical system, such as the shock absorbers of a car, critically damped behavior is generally the goal. In electrical circuits, underdamped responses are generally interpreted as an indication of possible trouble somewhere in the circuit.

The discussion above can be related to the discussion on resonant circuits by noting the relationship between the conditions for the various solutions and the definition of Q. Using the definition of Q, these conditions become

Overdamped:

$$\left(\frac{R}{2L}\right)^2 > \frac{1}{LC} \quad \text{or} \quad Q < \frac{1}{2} \tag{12-39}$$

Critically damped:

$$\left(\frac{R}{2L}\right)^2 = \frac{1}{LC} \quad \text{or} \quad Q = \frac{1}{2} \tag{12-40}$$

Underdamped:

$$\left(\frac{R}{2L}\right)^2 < \frac{1}{LC} \quad \text{or} \quad Q > \frac{1}{2} \quad\quad\quad (12\text{-}41)$$

SUMMARY

When a step function voltage is applied to an RC or RL circuit, the current changes smoothly from its initial value to its final value. When a step function is applied to an RCL circuit, the current often oscillates during its change from its initial to its final value. These oscillations, often called ringing, may be undesirable.

As stated earlier, all components (including real wires) have parasitic elements. Thus a real resistor has small parasitic inductive and capacitive elements associated with it. Even a short wire has a finite resistance, inductance, and capacitance. Thus any time a step function is applied to a circuit, oscillations or ringing may occur. In general, circuits involving only resistors and capacitors have such small parasitic inductances that they are usually overdamped and ringing does not develop. However, once an inductor has been added to a circuit, ringing becomes very common.

PROBLEMS

12-1 For the circuit of Figure 12-2, if V is 10 V, R is 10 kΩ, and C is 50 μF, what is the current the instant the switch is closed? How long does it take for the current to fall to 1 μA?

12-2 The dc voltage applied to an RC circuit is suddenly changed. How long does it take for the voltages and currents to settle to within $\frac{1}{2}$% of their final value if the value of R is 100 kΩ and C is 23 μF?

12-3 For the circuit of Figure 12-10, if V is 15 V, L is 10 H, and R is 100 Ω, what is the final current? How long after the switch is closed does it take the current to settle to within 1 mA of its final value?

12-4 In an RL circuit, it is observed that the current rises to one half its final value in 30 sec. If the resistor is 25 Ω, what is the value of the inductor?

12-5 For the RCL circuit of Figure 12-11, if $C = 10$ μF and $L = 50$ mH, what is the status of the circuit if $R = 1000$ Ω, 140 Ω, and 10 Ω?

12-6 For an RCL circuit with $C = 100$ pF and $L = 1$ μH, what is the minimum value R must have to prevent oscillations?

ANSWERS TO ODD-NUMBERED PROBLEMS

12-1 1 mA, 13.8 sec

12-3 150 mA, 50.1 sec

12-5 for 1000 Ω, circuit is overdamped; for 140 Ω, it is almost critically damped; for 10 Ω, it is underdamped

Chapter 13

CIRCUIT ANALYSIS— AN OVERVIEW

OBJECTIVES

1. To gain a perspective on the methods and uses of circuit analysis techniques.

2. To gain a perspective on the use of approximations to make circuit analysis is easier.

INTRODUCTION

The purpose of this chapter is to provide an overview and some perspective into the methods and uses of circuit analysis. Strategies for approaching problems and for approximating in problems are discussed. No new techniques are introduced, and no calculations are done in this chapter. The very nature of this chapter means that it will be more diffuse and rambling than previous chapters.

The goal of circuit analysis is to provide a framework for understanding electrical circuits. In using circuit analysis to study a circuit, a mathematical model of the circuit is created. This model can then be used to provide answers to questions such as "What would happen if this component were changed to ...?" The reason for using circuit analysis rather than building the circuit and measuring the results is the expectation that the circuit analysis route will provide more or better knowledge, besides being quicker, cheaper, and perhaps safer, than the experimental route.

The complexity of the mathematical model depends in part on the desired accuracy of the results. To obtain very accurate results frequently requires a mathematical model of considerable complexity. However, frequently only a rough understanding or an approximate result is needed. In this case, it is often possible to greatly simplify and approximate the mathematical analysis. This chapter is oriented toward these methods of simplification and approximation.

First, a distinction between transient and steady state situations is described. Then, a number of different classes of waveforms are described and some suggestions are made for dealing with each class. Finally, two simple circuits are discussed from several different points of view as an illustration.

TRANSIENT ANALYSIS

Whenever something in a circuit changes quickly compared to the natural time scale of the circuit, we say that a **transient** exists in that circuit. The step function responses studied in the previous chapter are examples of transient responses. These transients were caused by changing the switches or by changing the applied dc voltage. However, transients also occur when the ac voltage applied to a circuit suddenly changes in amplitude, frequency, or phase.

Implicit in this definition of a transient is an extended **steady state** period before and after the transient. The period of transition from the initial steady state to the final steady state is the transient response of the circuit. Note that the steady state solution does not necessarily refer to a dc situation; more frequently, it refers to the response to a steady ac signal.

If the nature of the transient can be described mathematically, then the usual circuit analysis laws can be applied; however, the differential equations that result will contain time-dependent coefficients. The solutions of these differential equations can be quite complex in even simple cases and impossible in many cases. Most often, transient problems cannot be solved exactly but must be approximated. One useful avenue for approximation is to replace the given transient with one of similar shape that is easier to analyze but that still is a reasonable approximation of the actual change—such as a series of small step function changes. Sometimes the effects of the transients are most easily described in terms of the transients being integrated or differentiated by the circuit.

POSSIBLE WAVEFORMS

A common approach to analyzing phenomenon is to sort them into a number of distinct classes and then to deal with each class separately. If done correctly, when all classes have been covered, all the possible phenomena have been covered; unfortunately, there is usually a leftover class. In this spirit, here is another classification of the possible types of voltages that can occur in a circuit. This is an expansion and refinement of the classification that was first advanced in Chapter 7, with the previous class of "other" much expanded and divided. Again, the justification of this classification is that it is often useful:

1. Direct current.
2. Very slowly varying dc.
3. Step function voltages.
4. Pulses.
5. Sinusoidal voltages: ac.
6. Sinusoidal plus dc.
7. Other periodic waveforms.
8. Others.

ANALYSIS

In this section, each of the voltage types listed above will be discussed. The discussion will include a characterization of the voltage, a description of what information is most often of interest, and what approximations are most often used to gain this information. Although this section is written in terms of voltages, it could just as well be written about currents, without significant changes.

Direct Current

A pure dc voltage is most easily characterized by the fact that its time derivative is zero at all times; that is, it is not changing at all. Direct current circuit analysis problems are relatively simple; there are no differential equations and at worst there are only large arrays of coupled linear equations.

Because of the relative simplicity of the circuit equations, the goals of analyzing dc circuits are usually quite comprehensive—find all the voltages and currents. The analysis of a dc circuit can often be greatly simplified by ignoring the small currents that flow through resistors that are very large compared to the other resistors in the circuit; this effectively means cutting any loop in which the current is very small. This quickly reduces the number of loops and the number of equations needed to be solved. If necessary, once the simplified problem has been solved, the information from the solution can be used to estimate the actual currents in loops that have been cut and to estimate the possible errors in the approximate solution.

Very Slowly Varying DC

This class is characterized by the fact that the voltage is changing with respect to time but changing so slowly that all reactive effects due to inductors and capacitors are very small compared to the resistive effects.

To the extent that all reactive effects are very small and can be ignored, this is a pure dc problem. In this approximation, the effect of the slowly varying voltage can be taken into account by solving the equivalent pure dc problem and then replacing the voltage V with the slowly varying voltage function. The errors caused by this approximation can then be estimated by using the V–I characteristics of the inductors and capacitors to estimate the voltage drops and currents flowing in the circuit because of the changing voltage. An even better approximate solution can be obtained by correcting the dc approximation with these estimated voltages and currents.

Step Functions

Step function voltages are characterized by relatively sudden changes in an otherwise stable dc voltage. The easiest step function to deal with is the idealized case considered in Chapter 12 and shown in Figure 13-1. Unfortunately, other shapes, such as those shown in Figure 13-2, occur commonly. In actual

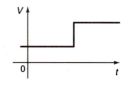
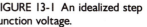

FIGURE 13-1 An idealized step function voltage.

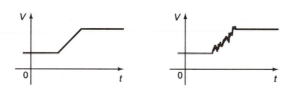

FIGURE 13-2 Two more realistic examples of step function voltages than that shown in Figure 13-1.

practice, a pure clean step function never exits; so to some approximation, all step function problems are messy.

Frequently, only either the steady state or the transient response of a circuit is of interest. To find the steady state solution, the detailed shape of the transition is not of interest, only the before- and after-solutions are. These are both pure dc problems, linked together by the transition. To find the transient response, it may be possible to simplify the circuit by eliminating those parts which affect only the steady state solutions.

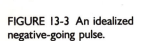

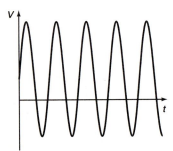

FIGURE 13-3 An idealized negative-going pulse.

Pulses

An ideal (negative going) voltage pulse is shown in Figure 13-3. In general, pulses never look as nice as these; more common shapes for pulses are shown in Figure 13-4.

FIGURE 13-4 More typical examples of negative-going pulses than that shown in Figure 13-3.

A pulse is characterized by its length (time), its amplitude and, if necessary, its shape. If the length of a pulse is much longer than the time constant(s) of the circuit, then pulse analysis involves the case of two step functions separated in time. If the length of the pulse is short compared to the time constants of the circuit, only a transient analysis is necessary. If the length of the pulse is comparable to the time constants of the circuit, the analysis is very hard. Pulsed circuit analysis is very hard when done in any detail. In general, the only easy ways are to use intuition and approximate the circuit in terms of integration, differentiation, and so on.

FIGURE 13-5 An ac voltage with a small dc bias or offset.

Sinusoidal Voltages

Alternating current circuit analysis has been extensively discussed in the previous chapters. In analyzing a circuit's response to ac voltages, most often the phase shift or gain as a function of frequency is desired. Much less often of interest is the ac transient response of a circuit.

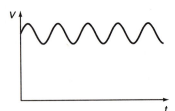

Sinusoidal Voltage Plus a DC Voltage

Perhaps the most common situation in electronics is a sinusoidal voltage superimposed on a dc voltage. There are two extreme cases that are often of interest. These are illustrated in Figures 13-5 and 13-6.

FIGURE 13-6 A dc voltage with a slight ac ripple.

In general, the first approximation to these circuits is to attempt to solve the ac and dc parts separately—it is not an approximation as long as all the circuit elements are linear. In cases where nonlinearities are important and simply separating the problem does not yield a good solution, concepts such as "dynamic" impedance can be used to give approximate solutions (see Chapter 14 for a brief discussion of dynamic impedance). Often the response of the circuit to one component of the voltage in terms of the variation of the other will be of interest, for instance, how the ac gain of the circuit varies as a function of the dc bias.

Other Periodic Waveforms

A periodic waveform is any waveform that repeats in time. The simplest example of a periodic waveform is ac. Another periodic waveform is shown in Figure 13-7. A periodic waveform is characterized by its period T, its frequency $1/T$, its amplitude, and its shape. The amplitude can be expressed in terms of peak values V_{max} and V_{min}, peak-to-peak value, average value, or rms value. It should be noted that, in general, the relationships among these various amplitudes are not the same as for the sine wave.

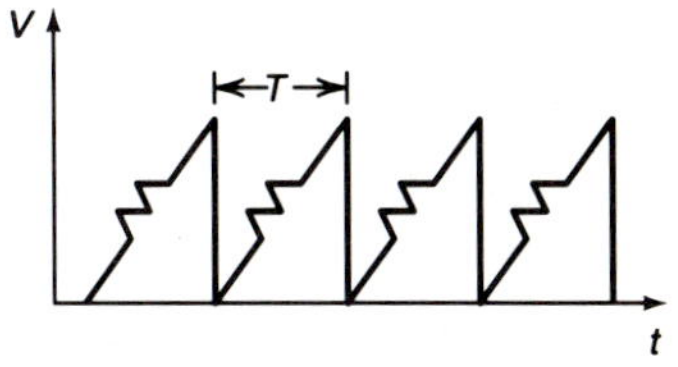

In general, the goal of analyzing a circuit subjected to a periodic waveform is the same as that for a sine wave—find the relationship between the input and the output under steady state conditions. However, in general, the direct analysis of a circuit subjected to a periodic waveform is very hard, because concepts such as impedance have no meaning. The solution to this problem is most easily obtained by decomposing the input waveform into a set of sine waves using the Fourier theorem, analyzing the response of the circuit to the various sine waves, and then synthesizing the output of the circuit. The use of Fourier analysis is far beyond the scope of an introduction to electronics; however, since it is so interesting and often so useful, a very brief introduction and example are given in Chapter 55 for those who are interested.

TWO CIRCUITS FROM MANY POINTS OF VIEW

To provide some perspective on all this material, many of the voltage waveforms discussed above will be considered with respect to the two circuits shown in Figure 13-8. Both of these circuits are simple voltage dividers. At any instant of time, the sum of the voltage drops across the two components equals the applied voltage. Furthermore, the voltage drop across the resistor is just iR. Thus all the analysis below is based on the v–i characteristics of the capacitor along with the variations in the applied voltage. Recall, however, the results of Chapter 9; the circuit in part (a) of the figure is a high-pass filter, whereas that in part (b) is a low-pass filter. Furthermore, from another point of view, the circuit in part (a) is a differentiator, whereas the circuit in part (b) is an integrator.

FIGURE 13-7 An example of a nonsinusoidal periodic voltage waveform.

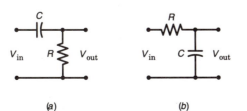

(a)   (b)

FIGURE 13-8 Two simple circuits for analysis with many different applied voltages. (See text.)

Direct Current

V_{out} for circuit A in Figure 13-8 is zero. That is, the dc voltage is blocked by the capacitor. For circuit B, $V_{\text{out}} = V_{\text{in}}$. The circuit has not changed the dc output other than adding a resistor in series with it.

Slowly Varying DC

The results are the same as those for the dc case as long as the rate of change is slow enough. As the rate increases, this case becomes the same as the step function case; the time scale for "slow enough" is set by the time constant of the circuit:

$$\tau = RC \qquad (13\text{-}1)$$

Step Function

For circuit A in Figure 13-8, the output is proportional to the derivative of the input voltage:

$$V_{\text{out}} = K\frac{dV}{dt} \qquad (13\text{-}2)$$

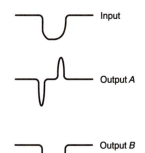

where K is some constant. For circuit B, the output is proportional to the integral of the input:

$$V_{\text{out}} = K' \int V dt \qquad (13\text{-}3)$$

Thus circuit A differentiates or picks out changes in the input voltage, whereas circuit B integrates or averages the input. Regarding circuit A as a **differentiator** and circuit B as an **integrator** is often a very productive point of view.

FIGURE 13-9 The input and output voltages for both circuits shown in Figure 13-8 in the case where the time constant of the circuit is very short compared to the length of the pulse.

Pulses

For the circumstance where the RC time constant of the circuit is very short compared to the pulse, circuit A of Figure 13-8 gives an output proportional to the derivative of the pulse, whereas the output of circuit B is proportional to the pulse. This is illustrated in Figure 13-9.

When the time constant of the circuit is long compared to the pulse, circuit A passes the pulse more or less unchanged, whereas circuit B integrates or averages it. These results are shown in Figure 13-10.

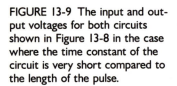

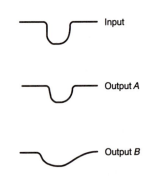

Alternating Current

Circuit A is a high-pass filter. $V_{\text{out}}/V_{\text{in}}$ is approximately 1 for high frequencies and about 0 for low frequencies. Circuit B is a low-pass filter. These cases were discussed extensively in Chapter 9. The frequency at which the period is equal to the RC time constant of the circuit is in the middle of the region where the response is changing rapidly as a function of the frequency.

FIGURE 13-10 The input and output voltages for both circuits shown in Figure 13-8 for the case where the pulse is long compared to the time constant of the circuit.

Biased AC

The response of the circuits can be determined from the discussions above. Thus circuit *A* blocks the dc and the very low frequencies and passes the high frequencies, *B* passes the dc and low frequencies and blocks the high frequencies.

Periodic Waveforms

By looking at circuit *A* as a differentiator and circuit *B* as an integrator, the effect of the circuits can often be predicted for periodic waveforms. The amplitude of the output will depend on the ratio of the time constant of the circuit to the period of the waveform. Two examples are shown in Figures 13-11 and 13-12.

Input

Output *B*

FIGURE 13-11 The response of circuit *B* in Figure 13-8 to a periodic input in a case that emphasizes the integrating nature of circuit *B*.

Input

Output *A*

FIGURE 13-12 The response of circuit *A* in Figure 13-8 to a periodic input in a case that emphasizes the differentiating nature of circuit *A*.

Chapter 14

DYNAMIC IMPEDANCE

OBJECTIVES

1. Know why the concepts of dynamic resistance and impedance are useful.

2. Know how dynamic resistance and dynamic impedance are defined.

INTRODUCTION

All circuit analysis done thus far has involved only linear devices, that is, devices for which the amplitude of the current is proportional to the amplitude of the voltage. At a fixed frequency, linear devices have one resistance or one impedance, no matter what the voltage or the current; and because of this, the circuit equations are always linear in the voltage and the current. But, as mentioned previously, electronics would not be interesting if there were only linear devices.

Unfortunately, once nonlinear devices are added to circuits, the resulting equations are no longer linear or simple differential equations, and they become difficult or impossible to solve. In this chapter, the problem of describing the resistance or impedance of a nonlinear device will be briefly discussed; then, an approximation that is often used to generate linear equations will be introduced.

THE BASIC PROBLEM

Consider the situation shown in Figure 14-1, where the voltage–current relationship for a nonlinear device is being measured. This is essentially the same situation as discussed in Chapter 2. Recall that resistance is defined as

$$R(I) = \frac{V}{I} \tag{14-1}$$

Thus for any given voltage–current point (V_p, I_p), the resistance can be calculated. Graphically, the resistance is the slope of the line through the particu-

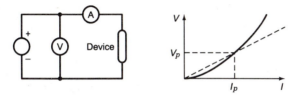

FIGURE 14-1 An experiment to measure the voltage–current relationship for a device and one possible result of this measurement.

lar voltage–current point and the origin of the graph; this line is shown as a dashed line on the graph in Figure 14-1.

For this particular device, the voltage–current graph is not a straight line; its voltage–current relationship is nonlinear. As discussed in Chapter 2, this means that this device does not obey Ohm's law. It also means that the resistance of the device is a function of the current passing through it.

If this device is put into a simple circuit, such as the one shown in Figure 14-2, then KVL gives

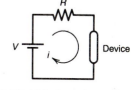

FIGURE 14-2 A nonlinear device used in a very simple circuit.

$$V - iR - iR_{device}(i) = 0 \qquad (14\text{-}2)$$

which is not a linear equation; depending on the nature of $R(i)$, it may be impossible to solve it analytically. This is the basic problem with nonlinear devices in circuit analysis.

A MORE COMPLEX PROBLEM

In Figure 14-3, this same nonlinear device is being used in a more complex circuit. A voltage consisting of a small ac ripple on a dc background is being generated by an ac voltage generator and a battery as shown. The capacitor and inductor are included in the circuit to make the situation as clear as possible. It is assumed that the inductor has a very large impedance at the frequency of the ac generator. In this situation, the battery and the signal generator will each feel that they are working into different impedances. To see this, consider the circuit from the point of view of each of the voltage generators.

Because of the capacitor, no dc current flows through the ac generator; and so the battery does not see the ac signal generator, it only sees the device. Thus it feels as if it were working into a load resistance given by its voltage

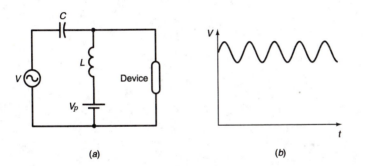

(a) (b)

FIGURE 14-3 A nonlinear device in a circuit (a) containing two voltage sources. The voltage applied to the device is shown in (b).

divided by the dc current flowing in the circuit; this is the slope of the dashed line shown in Figure 14-1 or Figure 14-4. As long as the ac wiggles are small compared to the dc voltage, there is no effect on this resistance due to the ac source.

But the ac signal source sees something entirely different. Because of the large inductor in series with the battery, essentially no ac current flows through the battery; and the generator does not see the battery, it sees only the load device in series with the coupling capacitor. However, if the capacitor is made very large, then its effects at the frequency of interest become small, and the ac signal source also sees only the device. It is only aware of the small changes in the current I_{ac}, that it is generating. Thus it feels that it is working into a resistance which is given by

$$R_{AC} = \frac{V_{AC}}{I_{AC}} = \frac{\Delta V_P}{\Delta I_P} \tag{14-3}$$

This corresponds to the slope of the voltage–current graph at the operating point (V_P, I_P), shown as a dotted line on the graph of Figure 14-4.

So here is a situation in which two different voltage sources are working into the same device at the same time, and each sees a different resistance. This is the result of the nonlinear nature of the device. The basic circuit analysis laws, KCL and KVL, applied to the situation involving nonlinear devices do not yield simple linear equations or linear differential equations. There is a solution to the equations. For some cases, there are numerical, graphic, or occasionally analytic methods for solving the nonlinear equations that arise. In general, it is much harder to solve the nonlinear equations than the linear equations discussed in previous chapters. However, the discussion above suggests an approximate method for obtaining linear equations for this situation.

AN APPROXIMATE METHOD

The advantages of linear equations and simple differential equations are so great that various approximations are used to generate these simple equations for nonlinear situations. A common situation is one where two voltage genera-

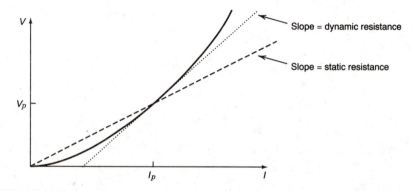

FIGURE 14-4 The voltage–current relationship for a nonlinear device, with the lines corresponding to the large-scale resistance (dashed line) and the dynamic resistance (dotted line) both indicated.

tors are present, one having a large amplitude, usually dc, and the other a small ac generator—this is exactly what was illustrated in Figure 14-3. In this case, there are two approximate resistances of interest. The **large-signal resistance**, or **static resistance**, is given by

$$R = \frac{V_P}{I_P} \tag{14-4}$$

where V_P and I_P are the **operating point** voltage and current, respectively. The **dynamic resistance**, or **small-signal resistance**, is given by the slope of the tangent to the voltage–current graph at the operating point:

$$R_S = \frac{\Delta V_P}{\Delta I_P} \tag{14-5}$$

In general, both the large-signal and the small-signal resistances depend on the large signal or the bias applied to the device. It is often the case that the small-signal resistance will be used in determining how a device will operate as an amplifier or similar circuit, whereas the large-signal resistance will be important in providing the necessary voltages and currents to make the device work at all. Thus the problem often divides up into two parts, each of which is a linear approximation. Using the large-signal resistance, pick the large-signal voltages and any other components necessary to obtain the desired operating point; then, use the small-signal resistance to determine how the circuit operates.

IMPEDANCE

In a similar manner, a **small-signal impedance**, or **dynamic impedance**, can be defined. When thinking about ac circuit analysis and impedances, the whole situation becomes much less clear, but nevertheless a distinction between large- and small-signal impedances remains. The large-signal impedance depends on the average slope of the v–i graph over the range of interest (which may not include the origin), whereas the small-signal impedance corresponds to the slope of the v–i curve at some particular point.

There is a distinction between the small-signal resistance and the small-signal impedance. It is, quite simply, that the small-signal impedance takes into account any phase shifts that occur between the applied small voltage and the resultant small currents, whereas the small-signal resistance defined above simply ignores these.

The approximate nature of the small-signal concepts becomes much clearer in the ac case, since if a pure sine wave voltage is applied to a nonlinear device, a sine wave current is not obtained. Thus because of Fourier's theorem, this is no longer a single pure frequency, and the whole concept of impedance begins to become fuzzy. However, as long as the approximate nature of the concept is kept clear, the concept remains very useful.

PART TWO

INSTRUMENTS

All electronic work requires making measurements. Although digital electronics may require only the simplest means of measuring voltages, any extensive effort of building, testing, or repairing equipment will require more extensive measurements of various electrical quantities. In fact, the dominant process in troubleshooting equipment is comparing the measured voltages, currents, or waveforms in a circuit with the expected values. To a considerable extent, doing electronics is making measurements.

Because measuring a quantity invariably changes the quantity being measured, it is generally useful to have some idea of the nature of the measuring instrument and how it can affect the circuit in which it is being used. The first three chapters of this part of the book deal with the most common measuring instruments—the multimeter, the digital multimeter, and the oscilloscope. Chapters 15, 16, and 17 cover the basic elements of these instruments and their use. Chapter 18 covers the most elementary features of the interaction between the measuring instrument and the quantity being measured.

Everyone should read these chapters at least once. For some, this will be enough of an introduction of these topics; others may want to master some of the details of the various circuits used in the instruments or explore in more detail the interactions between instrument and circuits.

PREREQUISITES

All four chapters assume a familiarity with elementary circuit analysis—Chapter 4.

CHAPTER 15 MULTIMETERS

1. In discussing the use of meters to measure ac voltages, some of the terminology used to describe ac voltages—Chapter 7—and diodes—Chapter 41—are used.

CHAPTER 16 DIGITAL MULTIMETERS

1. A familiarity with VOM circuits—Chapter 15.
2. References are made to simple operational amplifier circuits—Chapters 28 and 29—but no understanding of these circuits is needed here

CHAPTER 17 OSCILLOSCOPE

1. The opportunity to use an oscilloscope.

CHAPTER 18 MEASUREMENTS ERRORS

1. A familiarity with VOMs, DMMs, and oscilloscope circuits—Chapters 15 through 17.
2. Informal use of Norton's and Thevenin's theorems are made in the discussions—Chapter 6

Chapter 15

MULTIMETERS

OBJECTIVES

1. Become familiar with the D'Arsonval meter movement and the parameters used to describe any panel meter.

2. Become familiar with the circuits used to convert a panel meter into an ammeter, voltmeter, and ohmmeter.

3. Gain familiarity with the problems encountered when using a multimeter, or VOM.

INTRODUCTION

This chapter deals with the multimeter, or VOM. This is (or was) by far the most ubiquitous measuring instrument used in electronics laboratories. It is a relatively small box (usually black) having one or more switches, several inputs, and an electromechanical meter on which there are many scales. By setting the switches appropriately and then reading the correct scale, this instrument can be used to measure a wide range of ac or dc voltages and currents or to measure resistance. This functionality gives rise to its name, a **multimeter**, or a **volt-ohm-milliammeter,** or **VOM**. Although the digital multimeter is rapidly becoming more common than the VOM, this section begins with this older meter because it is conceptually simpler.

THE GALVANOMETER

The core of the VOM is some sort of moving pointer meter, sometimes called a **galvanometer**. Since galvanometers are usually intended to be mounted in a panel, the name **panel meter** is often used for a galvanometer. These two names will be used interchangeably here. From the user's point of view, the panel meter consists of a pointer and a scale. The most common type of galvanometer is based on a moving coil, or D'Arsonval meter movement. Figure 15-1 is a simplified sketch of the elements of this meter movement.

The functional parts of this meter movement are a small coil of light wire restrained by a spring and suspended on a very light bearing in a permanent

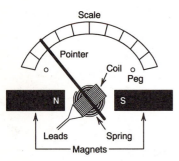

FIGURE 15-1 A simplified sketch of a D'Arsonval meter movement.

magnetic field. The operation is simple in principle; a current passing through the coil causes a magnetic field which is proportional to the current. The interaction of this magnetic field and the permanent magnetic field generates a torque that causes the coil to rotate. However, as the coil rotates, the spring is flexed and exerts a restoring torque on the coil. As a result, for any given value of the current, there is an equilibrium position of the coil, at which point the magnetic torque and the spring torque are equal. A pointer attached to the rotating coil moves along a scale which can be calibrated in terms of the current causing the deflection. By winding the coil on a light iron cylinder and by shaping the pole pieces, it is possible to make the meter so that the deflection of the pointer is linearly related to the current flowing in the coil.

Meter Parameters

A basic galvanometer can measure currents ranging from 0 to some maximum current or full-scale current, here called I_m. Further, the meter will have an internal resistance R_m due to the wire used in the coil and in the connections to the coil. Thus to obtain a full-scale deflection of the meter, a voltage of V_m must be applied to the galvanometer terminals. These quantities are all related by Ohm's law:

$$V_m = I_m R_m \qquad (15\text{-}1)$$

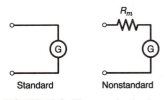

FIGURE 15-2 The standard schematic symbol for a galvanometer and the nonstandard symbol used here. The second symbol emphasizes the internal resistance of the meter movement.

When using the galvanometer, there is a tendency to ignore its internal resistance; in fact, the schematic for a meter—a simple circle—encourages this tendency. In this chapter, the nonstandard schematic symbol for the galvanometer in Figure 15-2 will sometimes be used to emphasize the internal resistance.

The smaller the value of I_m, the more sensitive is the meter. Typical values of I_m for normal meters range from 20 μA to 10 mA. The corresponding values for R_m range from about 3000 ohms for the most sensitive meters to a few ohms—see any electronics distributor's catalog for examples. A little reflection will show that the more sensitive a meter is, the more delicate it is; hence, the harder it is to use, the easier it is to damage, and the more expensive it is to make.

Meter Errors

Unfortunately, meters are not perfect. They exhibit various errors, such as slight nonlinearities, mismarked scales, incorrect calibrations, and other assorted errors. The accuracy of a meter is generally specified in terms of its full-scale deflection. For example, if a meter has a full-scale current of 1.0 mA and a quoted accuracy of 4% of full-scale value, then the error in any reading may be as large as 0.04 mA, even for a reading of 0.1 mA or smaller. Thus to obtain the smallest possible percentage error, always try to select a meter or meter range to obtain a reading as close to full scale as possible.

A meter can be calibrated by comparing its reading to the reading of a laboratory standard meter at a number of points spanning the full range of readings. In this way, a plot of the meter's error as a function of its reading is obtained. A sample of such an error calibration is presented in Figure 15-3.

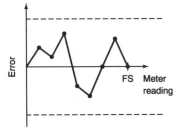

FIGURE 15-3 A typical error plot for a panel meter.

A normal panel meter has an accuracy of about 5%. Meters having accuracies of 1 or 2% are readily available. Of course, the abuse of any meter may make the quoted accuracy somewhat suspect.

Varying Current

Everything said above is based on the assumption that the current flowing through the galvanometer is constant. The meter is an example of a damped mechanical oscillator; it has a characteristic time constant that determines the time it takes to come to a new reading after the current has been changed. This time constant can be roughly estimated by applying a voltage to the meter and seeing how long it takes to come to its final deflection.

If the current in a galvanometer changes very slowly compared with the time constant of the meter, the meter's deflection simply changes as the current changes. If the current changes very rapidly, the deflection is the average of the current (this is a result of the impulse-momentum theorem). If the current changes at some intermediate rate comparable to the time constant of the meter, the deflection will represent some sort of smoothed function of the input current.

For mechanical reasons, meters with long time constants are more durable than meters that respond very rapidly. As can be easily verified, most of the meters in use have time constants that are in the range of seconds. Thus any current variations that are much shorter than a second will not be indicated by a normal meter. In particular, 60-Hz ac currents will be indicated as zero on a moving coil meter unless some additional circuitry is used (see below).

Other Meter Types

Today, there are a number of other mechanical meter designs on the market—taut band, hot wire meters, and so on. However, the basic parameters needed to describe these meters are the same. They differ only in the values of the parameters, in the mechanical details of construction, and in the other parameters such as cost, size, and ruggedness.

THE AMMETER

Current is measured with an ampere meter or, as it is usually called, an **ammeter**. The galvanometer is such an instrument. All that is needed is to put the meter in series in the circuit to be measured, that is, to somehow break the circuit, insert the meter, and read the current on the galvanometer. Figure 15-4 shows a galvanometer used to measure the current in the wire between points a and b in a complex circuit.

Larger Currents

But what if the current is too large to be measured with the galvanometer at hand? One solution is to find a less sensitive galvanometer; but a more economical way is to use a current divider to convert the galvanometer into an ammeter capable of measuring a larger current than I_m. This basic ammeter circuit is shown in Figure 15-5.

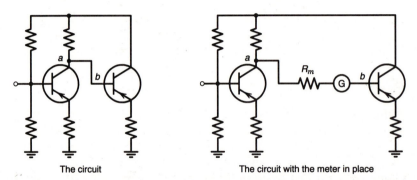

The circuit   The circuit with the meter in place

FIGURE 15-4 A galvanometer used to measure the current in the wire from points *a* to *b*. As the second part of the figure shows, adding the ammeter has added a resistance to the circuit.

A resistor R_s is placed in parallel with the galvanometer, R_s and R_m forming a current divider. R_s is called the **shunt** resistor because it shunts or bypasses some of the current around the galvanometer. The galvanometer measures the current actually flowing through itself, I_g. Using a knowledge of the current divider ratio, it is possible to calculate the total current I_x flowing in the ammeter circuit. The usual way of describing an ammeter circuit is in terms of a multiplier M, the ratio of the total current in the ammeter circuit to the current in the galvanometer. Thus

$$I_x = MI_g \qquad (15\text{-}2)$$

The current divider formula, Equation 4-16, applied to the circuit shown in Figure 15-5 gives

$$I_g = \frac{I_x}{M} = I_x \frac{R_s}{R_s + R_m} \qquad (15\text{-}3)$$

Hence

$$M = \frac{I_x}{I_g} = \frac{R_s + R_m}{R_s} \qquad (15\text{-}4)$$

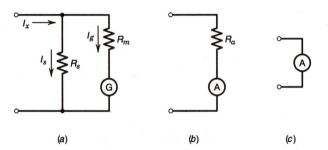

(a)   (b)   (c)

FIGURE 15-5 (*a*) An ammeter is a galvanometer with a shunt resistor. (*b*) A nonstandard schematic symbol for an ammeter which emphasizes the fact that the ammeter has an internal resistance. (*c*) The standard schematic symbol for an ammeter.

In general, the goal is to find the value of R_s needed to give the desired multiplier. Solving Equation 15-4 for R_s gives

$$R_s = \frac{R_m}{M - 1} \tag{15-5}$$

Finally, for future use, the overall resistance of the ammeter, R_a, looking into its terminals, is the parallel combination of R_s and R_m, or

$$\frac{1}{R_a} = \frac{1}{R_s} + \frac{1}{R_m} \tag{15-6}$$

Working this out gives

$$R_a = \frac{R_m}{M} \tag{15-7}$$

A certain amount of confusion sometimes arises when talking about the current flowing in an ammeter built from a galvanometer and a shunt resistor. The galvanometer and the shunt are treated as a single item called an ammeter. The ammeter has a certain full-scale current I_{fs} and resistance R_a. The confusion arises from the fact that there are two full-scale currents, the full-scale current of the ammeter circuit and the full-scale current of the galvanometer, which are related by the multiplier M. But the confusion is purely semantic. All that is needed is to be fairly careful about the terms being used.

Example of an Ammeter Design

As an example, assume there is a galvanometer with a full-scale current $I_m = 1$ mA and an internal resistance $R_m = 50\ \Omega$. Using this galvanometer, what shunt values are needed to make ammeters having full-scale currents of 10, 100, 200, and 1000 mA, and what are the resistances of the resulting ammeters?

Using the results above, Table 15-1 can be calculated very easily.

TABLE 15-1 Shunt Resistor Calculation

Full-Scale Current Desired (mA)	M	$R_s\ (\Omega)$	$R_a\ (\Omega)$
10	10	5.555	5.0
100	100	0.5050	0.5
200	200	0.2513	0.25
1000	1000	0.05005	0.05

VOM

By now it should be clear how the multiple ammeter ranges of a VOM are constructed. The switch selects which of the several different shunt resistors are used; this changes the current division ratio (the multiplier), and hence changes the range of the ammeter. To make the VOM more convenient to use, the manufacturer has had a number of scales printed on the meter. This way, no multiplication is needed—just find the correct scale, read it, and put the decimal point where it belongs.

Problem with Ammeters

There are two problems with using ammeters. The first is illustrated in the example shown in Figure 15-4. By inserting the ammeter into the circuit, you have changed the circuit being measured. Before the meter was inserted, there was a direct connection between points a and b; after insertion of the meter, there now is the resistance of the ammeter between these two points. Obviously, the smaller the value of this resistor, the smaller is the change in the circuit. This topic is discussed in more detail in Chapter 18.

The second problem with using an ammeter is purely practical. The ammeter must be placed in series in the circuit where the current is to be measured. In the laboratory, when constructing a circuit on some sort of breadboarding system, this is generally easy. Each wire can simply be pulled out and the meter put in its place. But in an existing circuit, it is often very difficult to break the circuit and insert the ammeter. Often, an indirect approach of measuring the voltage drop across an existing resistor and calculating the current is used rather than employing an ammeter to measure the current directly.

THE VOLTMETER

The galvanometer can be used to measure voltages. Since the galvanometer is ohmic, the voltage applied to it is to be calculated from the current flowing through it by using Ohm's law. Thus

$$V_g = I_g R_m \qquad (15\text{-}8)$$

Thus given a circuit such as the one shown in Figure 15-4, to measure the voltage between point a and ground, all that need be done is to connect the galvanometer between these two points, measure the current, and compute the voltage. To simplify the last step, it would not be too much trouble to print yet another scale on the meter face so that voltage can be read directly.

If the voltage to be measured is greater than V_m, the voltage required to produce a full-scale current in the galvanometer, the galvanometer cannot be used as a voltmeter as simply as described above. However, by adding a series resistor as shown in Figure 15-6, it is possible to create a meter with a higher full-scale voltage.

For the arrangement shown in Figure 15-6, it is obvious that the current flowing in the galvanometer is given by

$$I_g = \frac{V}{R_v + R_m} \qquad (15\text{-}9)$$

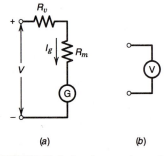

FIGURE 15-6 A voltmeter is a galvanometer with a series resistor. (a) The circuit; (b) the schematic symbol for a voltmeter.

Alternatively, the voltage applied to the terminals of the voltmeter (the entire circuit) can be calculated in terms of the current flowing in the galvanometer:

$$V = I_g(R_v + R_m) \tag{15-10}$$

Note the formal similarity of this case to the ammeter; in both cases, the quantity to be measured is simply a multiple of the current flowing in the galvanometer.

This problem can be approached from another point of view. Given a galvanometer having a full-scale current of I_m, the series resistor needed to make a voltmeter having a full-scale voltage of V_{max} can be calculated from

$$R_v = \frac{V_{max}}{I_m} - R_m \tag{15-11}$$

For the sample galvanometer used above, with $I_m = 1$ mA and $R_m = 50\ \Omega$, a voltmeter having a full-scale voltage of 1000 V requires a series resistor of

$$R_v = \frac{1000}{0.001} - 50$$

$$= 999{,}950\ \Omega$$

or almost 1 MΩ (megohm).

Voltmeter Sensitivity

To measure a voltage, the voltmeter is put across the voltage source. Since the voltmeter draws some current (less than I_m), there is an additional current flowing in the circuit that was not flowing there when the voltmeter was not in place. The resultant changes in the circuit being measured are said to be the result of the meter **loading** the circuit being measured. In Chapter 18, there is a discussion of how to calculate and/or estimate this loading effect. To do this, it is necessary to know the total resistance of the voltmeter. As can be seen from Equation 15-7, the total resistance is given by

$$R_t = \frac{V_{max}}{I_m} \tag{15-12}$$

A more sensitive voltmeter is made from a more sensitive galvanometer; it loads the circuit to be measured less, and it has a higher resistance. To make it easier to compare voltmeters, the sensitivity of a voltmeter is often stated in terms of **ohms per volt**. This is just $1/I_m$. Thus a voltmeter built with a 1-mA meter movement has a sensitivity of 1000 Ω/V, whereas one with a 50-μA meter has a sensitivity of 20,000 Ω/V. To find the load presented by a voltmeter, simply multiply the sensitivity in ohms per volt by the maximum voltage for the range being used.

VOM

Again, by now it should be clear how the voltmeter ranges of a VOM are constructed. The switch simply selects which of several different series resistors are used; this changes the full-scale voltage. Again, to make the use of the

VOM easier, the manufacturer has printed additional scales on the meter face to do all the multiplication for the users. All that must be done is to insert the decimal point in the reading. (Note that the same scale may be used to read either voltage or current.)

THE OHMMETER

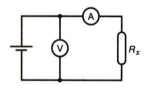

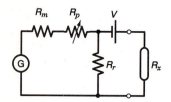

FIGURE 15-7 A poor ohmmeter.

FIGURE 15-8 A practical ohmmeter circuit.

The galvanometer can be converted to a device that will measure resistance in several ways. The simplest method is shown in Figure 15-7. The unknown device is hooked up to a battery and the galvanometer is used as a voltmeter to measure the voltage across the device and then as a ammeter to measure the current through it. The resistance can then be calculated from Ohm's law.

Although this method is very simple conceptually, it requires making two measurements and then doing a calculation. It may be necessary to take into account the effects of the voltmeter and the ammeter while making the measurement and calculation.

There are several other circuits that are more complex but that can measure resistance directly. One such circuit is shown in Figure 15-8. One way of looking at this circuit is that the unknown resistor R_x and the range resistor R_r form a voltage divider; the remainder of the circuit is a voltmeter measuring the voltage across R_r. From this arrangement, the unknown resistance can be calculated. To use this circuit, a two-step procedure must be followed:

1. After selecting the appropriate resistance range (this means selecting the value of R_r), the leads of the ohmmeter are shorted together ($R_x = 0$) and the variable resistor R_p is adjusted so that a full-scale reading I_m is obtained on the galvanometer.
2. The unknown resistor is connected to the leads and the current I_x flowing in the galvanometer is measured. After some tedious circuit analysis, it can be shown that the value of the unknown resistor is given by

$$R_x = \frac{R_r V}{R_r + (V/I_m)} \left(\frac{1}{I_x} - \frac{1}{I_m} \right)$$

(15-13)

If the meter is very sensitive, that is, if I_m is very small, R_x can be approximated as

$$R_x \approx R_r I_m \left(\frac{1}{I_x} - \frac{1}{I_m} \right) \approx R_r \left(\frac{I_m}{I_x} - 1 \right)$$

(15-14)

Ohmmeter Cautions

Several things must be remembered when using an ohmmeter. First, the circuit being tested must be off; if there are any batteries or power supplies operating in it, the readings may be incorrect. Second, you must be sure that you are measuring the device you want to measure and not some parallel path through the rest of the circuit—this often means that the device to be measured must be cut out of the circuit. Finally, the voltage applied to the device being measured may damage the device, so be careful.

VOM

To make a multirange ohmmeter, all that has to be done is to arrange the switching arrangement such that different values of the **range resistor** R_r, are switched into place. Again, the manufacturer has had additional scales printed on the meter face so that no calculations are needed to read resistance. To the extent that the approximation, Equation 15-14 above, holds true, the same scale can be used for several different ranges.

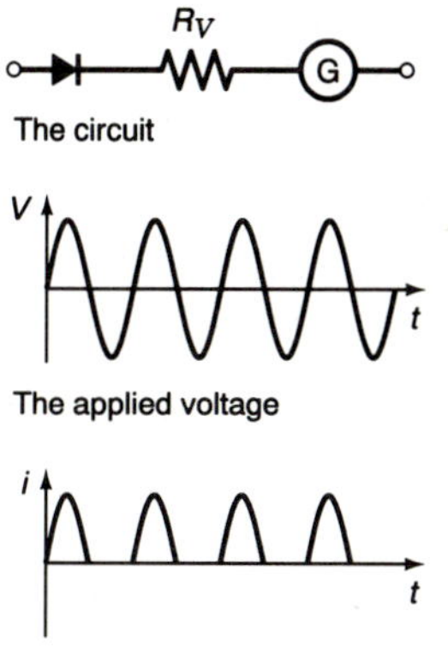

The circuit

The applied voltage

The current in the galvanometer

FIGURE 15-9 An ac voltmeter and some waveforms.

AN AC VOLTMETER

Thus far, we have referred to direct current measurements only. If 60-Hz ac is passed through a galvanometer, no deflection results. However, a diode can be used with a series resistor to make an ac voltmeter. The circuit and some waveforms are sketched in Figure 15-9. The meter movement itself takes an average of the current. From the figure, it can be seen that the average will be proportional to the peak value—the amplitude of the ac voltage. Again, by appropriate selection of the series resistor and printing of scales, the voltmeter can be made to directly read the peak value, the average value, or the rms value (all for sine waves only).

There are a number of problems with the ac ranges on VOMs. The most serious has to do with the turn-on voltage of the diode. Because the diode does not start conducting until about 0.6 V has been applied to it, this kind of instrument cannot be used to measure very small ac voltages. Furthermore, this turn-on voltage will cause a nonlinearity in the lower ac voltage ranges—look at the scales on a VOM. This same turn-on voltage makes the effect of an ac ammeter on the circuit being measured very hard to understand.

The calibration of the ac scales on a VOM is only correct when measuring pure sine waves, since for other waveforms the relationships between the peak values, average value, and rms value will be different. Finally, the instrument may respond differently as a function of frequency.

SOME PRACTICAL COMMENTS

Galvanometers and hence VOMs are delicate instruments. They can be very accurate, but they will not continue to be so if they are dropped on the floor or bounced along the workbench. The deflection of the meter pointer is limited by pegs at each end of its useful range. When a meter is deflected so that the pointer is pressing against the peg, it is said to be **pegged**. When a meter is pegged, its reading is worthless and it may soon burn out. A pegged instrument should always be disconnected immediately. It is, in fact, possible to peg a meter so violently that the pointer bends around the peg.

Most galvanometers are designed to deflect only the one way. Currents flowing the wrong way cause only a tiny deflection before the lower peg is reached. These small backward deflections are often overlooked. Often, when there is no apparent deflection, the meter has simply been plugged in backward; reversing the leads may give the desired reading. Many VOMs have a switch so that the polarity can be reversed without having to move any wires.

Some meters are made to be operated horizontally, other to be used vertically. To get the most accurate results from a meter, always use it in the correct position. Usually, the meter case will make clear which way the meter is to be used if this makes any real difference.

Most meters have a zero-adjust screw. This is used to set the meter pointer to zero when there is no current flowing in the meter. This adjustment is essential for accurate operation. This adjustment is often very sensitive to the position of the meter (horizontal/vertical), and it often has considerable mechanical backlash.

When using any meter, remember it is a delicate instrument. Before plugging it into the circuit, check over the circuit, make sure it looks right. If the meter has multiple ranges, start out with the least sensitive range (most volts or amperes) and progress to the most sensitive range you can use. Remember, a meter can be damaged far more rapidly than you can disconnect it from the circuit.

PROBLEMS

15-1 Given a panel meter with a resistance of 150 Ω and a full-scale current of 50 μA, calculate the shunt resistors needed to make ammeters having full-scale currents of 50, 100, and 500 μA and 1, 10, and 100 mA. Also calculate the series resistors needed to make voltmeters having full scale-voltages of 1, 10, and 50 V.

15-2 Given a panel meter with a full-scale current of 20 μA and a resistance of 1000 Ω, calculate the resistors needed for all the ranges listed in Problem 15-1.

ANSWERS TO ODD-NUMBERED PROBLEMS

15-1 no shunt required for the 50-μA current; shunts of 150, 16.67, 7.89, 0.754, and 0.075 Ω required for the 100- and 500-μA and the 1-, 10-, and 100-mA range, respectively; for the voltmeter, series resistors of 19,850 Ω, 199,850 Ω, and ~ 1 MΩ required.

Chapter 16

DIGITAL MULTIMETERS

OBJECTIVES

1. Become familiar with the circuits used to convert a digital panel meter into an ammeter, a voltmeter, or an ohmmeter.

2. Become familiar with some of the problems encountered when using a DMM.

INTRODUCTION

The **digital multimeter**, OR **DMM**, has the same functions as a volt-ohm-milliammeter, or VOM. However, it generally has more capabilities and greater accuracy, and frequently it is more convenient to use than a VOM. This improved performance coupled with rapidly declining prices has caused the DMM to become the dominant instrument in the electronics laboratory.

The DMM is similar to a VOM in many respects. The normal VOM consists of a panel meter and various simple circuits that convert it into an ammeter, a voltmeter, or an ohmmeter. Similarly, the DMM consists of a digital panel meter and various simple circuits that convert it into an ammeter, a voltmeter, or an ohmmeter. This chapter is a brief introduction to some of the circuits used in DVMs. The chapter follows the same pattern as the previous one for the VOM.

THE DIGITAL PANEL METER

For the purposes of this chapter, the **digital panel meter**, or **DPM**, is a black box. It has two input leads and a display; it also has either power connections or internal batteries to provide the power needed. The DPM measures the voltage at its input and displays the results of this measurement. For this chapter, the only DPM parameters of interest are its full-scale range, its accuracy, and its input impedance. It should be noted that there is nothing mysterious about the circuitry that makes up the digital panel meter. A number of different versions are covered in Chapter 36 on analog-to-digital converters.

The typical DMM is built around a digital panel meter that can measure voltages in the range of −100 to +100 mV, −150 to +150 mV, or −199 to

+199 mV. All of these are called 100-mV DPMs, but the one running to 150 mV is said to have **50% overrange** capability, and the one running to 199 mV is said to have **100% overrange** capability.

Most digital panel meters will have some way of indicating when the voltage applied to the input terminals is greater than what can be measured by the instrument. Some go blank, some flash random readings, some just blink, still others may display an error message. In any case, a totally unexpected display often means the voltage is out of range for the meter.

The typical digital panel meter has an input resistance of 10 to 100 megohms (MΩ). Units having input resistances of 1000 MΩ or more are not uncommon but cost slightly more. Today, the cost of the basic unit is in the range of $30 to $150, depending on overrange options, accuracy, technology, and packaging details.

Accuracy, Precision, and Resolution

Precision refers to how many significant digits there are in a measurement. **Resolution** refers to the smallest increment in the measurement that can be detected. **Accuracy** refers to the absolute error in the measurement compared to the true value. Thus for a DPM, precision refers to how many digits there are in the answer; resolution refers to the size of the least significant step; and accuracy refers to how well a measurement agrees with the results that would be obtained at the National Bureau of Standards. It is possible to have very precise measurements that have very poor accuracy; for example, a very finely ruled ruler could be made of rubber. With this ruler, several digits of precision can be obtained; but, by pulling on the ends, the length of the ruler might be changed by a factor of 2. Thus this ruler would have reasonable precision but terrible accuracy.

The precision of a digital panel meter is specified by stating how many digits of the results it can display. Thus a 3-digit panel meter can display results ranging from 000 to 999, whereas a $3\frac{1}{2}$-digit meter can display results from 000 to 1999. A 3-digit 100-mV meter would be able to measure and display results ranging from −99.9 to 99.9 mV; its resolution would be 0.1 mV. A $3\frac{1}{2}$-digit 100-mV meter would be able to display results ranging from −199.9 to 199.9 mV; again, its resolution would be 0.1 mV. A $4\frac{1}{2}$-digit 100-mV meter will be able to measure the same range of voltages, but its resolution would be 0.01 mV.

The accuracy of a digital panel meter is usually stated in terms of so many percent ± one least significant digit. A typical specification for a $3\frac{1}{2}$-digit meter might be 0.1% ± 1 digit. If this meter were used to measure a voltage of 23.4 mV, the error in the reading might be as large as 0.1 mV. For most units available today, the ±1-digit term dominates the error at all readings.

THE DIGITAL AMMETER

To make an ammeter with a digital panel meter is very easy. Simply pass the current through a known shunt resistor and measure the voltage drop across it—see Figure 16-1. To the extent that the input resistance of the DPM is much greater than R_s, all the current passes through the shunt resistor and the current can be directly calculated from the meter reading by using Ohm's

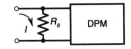

FIGURE 16-1 A digital ammeter.

law. As long as the shunt resistor is some integral power of 10 ohms (0.1, 1, 10, 100, etc.), the DPM will read the curent directly. All the user may have to do is supply the decimal point. On most modern DMM units, the manufacturer has arranged the switching network so that, as it selects the shunt resistor, it also puts the decimal point at the correct place in the display.

An Example

For all the examples in this chapter, unless specified otherwise, it is assumed that the DPM has an input impedance of 1000 MΩ, a full scale of 100 mV, 100% overrange capabilities, and an accuracy of 0.1% ± 1 digit.

To make a digital ammeter having full-scale current of 1.0 A requires a shunt resistor of 0.1 Ω (1.0 A flowing through a 0.1 Ω resistor generates a voltage drop of 0.1 V, or 100 mV). Clearly, the 1000-MΩ input resistance is very much greater than 0.1 Ω, so the loading effect of the DPM is negligible. To make an ammeter having a full-scale current of 100 mA would require a shunt resistor of 1.0 Ω.

Digital current meters are used the same way that VOM ammeters are used. The wire carrying the current to be measured is broken and the meter is inserted. As a result, a resistor and a voltage drop have been added to the circuit where formerly there were none. When the ammeter is reading full scale, this voltage drop is the full-scale voltage for the DPM. For the example used in this chapter, this is 100 mV. Similarly, when the VOM ammeter is reading full scale, the added voltage drop is V_m.

With today's techniques, it is often true that the V_m for a VOM is less than full-scale voltage of the popular DPMs. For the examples used in this book, V_m is 50 mV, whereas the full-scale voltage of the DPM is 100 mV. This means that the VOM will be a better ammeter on every range. Of course, it is possible to build a DPM that has a much smaller full-scale voltage than 100 mV, but the normal DVM's are not based on one of these. Nevertheless, the ease of reading a digital ammeter often makes it the more desirable instrument.

THE DIGITAL VOLTMETER

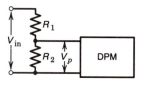

FIGURE 16-2 A digital voltmeter.

The DPM is a voltmeter, but it has only one range. If a DPM is to be used to measure a voltage higher than the full-scale value for the DPM, a voltage divider is needed, as shown in Figure 16-2. The voltage divider formula can be used either way to give

$$V_p = V_{in}\frac{R_2}{R_1 + R_2} \tag{16-1}$$

or

$$V_{in} = V_p\frac{R_1 + R_2}{R_2} \tag{16-2}$$

If the voltage divider ratio is picked to be some nice even number such as 10, 100, and so on, then, by simply moving the decimal point, the DPM will read the input voltages directly.

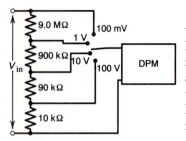

FIGURE 16-3 A multirange digital voltmeter.

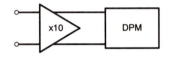

FIGURE 16-4 An amplifier used with a DPM to extend the range of the DPM.

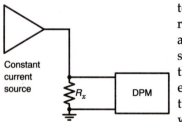

FIGURE 16-5 An ohmmeter made from a DPM and a constant current source.

A multirange voltmeter can easily be made by using a voltage divider with a multiple-position rotary switch, as shown in Figure 16-3. If the DPM has a full scale of 100 mV, with the values shown in Figure 16-3, there will be full-scale ranges of 100 mV, 1.0 V, 10.0 V, and 100 V. Additional ranges can be added in the obvious manner. In designing this multirange digital voltmeter, the loading effects of the DPM on the voltage divider chain have been ignored. For a DPM with a 1000-MΩ input resistance, this will cause only a 0.1% loading error on the 1.0-V range and less on the other scales.

But something has been lost in building this multirange voltmeter. The basic DPM had an input resistance of 1000 MΩ, whereas the four-range voltmeter has an input resistance of only 10 MΩ. This is still a rather high value; but if this instrument were being used to measure voltages in very-high-impedance circuits, this would be a serious decrease in performance. The DMM loads the circuit it is measuring just as the VOM does. However, the DMM presents a constant load—in the case considered here a load of 10 MΩ on all ranges. The VOM presents a variable load, one which changes as the range is changed.

A high-quality operational amplifier or instrumentation amplifier can be used to make a voltmeter having a full-scale value smaller than the full-scale voltage of the DPM. The circuit for making a 10-mV voltmeter from a 100-mV DPM is shown in Figure 16-4. The input resistance of this voltmeter will depend on the quality of the amplifier. Furthermore, the accuracy of the voltmeter will directly depend on the quality of the amplifier (see Chapters 29 and 32 for details).

An amplifier can also be used with a VOM to extend its range; there is nothing that restricts this trick to digital meters except tradition. VOMs have traditionally been small, simple, passive, low-technology instruments, whereas DVM's have used many electronic tricks. There have been a few VOMs manufactured including amplifiers. Some of these have apparent sensitivities of 100,000 ohms per volt or more.

THE OHMMETER

Digital ohmmeters generally take advantage of another electronic circuit. Using an op-amp, a source of a constant current can be constructed (see Chapter 29). If a constant current of known value is passed through the unknown resistor, the voltage developed across it can be measured directly by the DPM and the resistance computed from Ohm's law. If the current happens to be some power of 10, the DPM will directly read the value of the resistor after the decimal point has been put in place. A constant current of 1 mA will generate a voltage drop of 1 mV across a 1-Ω resistor, and a DPM will read 1.0—the resistance. Likewise, the same current passing through a 100 Ω resistor will result in 100 mV and a reading of 100 on the DPM. By using smaller currents, larger values of resistance can be measured (see Figure 16-5).

AC MEASUREMENTS

Alternating current measurements are made with a DMM the same as they are with a VOM. A diode changes the ac voltage into a pulsating dc voltage, and this is measured by the DPM. However, by using a number of techniques, most of the disadvantages of the VOM can be overcome. Using an op-amp in conjunction with the diode can eliminate the effects of the turn-on voltage of the diodes. Furthermore, amplifiers can be used to make a meter that reads the true peak value, the average value, or even the true RMS value. The possibilities are legion, and most of them have been used by one or another of the manufacturers. To meaningfully use a DPM on its ac ranges, the manual that comes with the instrument must be carefully read.

THE DMM

By combining a DPM and a suitable switching arrangement, an instrument can be created to measure resistance and ac or dc voltages and currents. As was true for the VOM, despite the apparent complexity of the instrument, once the range and function switches have been set, a conceptually simple instrument results.

There can be no doubt that DVMs are easier to use than are VOMs. Digital meters are easier to read, generally are more accurate, and give the user a feeling of confidence. However, they are somewhat seductive. As discussed above, these instruments can cause a circuit to change, and they can give incorrect readings just as a VOM can. While it is easy to remember that a VOM reading may be incorrect, the apparent accuracy and the assured nature of the DMM makes it easy to forget that its reading may be also incorrect.

PROBLEMS

16-1 What shunt resistor is reuired to make an ammeter having a full-scale current of 10.0 A from a DPM having a full-scale range of 100 mV? From one having a full-scale range of 10 mV?

16-2 What shunt resistor is required to make an ammeter having a full-scale current of 1.0 mA from a DPM having a full-scale range of 100 mV? From a DPM having a full-scale range of 10 mV?

16-3 What current is required to give 3-digit precision when measuring a 1.0-Ω resistor with a 100-mV DPM?

16-4 What current is required to give 3-digit precision when measuring a 10.0-MΩ resistor with a 100-mV DPM?

ANSWERS TO ODD-NUMBERED PROBLEMS

16-1 0.01 Ω and 0.001 Ω

16-3 100 mA

Chapter 17

THE OSCILLOSCOPE

OBJECTIVES

1. Become familiar with the purpose and use of the oscilloscope.

2. Know the purpose and function of each control on a typical simple oscilloscope.

3. Know some of the basic parameters used to describe the performance and input characteristics of an oscilloscope.

INTRODUCTION

This chapter is a brief introduction to the oscilloscope, its capabilities, and its use. The oscilloscope is the most versatile and useful of all electronic instruments. Almost any electronic measurement can be made with a good oscilloscope and a certain amount of technique. The oscilloscope is quite complex; it takes a long time and lots of practice to fully master its capabilities. This chapter is just a start. However, the first thing to learn is that oscilloscopes are almost universally called **scopes** in the laboratory.

Oscilloscopes come in many different sizes, shapes, capabilities, and price ranges. It is impossible to pick out one specific scope and describe it. What is described here is a hypothetical typical scope. Part of the process of learning how to use a scope is developing the ability to look at unfamiliar scope, understand its controls, and use it successfully.

The chapter begins with a brief summary of the purpose and operation of the scope, including a block diagram of the electronics that make up the scope. Then follows a discussion of the purpose and function of almost every control on the hypothetical scope. Since the scope is such a complex device, many of the controls interact. As a result, only repeated readings and much practice will make all this material understandable. If you do not have the opportunity to practice using a scope, much of this chapter will probably not be useful to you.

Finally, if there is one particular scope that you use more often than any other, make a determined effort to find the operating manual for that particular scope and read it carefully.

OSCILLOSCOPE BASICS

Expressed in the most primitive terms, the oscilloscope is a plotter. It plots points of light on the screen. The location of the point is determined by two voltages, one of which controls the horizontal position of the point and the other, the vertical position. A third voltage controls the intensity of the point.

In normal use, the oscilloscope is used to plot a voltage as a function of time. Vertical deflections are a measure of the input voltage and horizontal deflections are a measure of time, with later times displayed on the right of the screen.

The Cathode Ray Tube

The actual screen of the oscilloscope is the front of a **cathode ray tube**. This tube consists of an electron source, some grids that accelerate and focus the electron beam, two sets of plates that deflect the beam horizontally and vertically, and finally a screen coated with a phosphor that gives off light when the electron beam hits it. A schematic drawing of a cathode ray tube is shown in Figure 17-1.

The actual details of the cathode ray tube are not really suitable topics for an introduction to electronics. However, the cathode ray tube is extremely common in present-day society. Not only is it used in oscilloscopes, but the displays for computers and the picture tubes in television sets are variations on the basic design of the cathode ray tube.

A Block Diagram of An Oscilloscope

A block diagram of a simple complete oscilloscope is given in Figure 17-2. The **vertical section** of the scope electronics determines what input will be plotted, does a little filtering of the input, amplifies and sets the scale factor, and delivers the signal to the vertical deflection plates of the cathode ray tube.

The **horizontal section** is more complex. To have the trace sweep across the face of the scope at a uniform rate, a sawtooth voltage or a voltage ramp must be generated and applied to the horizontal deflection plates. The sweep speed (msec/cm) is determined by the slope (rise time) of this sawtooth voltage. Two examples are shown in Figure 17-3.

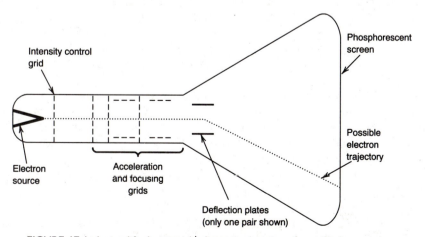

FIGURE 17-1 A simplified pictorial/schematic drawing of a cathode ray tube.

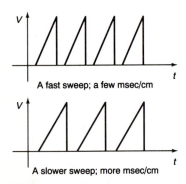

V

A fast sweep; a few msec/cm

V

A slower sweep; more msec/cm

FIGURE 17-3 Two different horizontal deflection sawtooth voltages causing different sweep speeds.

Input

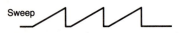

Sweep

FIGURE 17-4 An input voltage and a sweep voltage that are not synchronized.

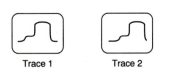

Trace 1   Trace 2

Trace 3

FIGURE 17-5 The display on the first three traces for the situation shown in Figure 17-4.

FIGURE 17-6 What the eye will see for the situation shown in Figure 17-4.

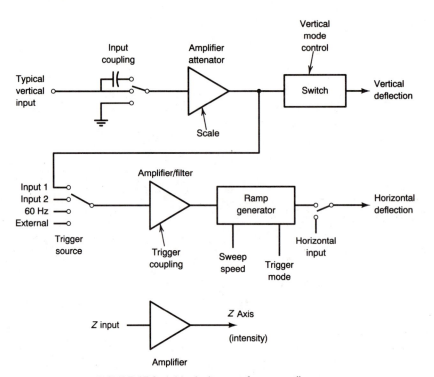

FIGURE 17-2 A block diagram for an oscilloscope.

For the trace to appear stationary on the scope, the horizontal voltage ramp must be synchronized with the vertical input signal. For instance, suppose a signal was connected to the vertical amplifier and the sweep circuit was not synchronized with this signal but rather ran as shown in Figure 17-4. Then, what would be seen on the first, second, and third traces is shown in Figure 17-5.

But if these traces are fast compared to the response of the eye (less than 30 msec or so), then the eye will see all three at the same time, and the result will be similar to that shown in Figure 17-6—or nonsense. In fact, if this series continued, the result would simply be a blur. What is needed is to somehow synchronize the starting of the trace with the input signal. This is the function of the triggering circuit.

The horizontal section of the oscilloscope electronics generates the sweep voltage. The trigger source switch determines what signal will trigger the sweep generator. The trigger coupling switch determines what part of the trigger input survives to the actual trigger circuit. The trigger mode switch sets the actual trigger condition and the sweep speed control determines the rise rate on the voltage ramp.

The **Z input** section of the scope can be used to allow an external voltage to control the intensity of the spot. There are usually no controls. The Z axis is rarely used in electronics, but it is essential in the use of a cathode ray tube in computer displays or in television.

CONTROLS

To fully explain the operation of each section of the scope, the function of each switch and/or control on the hypothetical typical scope will be explained.

Many of these controls interact with others, so several readings of this section will be required. The controls are grouped in what seem like logical groups to the author.

Basic Controls

Off-On

The oscilloscope is a complex electronic device. It must obtain power from some source, be it a battery pack or the normal ac power of the building. In any case, the off-on switch turns off or on the entire instrument. While the operation of this switch is fairly obvious, it is often true that the switch itself is hard to find.

Intensity

This controls the intensity or brightness of the spot light. (It actually controls the number of electrons per second that are hitting the screen, which determines the brightness of the spot.) In general, the spot should be no brighter than it needs to be, since high-intensity spots shorten the life of the cathode ray tube. Furthermore, allowing a bright spot to stand motionless for a long time can burn off the phosphore on the tube and produce a dead spot on the screen at that point.

Focus

This controls the sharpness or smallness of the spot. In general, a sharper spot is better.

Astigmatism

This control is present on many of the better oscilloscopes. It is really a second focus control. Because of the nature of the electron optics in the cathode ray tube, it is difficult to devise a simple system that will result in a very sharp spot both in the center and also at the edges of the screen. Adding extra focusing grids to the tube makes it possible to do a better job of maintaining a sharp focus over the entire screen. The focus and the astigmatism controls interact with each other (sometimes the intensity adjustment also interacts with these), so that it may be necessary to adjust both controls several times to get the best possible sharpness over the portion of the screen that is of interest.

Scale Illumination

Most oscilloscopes have some sort of plastic screen in front of the cathode ray tube. This screen usually has a grid engraved on it to facilitate making measurements on the scope. The scale illumination controls the ease with which these lines may be seen. Illuminating the lines makes the trace harder to see, but makes the divisions easier to see. This again is an adjustment which needs to be set to the best possible value for the current use.

Vertical Section

Most oscilloscope have at least two vertical inputs. These are usually called inputs 1 and 2 or A and B, although sometimes they are called X_1, X_2, and the like. In the following, the inputs will be called inputs 1 and 2. The following controls serve for each vertical input.

Scale

This control sets the vertical scale. That is, it determines how much the spot will be deflected by an input signal of a certain size. The usual units are either volts per centimeter or volts per division, where division refers to the grid marks on the screen. The actual input voltage can be determined by measuring the deflection and multiplying by the scale factor. Thus if the scale was set to 5 V/cm and the deflection is 1.2 cm, the input would be 6.0 V.

Attenuator (or Variable)

This is a knob that can be used to change the size of the input signal in a continuous manner—the scale switch changes the size in discrete steps. When the knob is used, the vertical scale is uncalibrated. There is always some way of locating the calibrated position on this knob. On better scopes, there will be a LED or some other indication that comes on when the scope is used in the **uncalibrated** mode. Be very careful about this knob. There is no faster way to make mistakes than to use a scope in the uncalibrated mode.

Invert

This inverts the input signal; that is, it multiplies it by -1. Thus positive input voltages become negative and cause downward deflections.

Position

This knob moves the trace up or down. It corresponds to adding a dc offset to the input signal.

×10

This makes the gain of the vertical amplifiers 10 times as great as normal; it changes the scale factor by a factor of 10. Thus if the ×10 switch is turned on and the scope is set on 0.05 V/cm, the actual scale factor is 0.005 V/cm, or 5 mv/cm. Usually, adding this extra factor of 10 of gain decreases the rise time of the vertical amplifier.

Vertical Coupling

This switch controls the coupling in the vertical amplifier. The usual choices are AC, DC, or Ground. The meanings of the various positions are as follows:

Ground. The input to the amplifier is grounded. This allows the user to find the vertical position corresponding to ground. It is useful for measuring voltages with respect to ground. On some scopes, this switch also grounds the signal source, which is plugged into the vertical input. In the author's opinion, this is a serious design error, so be careful of it.

DC. The vertical amplifier has dc coupling throughout, so that the deflection corresponds to both the ac and the dc components of the input.

AC. The vertical amplifier is ac coupled to the input. Thus the dc component of the input is blocked, and only the ac components cause deflections. This allows looking at small ac signals on large dc backgrounds. In the ac position, the scope usually does not respond to signals with frequencies below about 20 Hz.

The following controls serve for the vertical section of the scope as a whole.

Vertical Mode Control

This switch determines what is displayed on the trace. The usual choices are 1 Only, 2 Only, 1 + 2, 1 − 2, Alternate, and Chop. The meaning of each of these is briefly described below

1 Only. Only the signal at input 1 is displayed.

2 Only. Only the signal at input 2 is displayed.

1 + 2 Sum. The sum of the inputs 1 and 2 is displayed.

1 − 2 Difference. The difference between input 1 and input 2 is displayed.

Alternate. Input 1 is displayed on the first trace, then input 2 is displayed on the next trace, then input 1 again, and so on. By using the vertical position adjustment, the two traces can be separated vertically, and thus relations between the two signals can be studied. For this to be meaningful, the trigger circuit has to be set up correctly.

Chop. In this mode, first input 1 is displayed for a fraction of a microsecond, then input 2, then input 1 again, and so on. In this way, plots of both inputs can be drawn at the same time. The chop mode is useful with low repetition rate signals, while the alternate mode is useful with high repetition rate signals. Comparing the display for both chop and alternate will often show up triggering errors.

Horizontal Sweep Section

Time Base Control

Most of the horizontal section of the scope is devoted to generating a time base for the plots. That is, this switch controls the circuit which generates the voltage ramp which in turn **sweeps** the trace across the face of the scope at a constant rate. The time base control is calibrated in terms of time per centimeter (or time per division). A typical unit might be 0.1 msec/cm. The usual range on a scope is from about 0.1 sec/cm to 20 to 50 nsec/cm (remember, nsec is nanosecond, or 10^{-9} second).

One position of the time base control allows the scope to be used as an x–y plotter. For this position, a voltage applied to some input, usually the trigger input, is used to control the horizontal deflection.

Uncalibrated

The time base knob also usually has a central control that can be used to set the time base to intermediate uncalibrated values. All the comments above about using an uncalibrated setting apply. Be careful of this knob—its use can lead to serious errors.

Position

This knob can be used to shift the display left or right.

×10 (or Maybe ×5)

This switch makes the trace speed 10 times as fast and the trace 10 times as long; thus if the sweep speed is set at 10 cm/μsec and the ×10 switch is on, the functional sweep speed is 1 cm/μsec. Most of the trace will be off screen; but by using the position knob, the portion of interest can be brought into view. These two knobs can be used to get a more detailed look at some time interval well after the triggering signal.

Trigger Section

The trigger circuit generates a start signal at the time when the trigger input signal crosses some preset threshold in a preset direction. The various options are set by several switches. These are described below. The trigger circuit is the most difficult section of the oscilloscope for beginners to master. Only long practice and, unfortunately, many errors will give you the mastery you need.

Trigger Source Switch

This switch determines which signal is sent to the trigger circuit. The usual choices are Input 1, Input 2, Normal, Line, or External:

Input 1. The input to the trigger circuit comes from input 1. This is true even if only input 2 is being displayed.

Input 2. The same as above, only the names are reversed.

Normal. The channel being viewed provides the trigger signal. Thus if the alternate mode for the vertical channels is being used, the two inputs are alternately used for the trigger signal. In this case, there is no necessary time relationship between the two traces. To see time relationships between the two vertical inputs, the trigger signal must come from only one of the inputs or from the external trigger input.

Line. The input to the trigger circuit comes from the power line. This is very useful when looking for signals that are synchronized with the power line, such as hum in a hi-fi or a ripple in a power supply.

External. The input to the trigger circuit comes from the external trigger input. This setting allows an independent signal to start the sweep. Thus each signal being displayed is shown relative to the timing signal connected to the external input.

Trigger Mode Switch

The typical trigger section will have a number of additional controls. One is a switch which often has the choices Normal, Auto, or Single Sweep.

Normal. Only display a trace when the trigger condition is satisfied.

Auto. If no trigger occurs in 16 msec, generate a trace. This will keep the trace visible, but often it is meaningless.

Single Sweep (and Reset). This position results in a single trace after the reset button has been pushed and the trigger condition is met. This is useful for taking pictures of the scope trace.

Trigger Coupling Switch

This switch controls the coupling of the trigger amplifiers. There are many different settings on the various common scopes. Some of these are the following:

DC. The whole input signal, ac and dc combined, is used in the trigger logic.

AC. Only the ac component of the trigger input is used in the triggering decisions; that is, the trigger amplifier is ac coupled.

LF-Rej. The low-frequency parts of the trigger input are rejected, and only the high-frequency parts are used to trigger the sweep. The dividing line between low and high frequency differs from scope to scope; you must read the manual that comes with the scope to find this parameter.

HF-Rej. The high-frequency parts of the trigger input are rejected, and only the low-frequency parts are used to trigger the sweep. Obviously, this is the complement of the case above.

Triggering Mode Switches

This combination of switches and knobs sets the actual triggering condition. One knob sets the level at which the trigger occurs. This essentially sets the trigger level somewhere between the lowest and the highest position on the scope face. A second switch sets the direction of the trigger; this is usually marked as + and −. A plus setting means that the scope will trigger only when the trigger signal crosses the triggering level in the upward direction. A minus setting means that the scope will trigger only when the input crosses the level in the downward direction.

Hold/Stability

The knob controls the minimum delay between sweeps and also the jumpiness of the circuit. This is often a good knob to try when it is very difficult to get a stationary display.

Other Controls

There are many other controls on fancy scopes. Only experience and careful reading of the manuals for the scopes will enable you to use these controls. The next most common set of controls has to do with delayed sweep, but the use of this is far beyond the scope of a beginner. When you can meaningfully use this feature of a scope, you will not be reading this chapter.

OSCILLOSCOPE PARAMETERS

For normal scopes, there are two classes of parameters that differentiate between the cheapest and the most expensive scopes. These are the number of options and the speed of the scope.

Options

Extra options can be added to scopes until the front panels are great, huge masses of switches and knobs that remind one of an airplane control panel.

Obviously, every option adds to the cost and ultimate usefulness of the scope, but also makes it harder to use the scope. For a beginner, a simple scope is often the best one. In any case, a very brief list of some of the most common options is given below.

The trigger and sweep generation section of the scope can be made much more complex and flexible. The two most common options are to have more than one sweep generator, so that each trace may have its own independent time base; or to add a so-called **delayed sweep**, a circuit that allows the sweep to begin some adjustable delay after a trigger event, the delay being set by another knob on the scope. The typical use of a delayed sweep might be to look at some signal on a scale of 20 nsec/cm 50 msec after the trigger occurs. Obviously, either of these options makes it possible to make really horrible mistakes in using the scope.

The number of vertical inputs is another place where scopes differ greatly. The simplest scope has only one vertical input; the most common laboratory scope has two. However, scopes with four or more vertical inputs are not uncommon. Some scopes even have cathode ray tubes containing two or more complete electron beam circuits, so that multiple traces really can be drawn at the same time. Small four-beam scopes are often used with 4, 8, or 16 inputs to display multiple parameters about repetitive processes such as accelerators, power systems, or engine diagnostic systems.

For obvious reasons, it is much easier to learn how to use the simplest scope. Only when you have a considerable amount of experience should you attempt to use a more complex scope on a regular basis. With multiple inputs and delayed sweeps, it is possible to completely lose the trace and be unable to find it.

Speed

In simplest terms, the speed of a scope refers to how high a frequency the scope can display usefully. There are several ways in which this speed is measured. In general, fast scopes cost much more than do slower scopes.

Sweep Speed

Scopes that are designed for use with very fast signals will have sweep speeds as fast as 5 nsec/cm or more. Slower scopes may not go much faster than 1 μsec/cm. It should be noted that different phosphors are used for fast and slow scopes. For very slow scopes, a phosphor with long **persistence** is needed to prevent flicker in the trace; that is, the spot should continue to glow for a number of milliseconds after the electron beam has swept across it. For very fast scopes, very-short-persistence phosphors are needed to prevent blurring, but these phosphors must also produce very-high-intensity spots so that the trace can be seen.

Frequency

This is a measure of how high a frequency input signal can be displayed without loss of calibration. Slow scopes operate at up to 10 to 20 MHz, whereas the fastest scopes operate at up to 500 MHz or more.

Rise Time

This is a measure of how fast the trace will deflect if a step function input voltage is applied. The rise time is usually measured in nanoseconds. A slow scope may have a 20-nsec rise time, whereas very fast scopes will have rise times of 1 nsec or less.

Input Loading

Also of interests is the loading effect the scope has on the circuits being measured. This is similar to the loading effect either a VOM or a DMM has on the circuit it is measuring. Usually, specifications are quoted for both the horizontal and vertical inputs; the typical specification is so many megohms shunted by so much capacity. A typical scope might be 1 MΩ shunted by 25 pF. When working with either high-impedance circuits or with very high frequencies, the loading effects can be very serious.

Accuracy

The final thing to mention about an oscilloscope is the accuracy of measurements made with the scope. High-quality scopes have a number of calibration adjustments that can make the scope into a reliable measuring instrument. Because of the thickness of the scope traces, it is hard to measure things more accurately than 2% of the scope. For a typical uncalibrated laboratory scope, accuracies of better than 20% should not be expected.

ADDITIONAL OPTIONS

In their quest to produce more useful scopes, manufacturers have introduced many variations. A few of the most common of these are described in the discussion that follows.

One of the most common sets of options is a wide range of **probes**, which are used to actually connect the scope to the circuit to be measured. Almost all scopes come with some sort of a probe, a two-wire device that is designed to have one of the wires connected to ground and the other used as a probe to measure various points in the circuit. Most common probes have a relatively high impedance to minimize the loading on the circuit being measured—see Chapter 18. The calibration of the scope is usually based on the use of the normal probe.

One special probe which is often used is a **×10 probe**. This probe presents an increased load to the circuit being measured but results in a deflection 10 times as great. This probe is often useful for measurements of low-level signals in relatively low-impedance situations. Another special probe has a much higher input impedance, loads the circuit being measured much less, and attenuates the signal by a factor of 10 (makes the deflection 10 times as small). Unfortunately, these probes have also been called ×10 probes in the past. Before using any probe, check its effects on a signal of known amplitude.

Other probes are designed to be used with specific digital logic families. Such **logic probes** are designed to match the particular characteristics of one particular logic family and to facilitate the use of the scope with that logic family. The most common of these is designed to be used with the ECL family of logic.

Other scopes are provided with many different vertical inputs and a switching arrangement so that the various vertical inputs are shown at the same time on the scope. This is really an extension of the alternate mode on the vertical input to many different inputs. Such scopes are called **logic analyzers** and are useful for watching timing relationships in complex digital circuits.

Chapter 18

MEASUREMENT ERRORS

OBJECTIVES

1. Be familiar with the interaction between voltmeters and ammeters and the circuits being measured.

2. Be able to estimate or calculate the effects of these interactions on the circuits.

3. Know some of the problems with using an ohmmeter.

INTRODUCTION

It is a basic physical law that the act of measuring any quantity changes that quantity. In electronics, when the value a quantity had before it was measured is significantly different from the value it has while it is being measured, generally the value before the measurement, not the value during the measurement, is the quantity of interest. To obtain the desired result, either the effects of the measuring instrument on the quantity being measured must be reduced to the point where it is inconsequential, or the observed value must be corrected for the interaction. In either situation, it is necessary to understand how the measuring instrument interacts with the circuit being measured.

In this chapter, some simple examples of the effects of using VOMs, DMMs, or oscilloscopes to measure currents and voltages are explored. The purpose is to develop some general rules for minimizing these effects and to develop rules for estimating the magnitude of these effects.

There is a limit to how small the interaction between a circuit and an instrument can be made. This limit is stated by the Heisenberg Uncertainty Principle. But for common everyday electronics, this limit is so many orders of magnitude less than the level of the interaction caused by common instruments that it can be ignored. Only experts working in the extreme limits of microcircuits ever come close to the uncertainty principle limits.

139

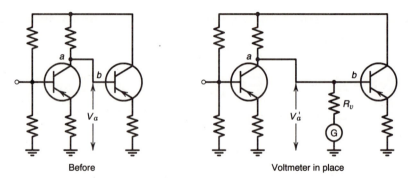

FIGURE 18-1 A circuit before and after a voltmeter has been connected between point *a* and ground.

VOLTMETERS

Figure 18-1 illustrates the basic problem in using a voltmeter. The first part of the figure shows a moderately complex circuit. To find the voltage between point *a* and ground, a voltmeter would be connected from *a* to ground. But this is equivalent to connecting a resistor from *a* to ground; the second part of the figure shows the circuit with this added resistor. Adding the resistor changes the circuit and may cause substantial changes in every voltage and current in the circuit. The problem is to be able to tell how much voltage V_a', which is what the meter will read, differs from the voltage V_a, which is the quantity of interest. (Note the general point that the meter reads the voltage between the two points **when it is the circuit**. No matter how accurate the meter, it will not read the desired voltage.)

One possible approach to this problem is to use circuit analysis and completely solve the circuit with and without the meter in place. Then, from the reading with the meter in place, it is possible to infer the value that would have occurred when the meter was not in the circuit. Obviously, this is not possible in most instances.

Since Thevenin's theorem says that an entire circuit can be replaced by a voltage source in series with a resistor, the entire circuit of Figure 18-1 can be replaced by the circuit shown in Figure 18-2. Now, the effects of the meter can be easily estimated or calculated. R_{Th} and R_v form a voltage divider. The voltmeter reads the voltage across R_v, and V_{Th} can be calculated from the equivalent circuit and the measured value. The algebra is trivial.

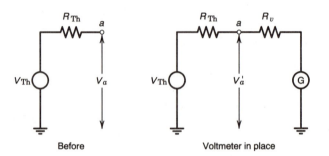

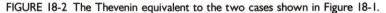

FIGURE 18-2 The Thevenin equivalent to the two cases shown in Figure 18-1.

If R_v is much larger than R_{Th}, then the voltage divider can be approximated and the result obtained with essentially no calculation. If R_v is 100 times the R_{Th}, essentially 99% of V_{Th} occurs across the voltmeter and 1% across R_{Th}; thus, the voltmeter reads about 1% low. If R_v is 1000 times the R_{Th}, then the voltmeter reads about 0.1% low.

Often, it is neither practical nor possible to calculate the Thevenin equivalent resistance for the circuit. But it can be estimated. If all the resistors in the circuit are in the range of hundreds to tens of thousands of ohms, and R_v is 10 MΩ. Then, R_v is probably at least a thousand times the R_{Th} and the meter reading is incorrect by less than 0.1%. If the resistors in the circuit are in the range of 1 MΩ and R_v is 10 MΩ, a safe assumption is that 10% errors are occurring.

From this discussion, it is easy to see that the higher the resistance of the voltmeter, the less is the influence on the circuit being measured. Thus the VOM with the most ohms per volt is the best VOM, and the DMM with the highest input resistance is the best DMM.

Finally, although all these comments have been about VOMs and DMMs, they also apply to the oscilloscope when it is being used to measure voltages. Because rise times are of interest with oscilloscopes, both the capacitive and resistive loading effects must be taken into account. Estimating these effects can be very difficult.

AMMETERS

Figure 18-3 illustrates the basic problem in using an ammeter. The first part of the figure shows a moderately complex circuit. To find the current between points a and b, an ammeter would be inserted into the wire connecting the two points. But this is equivalent to connecting a resistor between these two points. The second part of the figure shows the results of adding this resistor. An additional resistor again can cause substantial changes in the circuit. The problem is to be able to tell how much the current I'_{ab}, which is what the meter will read, differs from the current I_{ab}, which is the quantity of interest.

Just as the description of the problem is similar to the case for the voltmeter, so are the solutions. One possible approach is to use circuit analysis and completely solve the circuit with and without the meter in place. Again, from

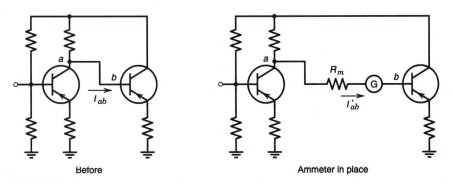

FIGURE 18-3 A circuit before and after an ammeter has been connected between points a and b.

this solution, it would be possible to infer the value of the current without the meter in place from the measured value. But again, this approach is not possible in most instances because of the complexity of the circuits involved.

Norton's theorem says that an entire circuit can be replaced by a current source and a shunt resistor. This is shown in Figure 18-4 for this case. From this equivalent circuit, the effects of the resistance of the ammeter can easily be estimated. If R_a is 1% of R_{Nor}, then the meter will read low by 1%, and so forth.

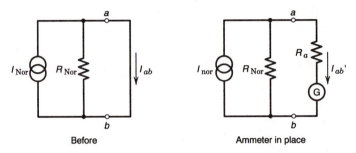

Before   Ammeter in place

FIGURE 18-4 The Norton equivalent to the two cases shown in Figure 18-3.

Finally, again for this case, the effects of the meter can be estimated by simply estimating the size of the Norton equivalent resistor. If all the resistors in the circuit have resistances of hundreds or more ohms and the ammeter has a resistance of 1 ohm, then the errors will be of the order of 1% or less. If the ammeter has a resistance that is 10% or more of the resistances in the circuit, then the errors will be 10% or greater.

From the preceding discussion, it should be very clear that the better ammeter is the one that has a lower resistance. The lower the resistance that is inserted into a circuit, the less interaction there is between the meter and the circuit being measured.

OHMMETER

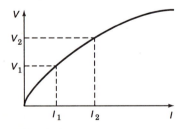

FIGURE 18-5 The voltage–current graph for a non-ohmic device.

There are three common ways in which to make mistakes in using a ohmmeter. The first way is to measure a device that is connected to a power source. this results in a measurement that is unrelated to the resistance of the device being measured. The second way is to attempt to measure a device in a complex circuit without isolating it from the rest of the circuit. This can result in a measurement of the resistance of other parts of the circuit and not of the device of interest.

The third way is to use an ohmmeter to measure a non-ohmic device. This is illustrated in Figure 18-5. In this case, the measured value of the resistance may depend on which range of the ohmmeter is being used. It may vary substantially from range to range or if different ohmmeters are used. This is because the voltage–current relationship for a non-ohmic device is not a straight line. The figure shows this. If V_1 is applied to the device, the current I_1 will result, and one particular value of the resistance will be calculated. However, if V_2 is applied, then I_2 will result, and another value of the resistance will be computed. This is not really a fault of the ohmmeter, for it is not possible to assign one resistance value to a non-ohmic device. Instead, this is a case where a nonsense measurement is being made.

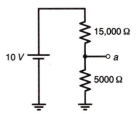
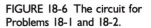
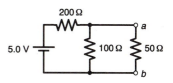

FIGURE 18-6 The circuit for Problems 18-1 and 18-2.

PROBLEMS

18-1 For the circuit in Figure 18-6, what is the voltage between point *a* and ground? What will be measured with a voltmeter on the 5-V range if the meter has a sensitivity of 10,000 Ω/V?

18-2 The same as Problem 18-1, only this time the voltmeter has an input resistance of 10 MΩ on all ranges.

18-3 For Figure 18-7, what is the current between points *a* and *b*? What will be the reading of an ammeter having a resistance of 5 Ω?

18-4 What will be the reading of the ammeter in the circuit in Figure 18-7 if it has a resistance of 0.05 Ω?

FIGURE 18-7 The circuit for Problems 18-3 and 18-4.

ANSWERS TO ODD-NUMBERED PROBLEMS

18-1 estimated errors will be 10% or less: exact value is 2.5 V; voltmeter reads 2.326 V

18-3 estimated errors will be 10% or less: exact value is 14.286 mA; meter reads 13.699 mA

PART THREE

DIGITAL ELECTRONICS

Electronics can be divided into two broad classes: analog and digital. In analog electronics, the voltages and currents can take on essentially any value within the limits set by the nature of the circuits; in digital electronics, the voltages and currents generally take on only two values. In recent years, great progress has been made in both areas, but the most spectacular advances have been in digital electronics. Techniques have been developed for using digital electronics in more and more situations that formerly were exclusively analog. Because digital technology is rapidly outrunning analog technology, at present, a general rule is: "If you can do it digitally, you can do it better and easier." Well-known examples of this progress involve computers, calculators, and TV games.

As with any other simple distinction, the division of electronics into two distinct types is imperfect. Most useful devices involve both analog and digital circuits; digital record playback systems are an excellent example of this combination. However, as a device for making learning easier, this distinction is quite useful. Furthermore, since digital electronics often appears to be somewhat simpler and more interesting, this next part of the book is an introduction to digital electronics. Part Four is an introduction to analog electronics, and subsequent parts deal with complete systems that combine both analog and digital electronics.

Again, it should be stressed that this entire part should be read more than once. No matter in what order things are explained, it is always necessary to refer to topics that have not yet been introduced to fully understand the present topic. The order in which the material is presented here minimizes the number of these forward references, but still repeated readings will help. The first three chapters of this section deal with the fundamentals of digital logic, the basic terminology, and the basic building blocks. Chapter 22 deals with practical details about packaging, families of gates, interconnections of gates, and types of outputs. The next four chapters deal with more complicated circuits made from the basic building blocks. The last chapter, Chapter 27, presents a simple outline of a computer design which illustrates the use of many of these circuits in one functioning unit. Chapter 39, which presents a schematic design of an analog-to-digital converter interfaced to a computer system, also illustrates the use of many of these circuits in one functioning unit.

PREREQUISITES

CHAPTER 19 INTRODUCTION TO DIGITAL ELECTRONICS
1. A familiarity with the concepts of voltage and current—Chapter 2—and with the conventions used in schematic diagrams—Chapter 3.

CHAPTER 20 DESIGNING LOGICAL NETWORKS
1. A knowledge of the basics of digital logic—Chapter 19.
2. A knowledge of binary numbers is needed for the example and the problem—Chapter 54.
3. Some of the terminology and ideas used in this chapter are more fully developed in the following chapters in this part of the book.

CHAPTER 21 FLIP-FLOPS
1. A knowledge of basic logic and logical gates—Chapter 19.
2. Some of the terminology used in this chapter is more fully explained in Chapter 22.

CHAPTER 22 DIGITAL LOGIC PRACTICAL DETAILS
1. A knowledge of the basic concepts of digital logic—Chapter 19.

CHAPTER 23 CLOCKS, TIMERS, AND ONE-SHOTS
1. A knowledge of *RC* circuits—Chapter 12.
2. A knowledge of simple logic—Chapter 19—and of flip-flops—Chapter 21.
3. To fully understand the internal operations of some of these circuits, a familiarity with comparators and Schmitt triggers is required—Chapter 34.

CHAPTER 24 REGISTERS, SHIFT REGISTERS, AND COUNTERS
1. A general familiarity with digital logic, with particular emphasis on flip-flops—Chapters 19–22.
2. A familiarity with binary numbers—Chapter 54.

CHAPTER 25 MORE COMPLEX COMBINATIONAL LOGIC
1. A knowledge of the basics of digital logic—Chapters 19–22.
2. A knowledge of the binary number system—Chapter 54.

CHAPTER 26 MEMORIES
1. A familiarity with digital electronics—Chapters 19 through 22.
2. A familiarity with a number of the special circuits described in Chapter 25.
3. Some familiarity with modern computer systems and terminology.

CHAPTER 27 A SIMPLE CONCEPTUAL COMPUTER

1. A familiarity with registers, shift registers, and counters—Chapter 24—data buses—Chapter 25—and memories—Chapter 26.
2. A familiarity with some computer terminology.

Chapter 19

INTRODUCTION TO DIGITAL LOGIC

OBJECTIVES

1. Know the definitions of the basic logical operations.

2. Know the basic rules for Boolean algebra and its correspondence with logic.

3. Be able to prove simple logical relationships using either truth tables or Boolean algebra.

4. Know the relationship between the operations of logical gates and the basic logical operations.

5. Know the schematic symbols for the logical gates.

6. Be able to work out the truth tables for simple logical circuits.

7. Become familiar with the packaging of gates in integrated packages.

INTRODUCTION

In analog electronics, the magnitude of the voltages and currents carry information directly; for example, a voltage may be proportional to the light intensity at some sensor. In digital electronics, on the other hand, the voltages and currents can take on only two values; any information being carried by the voltage, for example, has to be encoded in terms of high voltages and low voltages, corresponding to True and False. This means that logic, binary numbers, and many related topics form the background for an understanding of digital electronics.

In this chapter, the elements of logic and the related electronic circuits, their terminology, and their schematic symbols are introduced. Subsequent chapters deal with assembling these logical units into more complex useful circuits. The first sections of this chapter are pure mathematics; but since the rest of the chapter is electronics, this information is included here rather than in a separate mathematical chapter.

AN INTRODUCTION TO LOGIC

Logic deals with variables that are either **True (T)** or **False (F)**. No other values are permitted. Although there may be other meanings associated with the variables, within the context of logic, only the values of True and False are considered. Let's explain this in a little more detail. Assume there is a variable which, by agreement, has the following meaning: True means the sun is shining and False means the sun is not shining. However, within the context of logic, only the values of True and False are considered; the actual meaning associated with this variable is outside the context of logic. In electronics, a logical circuit deals only with the voltages corresponding to True or False, not the meaning associated with these terms. That is, a circuit responds to True or False inputs, not the statement "The sun is shining."

The ancient Greeks started the formal study of logic, and it has been going strong ever since. Many textbooks are devoted to the subject, and there are many symbolic systems for indicating logical variables and operations. See Appendix A for several references. In this text, only the simplest notation system will be used; it is the one used in most of the literature provided by the digital circuit manufacturers. Variables will be indicated by capital letters, A, B, C, and so on, and the operations discussed will be indicated in boldface letters at first. The use of boldface will quickly be phased out, except where the context demands it for clarity.

There are several logical operations that can be carried out on one or more logical variables. The most basic of these operations are **AND**, **OR**, and **NOT**. The meaning of the basic logical operations is demonstrated by means of **truth tables**. These are tables where the resultant output is indicated for every possible combination of input variables. Since each logical variable can take on only two values, True or False, the number of lines in a truth table will be some power of 2. Thus if there is only one input variable, the truth table has only two lines; if there are two input variables, the truth table has $2^2 = 4$ lines. In general, if there are N input variables, the truth table has 2^N lines. The definitions of the basic logical operations follow.

AND

The quantity (A **AND** B) is true if and only if (iff) both A and B are true. This is illustrated by the following truth table:

A	B	A **AND** B
T	T	T
T	F	F
F	T	F
F	F	F

OR

(A **OR** B) is true if either A or B is true. The truth table is

A	B	A **OR** B
T	T	T
T	F	T
F	T	T
F	F	F

NOT

The **NOT** operation converts a True into a False, and vice versa. The truth table for a **NOT** is particularly simple:

A	NOT A
T	F
F	T

Other Logical Operations

Many other logical operations can be constructed from these three basic ones. Some are so common and useful that they have been given names and symbols, as follows:

$$\text{NAND} = \text{NOT (A AND B)} \qquad \text{NAND is NOT AND}$$
$$\text{NOR} = \text{NOT (A OR B)} \qquad \text{NOR is NOT OR}$$
$$\text{XOR} \qquad \text{an exclusive OR}$$
$$\text{XNOR} = \text{NOT XOR}$$

These operations will be fully defined later in this chapter.

BOOLEAN ALGEBRA

Boolean algebra, first studied by George Boole, is the algebra of a two-state variable. In Boolean algebra, the variables can be only 0 or 1. The basic operations in the algebra are **addition**, **multiplication**, and **negation** (sometimes called **complementation**). These operations are defined by the following addition and multiplication tables:

A	B	A + B
1	1	1
1	0	1
0	1	1
0	0	0

A	B	A * B
1	1	1
1	0	0
0	1	0
0	0	0

The negation operation, indicated by a bar over the variable, changes a 0 to a 1, and vice versa. Thus

A	$\overline{A}$
0	1
1	0

A comparison of the truth tables for the basic logical operators and the multiplication and addition tables above shows that the two systems are identical once the equivalent operations and states are identified. The equivalent operations and states are

Boolean Notation	Logical Notation
1	T
0	F
+	OR
*	AND
$\overline{A}$	NOT A

With these identifications, the two systems are identical, and the two systems of notation and proof can be used interchangeably. (Once again, remember that many different notation schemes are used in the literature. Although the logical systems are always the same, the notation often looks quite different. The notation used here is consistent with that used in most electronics data books.)

LOGICAL THEOREMS

Many logical theorems can be proved by using Boolean algebra. A few of these theorems are useful in electronic design, but all of them and the methods of their proof form part of the background of the language of digital logic.

An equal sign (=) used in a logical equation means that both sides of the equation have the same value of True or False for all lines of the associated truth table. This is illustrated below.

The definitions of the basic operations, **AND, OR,** and **NOT,** are such that associativity and commutativity are built into the operations. For those students who are mathematically inclined, this means that the logical system forms an **abelian** group.

The following list contains the most important fundamental logic theorems:

(a). $A + A = A$
(b). $A * A = A$
(c). $A + 1 = 1$
(d). $A * 1 = A$
(e). $A + 0 = A$
(f). $A * 0 = 0$
(g). $A + \overline{A} = 1$

(h). $A * \overline{A} = 0$

(i). **NOT** (**NOT** A) = A

(j). $A + (A * B) = A$

(k). $A * (A + B) = A$

(l). $A * (\overline{A} + B) = A * B$

(m). $A + (\overline{A} * B) = A + B$

(n). $\overline{A} + A \cdot B = \overline{A} + B$

(o). $\overline{A} + A \cdot \overline{B} = \overline{A} + \overline{B}$

(p). $\overline{A * B} = \overline{A} + \overline{B}$

(q). $\overline{A + B} = \overline{A} * \overline{B}$

These theorems can be proved in one of two ways: either by testing all possible combinations by means of a truth table or the addition/multiplication tables, or by using theorems that have already been proved. Both these methods are illustrated here.

Theorem (l) will be proved algebraically assuming that all the theorems before (l) in the list are proved:

$$A * (\overline{A} + B) = A * \overline{A} + A * B \qquad \text{using associativity}$$

$$= 0 + A * B \qquad \text{using theorem (h)}$$

$$= A * B \qquad \text{using theorem (e)}$$

and the proof is completed.

Theorem (p) will be proved using the following truth table

A	B	A * B	$\overline{A * B}$	$\overline{A}$	$\overline{B}$	$\overline{A} + \overline{B}$
T	T	T	F	F	F	F
T	F	F	T	F	T	T
F	T	F	T	T	F	T
F	F	F	T	T	T	T
			*			*

The two columns linked together are identical, that is, they have the same values of True or False on all lines, and all combinations of the input variables are included in the truth table. Thus the proof is completed.

The last two theorems, (p) and (q), are known as **DeMorgan's laws**. They say that all logical operations can be constructed from a collection of either **NOT**s and **OR**s or **NOT**s and **AND**s. Thus if an electronic equivalent of a **NOT** and an **AND** can be constructed, these units can be combined together to form all of the basic logical operations. This fact was often used in the construction of early electronic logic.

LOGICAL GATES

The electronic circuits that carry out the basic logical operations are called **gates**, or **logical gates**. The schematic symbols for the basic logical gates are given in Figure 19-1.

AND

OR

NOT

FIGURE 19-1 The schematic symbols for the basic two-input logical gates.

Logical gates are designed to have only two possible input and output levels. One of these two levels is more positive than the other. The more positive level is indicated as **High (H)**, the other level is indicated as **Low (L)**. There are two possible choices of the way in which these output levels can be related to the logical values of True and False. The most common choice is for the High level to correspond to a True. This is called **positive logic**. The complete correspondence for positive logic is

$$H = T = 1$$
$$L = F = 0$$

However, there are situations when it is useful to employ the other possible correspondence, called **negative logic**. This correspondence is

$$H = F = 0$$
$$L = T = 1$$

The output of a logical gate is determined by the nature of the circuit and the combination of the highs and lows at the input. The name and the schematic symbol used for this circuit depends on the choice of positive or negative logic. For instance, the same circuit may be either a positive logic NAND or a negative logic NOR—check the truth tables in the next section. In general, all symbols and names used throughout this book and most of the electronics literature correspond to the positive logic choice.

The following pages summarize all of the above discussion for the common logical gates. For each gate, the definition, the Boolean symbol, the schematic symbol, and the truth tables written in three different notations are given. There is no choice but to memorize this material. The truth table in terms of Hs and Ls is the critical table; positive logic is assumed throughout this section.

NOT

FIGURE 19-2 The schematic symbol for a **NOT** gate.

This is indicated as a bar above the symbol (⁻) in Boolean algebra. It should be noted that **NOT** is sometimes called an **inverter** or referred to as **complementation**.

The symbol for the logic **NOT** gate is given in Figure 19-2.
The truth tables for a **NOT** written three ways are as follows:

A	$\overline{A}$	A	$\overline{A}$	A	$\overline{A}$
T	F	1	0	H	L
F	T	0	1	L	H

AND

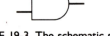

FIGURE 19-3 The schematic symbol for a two-input **AND** gate.

This is indicated as * in Boolean algebra. The symbol for the **AND** gate is shown in Figure 19-3

The truth tables for the **AND** gate are as follows:

A	B	A **AND** B	A	B	A * B	A	B	A * B
T	T	T	1	1	1	H	H	H
T	F	F	1	0	0	H	L	L
F	T	F	0	1	0	L	H	L
F	F	F	0	0	0	L	L	L

OR

FIGURE 19-4 The schematic symbol for a two-input **OR** gate.

This is indicated as + in Boolean algebra. The symbol for the **OR** gate is shown in Figure 19-4.

The truth table for the **OR** gate written three ways is as follows:

A	B	A **OR** B	A	B	A + B	A	B	A + B
T	T	T	1	1	1	H	H	H
T	F	T	1	0	1	H	L	H
F	T	T	0	1	1	L	H	H
F	F	F	0	0	0	L	L	L

NAND

FIGURE 19-5 The schematic symbol for a two-input **NAND** gate.

This is indicated as $\overline{*}$ in Boolean algebra. The schematic symbol for the **NAND** gate is shown in Figure 19-5.

The truth tables for the **NAND** gate are as follows:

A	B	A **NAND** B	A	B	$\overline{A * B}$	A	B	$\overline{A * B}$
T	T	F	1	1	0	H	H	L
T	F	T	1	0	1	H	L	H
F	T	T	0	1	1	L	H	H
F	F	T	0	0	1	L	L	H

NOR

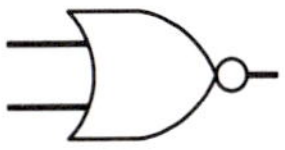

FIGURE 19-6 The schematic symbol for a two-input **NOR** gate.

This is indicated as $\overline{+}$ in Boolean algebra. The schematic symbol for the **NOR** gate is in Figure 19-6.

The truth tables for the **NOR** gate are as follows:

A	B	A **NOR** B	A	B	$\overline{A + B}$	A	B	$\overline{A + B}$
T	T	F	1	1	0	H	H	L
T	F	F	1	0	0	H	L	L
F	T	F	0	1	0	L	H	L
F	F	T	0	0	1	L	L	H

XOR

This is the exclusive OR. The Boolean algebra equivalent is $(A + B) * (\overline{A * B})$. The English language equivalent is: A or B, but not both.

FIGURE 19-7 The schematic symbol for a two-input **XOR** gate.

The schematic symbol for the **XOR** gate is shown in Figure 19-7.
The truth tables for the **XOR** gate are as follows:

A	B	A **XOR** B		A	B	A **XOR** B		A	B	A **XOR** B
T	T	F		1	1	0		H	H	L
T	F	T		1	0	1		H	L	H
F	T	T		0	1	1		L	H	H
F	F	F		0	0	0		L	L	L

XNOR

FIGURE 19-8 The schematic symbol for a two-input **XNOR** gate.

This is the exclusive NOR. The Boolean equivalent is $(A * B) + (\overline{A} * \overline{B})$. The English language equivalent is: This is an equality detector, that is, it is true if and only if A = B.

The schematic symbol for the **XNOR** gate is shown in Figure 19-8.
The truth tables for the **XNOR** gate are as follows:

A	B	A **XNOR** B		A	B	A **XNOR** B		A	B	A **XNOR** B
T	T	T		1	1	1		H	H	H
T	F	F		1	0	0		H	L	L
F	T	F		0	1	0		L	H	L
F	F	T		0	0	1		L	L	H

MULTIPLE-INPUT GATES

The definitions for the basic gates can be easily extended to cases with more than two inputs. For instance, a three-input **AND** can be thought of as two consecutive **AND**s as follows:

$$A * B * C = A * (B * C) = (A * B) * C$$

Its meaning can be illustrated by the truth table below:

A	B	C	A · B · C
T	T	T	T
T	T	F	F
T	F	T	F
T	F	F	F
F	T	T	F
F	T	F	F
F	F	T	F
F	F	F	F

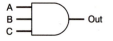

FIGURE 19-9 The schematic symbol for a three-input **AND** gate.

Finally, the symbol for a three-input **AND** is given in Figure 19-9. In a similar manner, the meanings of multiple-input **NAND**s, **OR**s, and **NOR**s and their associated schematic symbols should now be clear.

INTEGRATED CIRCUIT LOGICAL GATES

(a)

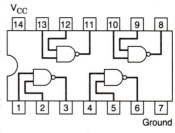

(b)

FIGURE 19-10 (*a*) A life size drawing of a 14-pin DIP package. (*b*) The pinning of a common quadruple two-input NAND integrated circuit. Both are top views.

The primary cause for the current electronics revolution is the development of the **integrated circuit**—small pieces of silicon crystals containing complete circuits (see Chapter 46 for details about these circuits). The feature which has made digital electronics so powerful is that the basic logical gates can be implemented in truly microscopic areas in an integrated circuit. Today, more than 100,000 gates can be packed into one integrated circuit, allowing circuits of immense complexity to be implemented in one integrated package. The actual integrated circuits, often called **chips**, are tiny fragile objects that must be carefully protected from air, moisture, and other environmental features. To do this, the chips are mounted in various standardized packages. The package most commonly used in the laboratory is called a **DIP** package, standing for **D**ual **I**n-line **P**ackage (see Figure 19-10 for a pictorial drawing). The normal packages for the simple logical gates are smaller than a postage stamp and have two rows of connecting **leads**, or **pins**, having a 0.1-inch spacing between the leads. These two rows of leads are what gives the package its name. DIP packages with 14 and 16 pins are the most common ones for simple logical gates, although many others exist.

For the basic logical gates, the limitations on what is available is the number of pins on the circuit packages. For instance, four independent **NAND**s require 12 leads; thus four **NAND**s and the two power leads necessary to make the circuit operate can fit into a 14-pin DIP package. For illustrative purposes only, Figure 19-10 shows an approximately life-sized drawing of a 14-pin DIP package and the pinning of the most common integrated quad-NAND gate. Note that when viewed from above, the pins, or leads, are numbered anticlockwise. Of course, when viewed from below, the pins are numbered clockwise. TABC is a good mnemonic for remembering this; it stands for "Top—Anticlockwise, Bottom—Clockwise." All DIP packages will have some indication of which end corresponds to pin 1, either a notch as shown in Figure 19-10 or a small dot near pin 1. Chapter 22 contains further information about the details of actual gates, and there is more information about this particular quad-NAND gate and others in Appendix B.

PROPAGATION DELAY

Although most technical details concerning logical gates are discussed in Chapter 22, there is one detail that needs to be kept in mind from the very beginning. This is the concept of the **gate propagation delay**. Electronic logical gates do not generate their outputs instantaneously, but rather there is a time delay before their outputs reflect the correct value based on their inputs. For instance, if both inputs of an **AND** go from Low to High at some instant of time, there is a characteristic time delay before the output goes from Low to High—the gate propagation delay. Whereas the gate delay depends on many details—such as the direction of the input transition, how many gates and how much capacitance is connected to the output of the gate, the gate type, and so on—to a good approximation, the gate propagation delay depends only on the type of gate being used. Typical propagation delays range from about 1 nsec for the fastest logic in use to a little less than 100 nsec for the slow-

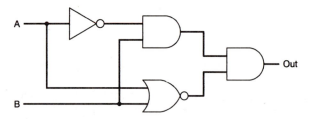

FIGURE 19-11 An example of a reverse design problem.

est logic still in use. While these times may seem incredibly short, it should be remembered that in complex situations, signals may have to propagate through 20 to 50 layers of gates. In computer design, gate delays are one of the dominant elements that limit operational speed.

REVERSE DESIGN

There are two basic types of design problems that can be posed with respect to digital circuits. The first one is, "Design a circuit whose output corresponds to a certain truth table or Boolean expression." This is discussed in the following chapter. The second question generally is asked when one tries to modify or fix existing circuits: "What does this circuit do?"; or a truth table or a Boolean expression for the output of the circuit must be worked out.

As an example of this second question, find the truth table and the Boolean expression for the output of the circuit in Figure 19-11.

A very systematic approach to the problem of reverse design is to simply build up the truth table for the output or the Boolean expression for the output one section at a time. To aid in this process, several intermediate points in the sample circuit are given names; these are shown in Figure 19-12.

Since there are only two inputs to this circuit, the truth table will have only four lines. The truth table for the intermediate result X can be written directly—an extra column for $\overline{A}$ is included for clarity:

A	B	$\overline{A}$	X
T	T	F	F
T	F	F	F
F	T	T	T
F	F	T	F

In a similar manner, the truth table for the intermediate result Y can be constructed:

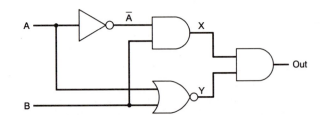

FIGURE 19-12 The circuit of Figure 19-11 with intermediate points given names.

A	B	Y
T	T	F
T	F	F
F	T	F
F	F	T

Finally, the truth table for **Out** can be constructed from these two, since **Out** is simply the **AND** of X and Y. Putting these two tables together to do this final step yields the following:

A	B	X	Y	Out
T	T	F	F	F
T	F	F	F	F
F	T	T	F	F
F	F	F	T	F

The Boolean expression for **Out** can be constructed in a similar step-by-step process. Thus

$$Out = X * Y$$
$$X = \overline{A} * B$$
$$Y = A \textbf{ NOR } B = \textbf{NOT}(A + B)$$
$$= \overline{A} * \overline{B}$$

And finally, by combining these,

$$Out = (\overline{A} * B) * (\overline{A} * \overline{B})$$
$$= (\overline{A} * \overline{A}) * (B * \overline{B})$$
$$= \overline{A} * 0 = 0$$

Obviously, this particular circuit, Figure 19-11, is not very useful. However, as an actual mistake in a design, it is a nice illustration of the use of reverse design.

PROBLEMS

19-1 Using a truth table, demonstrate that

$$A \textbf{ XOR } B = (A + B) * \overline{(A * B)}$$

19-2 Show that the following relationships are true; these are the usual associativity relationships:

$$A * (B + C) = (A * B) + (A * C)$$
$$A + (B * C) = (A + B) * (A + C)$$

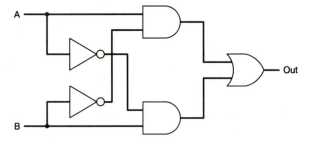

FIGURE 19-13 The circuit for Problem 19-3.

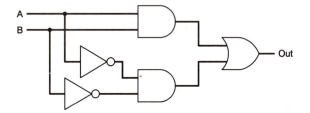

FIGURE 19-14 The circuit for Problem 19-4.

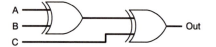

FIGURE 19-15 The circuit for Problem 19-5.

19-3 Work out the truth table for the circuit in Figure 19-13. What is this circuit?

19-4 Work out the truth table for the circuit in Figure 19-14. What is this circuit?

19-5 Work out the truth table for the circuit in Figure 19-15. State in words what this circuit does. Note that this is not a three-input exclusive OR.

ANSWERS TO ODD-NUMBERED PROBLEMS

19-3

A	B	X	Y	Out
T	T	F	F	F
T	F	T	F	T
F	T	F	T	T
F	F	F	F	F

This is the truth table of an XOR.

19-5

A	B	C	A **XOR** B	Out
T	T	T	F	T
T	T	F	F	F
T	F	T	T	F
T	F	F	T	T
F	T	T	T	F
F	T	F	T	T
F	F	T	F	T
F	F	F	F	F

The output is true when an odd number of inputs are true. This is called a parity generator. It can be extended to any number of bits.

Chapter 20

DESIGNING LOGICAL NETWORKS

OBJECTIVES

1. Be able to design a circuit to implement a given truth table.
2. Be familiar with several techniques for simplifying logical circuits.

INTRODUCTION

As stated in the last chapter, there are only two basic problems which can be posed about logical networks. One is "what does this circuit do" and the other is "design a circuit which corresponds to a certain truth table." The first of these questions, called reverse design, was discussed in the previous chapter. The second of these processes, designing a circuit corresponding to an arbitrary truth table, will be discussed here.

The process of designing a digital logical circuit for a specific application involves several steps. First the situation has to be described in terms of logical variables, then the truth table for the situation has to be filled in, and finally a circuit which will reproduce this truth table has to be designed. While strictly speaking, only the last step of this process is electronics, several problems which involve the full process are included in the chapter.

A SIMPLE EXAMPLE

As a simple example, let's design a digital logical network to implement a standard two-way light switch. The nature of a two-way light switch is as follows. There are two switches controlling one light. If the light is on, changing the position of either switch will turn the light off. Similarly, if the light is off, changing the position of either switch will turn it on.

The usual way of designating the position of a light switch is in terms of its being in the up or down position. In these terms, the way of wiring up a two-way switch described in Table 20-1 will result in the desired operation.

162

TABLE 20-1 One Way of Implementing a Two-Way Light Switch

Switch		Light
A	B	
up	up	on
up	down	off
down	up	off
down	down	on

It is easy to verify that this is a suitable truth table. If the light is off, one of the switches must be up and the other must be down. Changing either switch will result in both being up or both being down; and in either case, the light will be on. Similarly, if the light is on, both switches are up or both switches are down; changing either switch will result in the light going off.

To make this look more like a logic problem, a switch being in the up position will be called a True, and a switch being down will be a False. Furthermore, the light will be on when the output of the circuit is True. With these definitions, Table 20-1 can be written in a more standard form (with the lines numbered for convenience), as shown in Table 20-2.

TABLE 20-2 The Two-Way Switch Described by a Truth Table

Line	A	B	Light
1	T	T	T
2	T	F	F
3	F	T	F
4	F	F	T

The next step of the process is to design a circuit corresponding to Table 20-2. The most straightforward way to do this takes advantage of the fact that an **AND** has a True output on only one line of its truth table. A circuit that is an **OR** of the output of a number of **AND**s having different combinations of inputs will have a True output only for those lines of the truth table for which one of the **AND**s has a true output. By picking the inputs of the individual **AND**s suitably, the desired truth table can be implemented. This is called a sum of products solution because the Boolean expression for the output is a sum (**OR**) of products (**AND**s). Each line for which the truth table has a True output is called a miniterm. This process will be illustrated in the following discussion.

For the example being considered here, the solution is the **OR** of the miniterms corresponding to lines 1 and 4 of the truth table. The circuit corresponding to the miniterm for line 1 is shown in Figure 20-1, the circuit corresponding to the miniterm for line 4 is shown in Figure 20-2, and the complete solution to the example is given in Figure 20-3.

This solution can also be written in Boolean algebra terms. The miniterm corresponding to line 1 of the truth tables is A * B, while the miniterm corresponding to line 4 is $\overline{A}$ * $\overline{B}$. The total solution to the problem is then

FIGURE 20-1 The circuit corresponding to the miniterm for line 1 of the truth table (Table 20-2) for the two-way switch circuit.

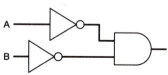

FIGURE 20-2 The circuit corresponding to the miniterm for line 4 of the truth table for the two-way switch circuit.

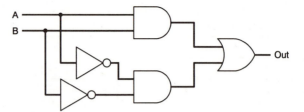

FIGURE 20-3 The complete circuit for the two-way switch example discussed in the text.

$$Out = (A * B) + (\overline{A} * \overline{B})$$

which illustrates the sum of products notation. Finally, it should be noted that the solution of this problem is the equality detector of Problem 19-4.

A MORE COMPLEX EXAMPLE

The method used above may seem much too complicated for the difficulty of the problem. However, the same method will work on a problem of greater complexity. Before discussing further techniques, a slightly more complex example is offered as an illustration. The goal is to design a circuit that will implement the truth table, Table 20-3.

TABLE 20-3 A Truth Table To Be Reproduced by a Logical Circuit

Line	A	B	C	D	Output
1	T	T	T	T	F
2	T	T	T	F	F
3	T	T	F	T	T
4	T	T	F	F	F
5	T	F	T	T	F
6	T	F	T	F	T
7	T	F	F	T	T
8	T	F	F	F	F
9	F	T	T	T	F
10	F	T	T	F	F
11	F	T	F	T	F
12	F	T	F	F	F
13	F	F	T	T	F
14	F	F	T	F	F
15	F	F	F	T	F
16	F	F	F	F	T

Again, a sum of products solution will be used. This time the miniterms will have four inputs. For instance, the circuit drawn two ways in Figure 20-4 corresponds to the miniterm for line 3 of the truth table, Table 20-3. (Remember, an AND has a True output for only one line of its truth table.) This particular miniterm can be written in Boolean terms as $(A * B * \overline{C} * D)$.

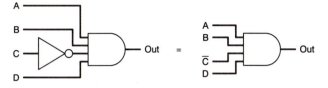

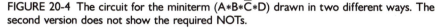

FIGURE 20-4 The circuit for the miniterm $(A*B*\overline{C}*D)$ drawn in two different ways. The second version does not show the required NOTs.

For the truth table, Table 20-3, the complete logical solution involves four miniterms. The solution in Boolean terms is

$$\text{Output} = (A * B * \overline{C} * D) + (A * \overline{B} * C * \overline{D}) + (A * \overline{B} * \overline{C} * D)$$
$$+ (\overline{A} * \overline{B} * \overline{C} * \overline{D})$$

A circuit that implements this equation is shown in Figure 20-5. Two different ways of drawing this circuit are used to illustrate the two most common techniques for drawing complex logical circuits. The author prefers the first technique, which involves fewer connections and uses symbols or names on the inputs to replace the connections. The second technique is more common in computer-drawn diagrams.

ALTERNATIVE SOLUTIONS

The circuit in Figure 20-5 is not the only circuit that will have Table 20-3 as its truth table. Other circuits can have the same truth table. Furthermore, this is not necessarily the simplest circuit that has this truth table. However, for building one-of-a-kind circuits for laboratory use, the design technique presented above is quite acceptable and perhaps even the recommended technique. The circuit can be worked out directly from the truth table; once the technique is mastered, it is very fast and straight-forward to design and implement. Also, if it should turn out that a mistake has been made in the truth table, miniterms can be added or removed easily.

There are several other methods of building circuits to represent arbitrary truth tables; one of these is very simple and elegant and is suitable for a beginner. A circuit called a multiplexer can be used to represent an arbitrary truth table that is not too large — say, up to five input lines. Multiplexers are discussed in Chapter 25, and a circuit using a multiplexer that reproduces Table 20-3 is given in Figure 25-5. Just looking at this figure, however, shows that this solution is much simpler to wire than the one given in Figure 20-5.

This method of creating circuits that represent arbitrary truth tables becomes very tedious when there are many inputs — say, more than five. However, computer design and similar situations often require creating circuits that have 8 to 16 inputs. Two techniques can be used in these situations. Although these methods are too complex for use in simple situations, they are mentioned so often in the literature and in product descriptions that a very brief description of them will be given here.

Programmable logic arrays, or **PLA**s, are circuits that can be made into arbitrary sum of products circuits. These circuits consist of arrays of NOTs,

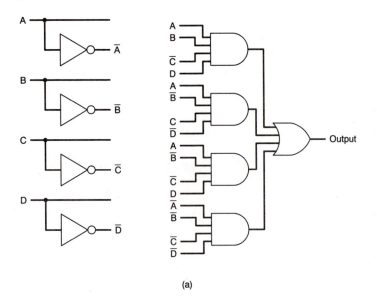

(a)

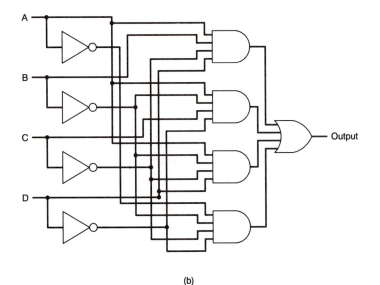

(b)

FIGURE 20-5 The circuit that reproduces the truth table, Table 20-3, drawn in two different ways. The first way (a) is clearer, the second (b) is more typical of a computer-drawn diagram.

ANDs, and ORs, without all the interconnections. The user essentially completes the manufacturing process by making these internal interconnections. This final step is done with an electronic unit controlled by a computer, most often a PC. The truth table is input to the program, either as a table or in some mathematical form so that the program can work out the truth table. Then, the electronic unit under the control of the computer "burns in" the appropriate connections inside the integrated circuit so that the circuit now represents the desired truth table. The advantages of these units is that a very large sum of products circuits can be put on one integrated circuit. For instance, the PLS153 made by Signetics can have up to 18 inputs and can generate up to 10 independent sum of products outputs from these inputs. Thus this one integrated circuit could be used to implement the entire solution to the "Final Example" worked out at the end of this chapter.

Memories, discussed in Chapter 26, can be used to implement arbitrary truth tables. The input variables are used as the address lines for the memory, and the appropriate True or False value is loaded into the memory cell. In this way, circuits corresponding to truth tables with 16 to 20 inputs can be constructed. This is a particularly useful technique in situations in which the truth table changes during the operation of the circuit.

SIMPLIFICATION OF LOGICAL NETWORKS

For a design of a commercial product where price is a consideration, it is of considerable importance to simplify a logical network to the greatest extent possible, since every component and/or connection increases the price of the product. There are a number of techniques for doing this, ranging from very informal to very formal. These techniques are extensively treated in the standard textbooks on Digital Logic Design—see Appendix A for references. Two of these techniques are discussed here. However, it should be remembered that frequently, when designing, building, and debugging a circuit, it turns out that the initial truth table needs to be changed. A circuit that has been simplified is harder to change than a simple sum of products circuit. Furthermore, in general simplified circuits are harder for other people to understand than are straightforward sum of products circuits. With this warning in mind, let us consider two techniques for simplifying circuits.

The first, easiest, and least formal way to simplify a logical network is simply to inspect the Boolean equivalent of the network for ways in which it can be simplified. The primary simplifications are cases where several miniterms can be grouped together so that they do not involve all the variables. For instance, the following miniterms can be simplified as shown:

$$(A * \overline{B} * C) + (A * B * C) = (A * C) * (\overline{B} * B) \qquad \text{due to associativity}$$

$$= (A * C) \qquad \text{theorems (g) and (d) of Chapter 19}$$

In doing this, one or more miniterms can be used several times. For instance, by using the groupings indicated, the expression below can be simplified as shown:

$$(A * B * C) + (\overline{A} * B * C) + (A * \overline{B} * C) + (A * B * \overline{C}) = (A * B) + (A * C) + (B * C)$$

Finally, this technique can be applied to the example done in the previous section. The solution worked out there can be simplified, by inspection, to

$$\text{Output} = (A * B * \overline{C} * D) + (A * \overline{B} * C * \overline{D}) + (A * \overline{B} * \overline{C} * D)$$
$$+ (\overline{A} * \overline{B} * \overline{C} * \overline{D})$$

$$= (A * \overline{C} * D) + (A * \overline{B} * C * \overline{D}) + (\overline{A} * \overline{B} * \overline{C} * \overline{D})$$

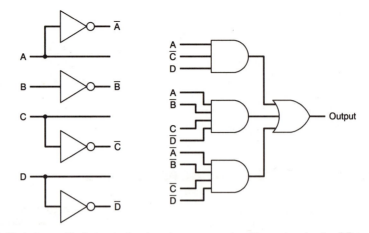

FIGURE 20-6 A simplified circuit that has the same truth table as the circuit of Figure 20-5.

This solution is illustrated in Figure 20-6; it is a clear simplification of the result shown in Figure 20-5.

Another simplification of this solution can be obtained by writing it in the form

$$\text{Output} = (A * \overline{C} * D) + (\overline{B} * \overline{D}) * (A * C + \overline{A} * \overline{C})$$

However, this involves more gates and more connections than the circuit in Figure 20-6, so it represents no further simplification.

KARNAUGH MAPS

A very elegant way for finding a simple circuit to represent a given truth table involves using **Karnaugh maps**. This is a technique that works well for systems with two, three, or four input variables. The truth table is redrawn as a two-dimensional array, the Karnaugh map. Usually, Boolean algebra notation (0s and 1s) is used. The variables are represented along the axes of the array. The map is drawn in such a way that only one variable changes when moving to an adjacent cell either horizontally or vertically on the map. Once the graph is drawn, it is inspected for contiguous regions where the cells are all True (containing 1). The regions can be represented by simple Boolean expressions. Any set of these expressions that covers all the cells containing 1s is a complete solution to the problem.

This process will be illustrated for two examples, one having three input variables and one having four input variables. Table 20-4 is a truth table for a system having three input variables. For convenience, the truth table is written in terms of 0s and 1s.

Figure 20-7 is the Karnaugh map for this problem. Note the order in which the variables change along the top axis; this way, only one variable changes when moving from cell to cell horizontally. Figure 20-8 is the same as Figure 20-7, except that the areas containing 1s are indicated, and the appropriate logical equation for these areas are also shown on the map. There are three

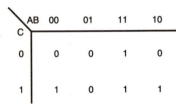

FIGURE 20-7 The Karnaugh map for Table 20-4. Note the order of the variables on the axes.

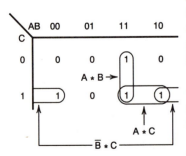

FIGURE 20-8 The same Karnaugh map as in Figure 20-7, indicating the areas containing 1s and the logical equations for these areas.

TABLE 20-4 A Truth Table To Be
Reproduced by a Logical Circuit Designed
Using a Karnaugh Map

A	B	C	Output
1	1	1	1
1	1	0	1
1	0	1	1
1	0	0	0
0	1	1	0
0	1	0	0
0	0	1	1
0	0	0	0

of these areas. The drawing indicates the important fact that the map must
be thought of as if it were drawn on a cylinder. Thus the right-hand element
is adjacent to the left-hand element on the same row. A similar statement could
be made about the elements on the top and bottom of some column if the
map had more than two rows.

After identifying all of the areas containing 1s, the logical equation for this
map can be written down. For this case, there are two possible expressions:

$$\text{Output} = (A * B) + (A * C) + (\bar{B} * C)$$

and

$$\text{Output} = (A * B) + (\bar{B} * C)$$

FIGURE 20-9 The Karnaugh map
for Table 20-3.

Of these, the second is clearly the simplest.

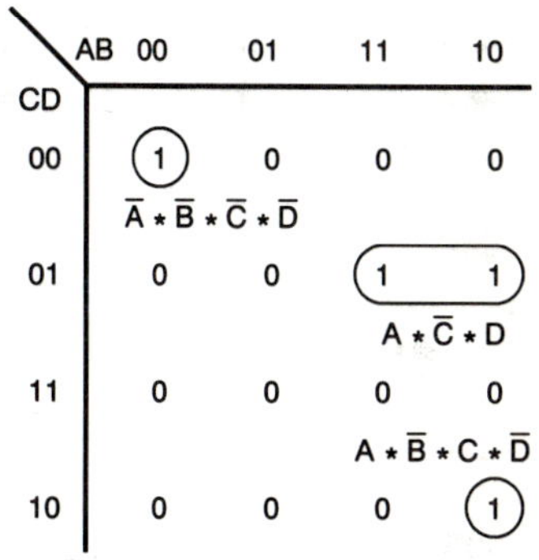

As a second example, Figure 20-9, the Karnaugh map for the truth table,
Table 20-3, is presented. Again, note the order in which the variables change
along the axes. The regions containing 1s are indicated on this map. From the
map, it is easy to verify that the logical equation for this truth table is

$$\text{Output} = (A * \bar{C} * D) + (A * \bar{B} * C * \bar{D}) + (\bar{A} * \bar{B} * \bar{C} * \bar{D})$$

Which is, of course, exactly the same result as that obtained by inspection of
the original truth table. This should come as no surprise.

A FINAL EXAMPLE

This chapter ends with a moderately complicated example of a complete design problem, starting with the description of the situation and ending with
a circuit drawing. The problem is to design a circuit whose output represents
the four-bit binary product of two two-bit binary input numbers. (See Chapter 54 for a discussion of binary arithmetic.)

The goal is to produce a circuit that has four inputs corresponding to the
two two-bit binary numbers which are to be multiplied together and four
outputs corresponding to the four-bit product. If the first input number is AB

and the second input is CD and the output is WXYZ, then the truth table must correspond to the multiplication table of

$$\begin{array}{r} AB \\ \underline{CD} \\ WXYZ \end{array}$$

For instance, if AB = 10 and CD = 11, then the product is

$$\begin{array}{r} 10 \\ \underline{\times 11} \\ 0110 \end{array}$$

As an aid to preparing the truth table, the multiplication table for two-bit binary numbers is shown as Table 20-5 here. (It includes decimal subtitles for those who do not know the binary arithmetic tables.) The truth table can be read directly from the multiplication table. It is shown as Table 20-6.

TABLE 20-5 The Two-Bit Binary Multiplication Tables[a]

CD AB	00 (0)	01 (1)	10 (2)	11 (3)
00 (0)	0000 (0)	0000 (0)	0000 (0)	0000 (0)
01 (1)	0000 (0)	0001 (1)	0010 (2)	0011 (3)
10 (2)	0000 (0)	0010 (2)	0100 (4)	0110 (6)
11 (3)	0000 (0)	0011 (3)	0110 (6)	1001 (9)

[a]Values given in parenthesis ( ) are the decimal equivalent.

TABLE 20-6 The Truth Table for the Two-Bit Binary Multiplier Described in Table 20-5

A	B	C	D	W	X	Y	Z
F	F	F	F	F	F	F	F
F	F	F	T	F	F	F	F
F	F	T	F	F	F	F	F
F	F	T	T	F	F	F	F
F	T	F	F	F	F	F	F
F	T	F	T	F	F	F	T
F	T	T	F	F	F	T	F
F	T	T	T	F	F	T	T
T	F	F	F	F	F	F	F
T	F	F	T	F	F	T	F
T	F	T	F	F	T	F	F
T	F	T	T	F	T	T	F
T	T	F	F	F	F	F	F
T	T	F	T	F	F	T	T
T	T	T	F	F	T	T	F
T	T	T	T	T	F	F	T

From this truth table, the four Boolean expressions for W, X, Y, and Z can be read off directly. They are, after some simplification,

$$W = A * B * C * D$$

$$X = A * \overline{B} * C * \overline{D} + A * \overline{B} * C * D + A * B * C * \overline{D}$$
$$= A * \overline{B} * C + A * C * \overline{D}$$

$$Y = \overline{A} * B * C * \overline{D} + \overline{A} * B * C * D + A * \overline{B} * \overline{C} * D + A * \overline{B} * C * D$$
$$\quad + A * B * \overline{C} * D + A * B * C * \overline{D}$$
$$= B * C * \overline{D} + \overline{A} * B * C + A * \overline{B} * D + A * \overline{C} * D$$

$$Z = \overline{A} * B * \overline{C} * D + \overline{A} * B * C * D + A * B * \overline{C} * D + A * B * C * D$$
$$= \overline{A} * B * D + A * B * D$$
$$= B * D$$

Finally, the circuits corresponding to these terms are all shown in Figure 20-10. Although this problem may seem like an exercise in tedium, it does illustrate one of the reasons why modern digital computers are much more powerful than those of just a few years ago. In early computers, multiplication was done by a shift-and-add algorithm—See Chapter 54. As a result, multiply instructions took a great deal of time. However, as the exercise just done shows, two two-bit binary numbers can be multiplied together in no more time than it takes for a signal to propagate through three gates. Furthermore, there is no reason why this process cannot be extended to any number of bits other than the fact that the truth table becomes prohibitively long. However, when PLAs and computer programs to produce the truth table and burn in the PLA connections are used, there is no reason why a circuit that produces the product of two eight-bit numbers with only three gate delays cannot be built; and, in fact, such circuits exist.

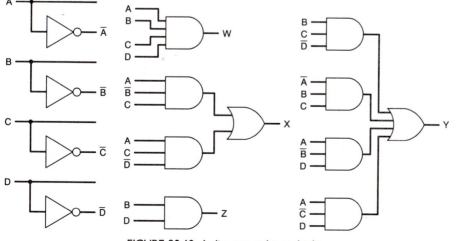

FIGURE 20-10 A direct two-bit multiplier.

PROBLEMS

20-1 Design three circuits to produce outputs corresponding to the following truth table, where X, Y, and Z are three independent outputs. Do any obvious simplifying:

A	B	C	X	Y	Z
T	T	T	F	F	F
T	T	F	F	T	T
T	F	T	F	T	T
T	F	F	T	F	F
F	T	T	F	F	T
F	T	F	T	F	F
F	F	T	T	T	F
F	F	F	F	T	T

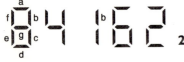

FIGURE 20-11 A seven-segment display with the various segments given names and several sample digits that can be displayed by the seven segment display.

20-2 Four binary bits can be used to count from 0 to 9 using the standard BCD coding (see Chapter 54). An array of seven lights (a seven-segment display) laid out and labeled as shown in Figure 20-11 can be used to display the ten digits 0 to 9. Several sample digits are also shown in the figure. Design circuits to control segments *b* and *e* given the four BCD input bits.

20-3 Design a four-bit digital comparator, that is, a circuit which has two two-bit binary inputs and outputs corresponding to >, =, and <, which are True if the first number is greater than, equal to, or less than the second number. Once the truth table is constructed, complete the design process only for the > output to save pointless work.

ANSWERS TO ODD-NUMBERED PROBLEMS

20-1

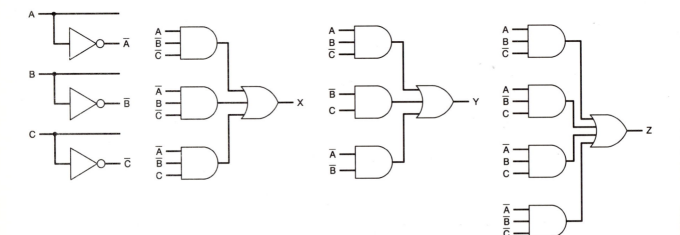

FIGURE 20-12 The solution to Problem 20-1.

20-3

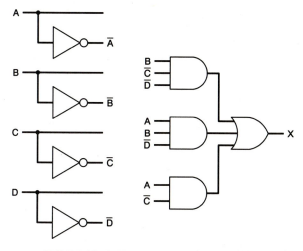

FIGURE 20-13 The solution to Problem 20-3.

Chapter 21

FLIP-FLOPS

OBJECTIVES

1. Understand the operation of the RS flip-flop.

2. Be familiar with the different types of flip-flops and know the characteristic or truth tables for each type.

3. Know the terminology used to describe digital pulses.

4. Be familiar with the different clocking schemes used in flip-flops.

INTRODUCTION

In the two previous chapters, combinational logic was considered. In this chapter, sequential logic will be presented. In **combinational logic**, the outputs depend only on the current status of the input signals (after allowing for any gate propagation delays), whereas in **sequential logic** the outputs may depend on the status of the inputs at earlier times. For combinational logic, a change in the inputs will generally result in a change in the output; this is not necessarily true in sequential logic. Thus sequential logic involves some sort of memory of the past conditions in the circuit. This memory is provided by flip-flops.

This chapter is an introduction to flip-flops, how they are constructed, the various types, the terminology used to describe them, and some of the more common uses of flip-flops. However, before actually considering flip-flops, some terminology concerning time and pulses must be introduced.

PULSES

Common to all discussions of sequential logic circuits are references to time such as, "the state of a circuit before some particular time," or "the state after ...," or "the state at a certain time." Pulses provides the means for distinguishing before from after.

Up to this point, there has been no discussion of changes in logical signals. Obviously, logical signals must change or else logic would be totally static and uninteresting. Logical signals that only change rarely are some-times called **logical levels**, whereas signals that change often are sometimes called **logical pulses**. Although this terminology may appear strange in circuits such as a computer, where most if not all the signals are frequently changing, it is still useful. Signals for which the value True or False is the important feature are called levels; signals for which the transitions are the important items are called pulses. Thus the names focus attention on different features of the signal.

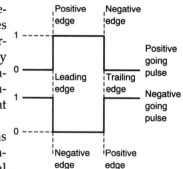

FIGURE 21-1 A positive and a negative going pulse, with the various edges labeled.

At least for conceptional purposes, it is convenient to think of a pulse as having an inactive, or False, state and an active, or True, state and, at least conceptually, most of the time being in the inactive state, with only occasional short intervals in the active state. Figure 21-1 illustrates much of the terminology used with respect to pulses. Clearly, a pulse, which is in its active state for only a very brief period of time, furnishes a fairly unambiguous definition of the periods of time designated as before, after, and during the pulse.

There is a great deal of terminology associated with pulses. Signals that consist of a series of pulses are sometimes called **pulse trains**. Signals consisting of pulses at regularly spaced time intervals are often called **clock pulses**, or even just **clocks**. As indicated in Figure 21-1, pulses can be either **positive going** or **negative going**. Since a simple inverter (NOT) will convert one into the other, there cannot be any critical differences between the two. For most of this chapter, a positive going clock pulse will be assumed. However, many common integrated flip-flops are made to work with negative clocks, so be careful when reading data books. The terminology **positive rail** for a positive going clock and **negative rail** for a negative going clock is sometimes used with flip-flops.

THE RS FLIP-FLOP

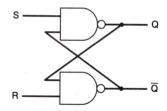

FIGURE 21-2 An RS flip-flop. The R input is the reset input S is the set input, and Q is the output.

The central element in all flip-flop circuits is the RS flip-flop shown in Figure 21-2. This is a simple circuit constructed from two two-input NANDS. The key feature is the connection from the output of one gate back to the input of the other gate. This circuit has two inputs and two outputs. The inputs are called R and S, which stand for Reset and Set; the discussion of the operation of the circuit that follows will make clear the reason for these names. The outputs are Q and $\overline{Q}$; in normal operation, one output is the complement of the other; as a result, they are given the names of Q and $\overline{Q}$ to emphasize their normal complementary nature. Further, the state of a flip-flop is usually indicated by the state of the Q output. Thus saying that a certain flip-flop is High or True means that the Q output is High and the $\overline{Q}$ output is Low.

It will be shown that this circuit has two stable states; once it is set in one of these states, it will remain unchanged until something happens at the inputs to cause it to change. Both the R and S inputs being High corresponds to the normal or unchanging state of an RS flip-flop. A low level or a negative going pulse at either input may cause the outputs to change. A Low at the S input causes Q to go High. A Low at the R input causes $\overline{Q}$ to go High. All of these statements are demonstrated in the following analysis of the RS flip-flop.

The operation of this RS flip-flop will be examined by carrying out a rather lengthy and verbose analysis of the circuit. The analysis scheme used here can be used to understand any flip-flop circuit. Thus this section serves two functions: it explores the operation of the RS flip-flop and it is a model approach to an unknown flip-flop circuit. The basic analysis scheme is to consider essentially all possible combinations of inputs and outputs.

First, all the possible states with S and R both High are examined. It will be shown that with both S and R High, Q and $\bar{Q}$ cannot both be High. To see this, assume that Q and $\bar{Q}$ are both High. Then, consider the top NAND in Figure 22-2; both of its inputs are High so its output must be Low—which contradicts the assumptions.

It is also true that with both S and R High, both Q and $\bar{Q}$ cannot be Low. To see this, assume that Q and $\bar{Q}$ are both Low. Consider the upper NAND. It has one High and one Low as its input. The truth table for a NAND says that the output is High in this condition—again this contradicts the assumptions.

With both S and R High, it is possible for Q to be High and $\bar{Q}$ to be Low. With this combination, consider the upper NAND. It has one High and one Low as its inputs, thus its output is High as assumed. The lower NAND has two Highs for its inputs, thus its output is Low, again as assumed. Thus this particular combination of inputs and outputs is acceptable and stable; that is, it does not change.

In a similar manner, the situation with S and R both High, Q Low, and $\bar{Q}$ High is also acceptable and stable.

Conclusion 1

As long as both S and R are High, the state of the flip-flop does not change and one and only one of the two outputs can be High.

Thus as promised above, one output is the complement of the other during normal operation. This is the stable or **quiescent state** of the flip-flop.

The next step in the analysis of the circuit is to investigate what happens when one or the other of the inputs receives a negative going pulse. First, the case where a negative going pulse occurs on the R input is considered. This involves looking at both possible initial states and determining what happens at the leading edge and trailing edge as well as during the pulse.

The flip-flop starts in one of its two stable states, with both R and S High and only one of Q and $\bar{Q}$ High. First, assume that Q is High and $\bar{Q}$ is Low. As shown above, this is a stable state; that is, nothing happens.

At the leading edge of the pulse, the R input goes negative. Because of this, the lower NAND now has one High and one Low input. Hence, the output of the lower NAND ($\bar{Q}$) must become High. After this change, the upper NAND has two Highs as input, hence its output (Q) must become Low. This change means that the lower NAND then has two Lows as its input, but also that its output must be High, so no further change takes place in either NAND as long as the R input remains low. A brief summary of this, because the lower input changed, the output of the lower NAND changed; because of this change, the output of the upper NAND changed. This last change caused no further changes.

Finally, at the trailing edge of the negative going pulse on the R, the R input returns to the quiescent state of High, and nothing happens. The outputs remain unchanged in the state they were in while the R input was Low. Thus Q is Low and $\bar{Q}$ is High.

This can be summarized as follows:

Initial state (before the pulse)   Q is High
Input   Negative going pulse on R
Final state (after the pulse)   Q is Low

Next, consider what happens if Q were initially Low and a negative going pulse occurred on the R input. The analysis proceeds similarly to the case above, but it is much shorter.

Because Q is Low, this means that $\bar{Q}$ is initially High. When R goes Low, the lower NAND gate now has two Lows as its inputs. But this also means that its output should be High; thus there is no change.

This can be summarized as follows:

Initial state   Q is Low
Input   Negative going pulse on R
Final state   Q is Low

Thus no matter what the initial state, a negative going pulse on the R input will result in the flip-flop being Low after the pulse—remember the state of the flip-flop is often specified by the state of the Q output.

Because of the symmetry of the circuit, exactly the same arguments will apply to the case of a negative going pulse on the S input, as long as all references to Qs are replaced by $\bar{Q}$s and all references to lower NANDs by upper NANDs, and so on. The conclusion of this is that no matter what the initial state, a negative going pulse on the S input will result in the state of the flip-flop being High after the pulse.

Conclusion 2

No matter what the initial state of the flip-flop, a negative going pulse on the S line results in the flip-flop being **set (S)** to the High state, whereas a negative going pulse on the R line results in the flip-flop being set to the Low state; this is often called **reset (R)**—hence the terminology R and S.

The final condition to be investigated is what happens when negative going pulses occur on both the R and S lines at the same time.

If both R and S are Low, then both the Q and $\bar{Q}$ outputs are High. There is absolutely no ambiguity about this state. However, when both R and S go back High, then only one or the other of Q and $\bar{Q}$ can remain High. If one of the two inputs goes back to the quiescent High state before the other, the one that goes High last determines what the final state of the flip-flop will be. If both the R and S go High at the same time, the final state will depend on random features of the circuit (which gate is slower), and the final state is unpredictable. The usual terminology used for this situation is to say that the final state is **indeterminate**.

Because the inputs of the flip-flop discussed above respond to negative going inputs, this circuit is more properly called an $\overline{RS}$ flip-flop. The circuit diagram is given in Figure 21-3. However, the terminology used in this book

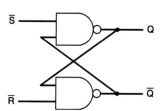

FIGURE 21-3 An $\overline{RS}$ flip-flop.

and in most other sources is quite imprecise in this respect; this flip-flop will simply be called an RS flip-flop.

There are several different ways to summarize the operation of the flip-flop described above, two of which are shown below (Tables 21-1 and 21-2). These tables give the state of the output of the flip-flop after the changes at the input indicated in the table have taken place. In these tables, the symbol Q_0 means "the same as the initial state" or "no change." Table 21-1, for the $\overline{RS}$ flip-flop, is similar to a truth table for a normal gate. The indeterminate in the final state refers to what will happen when the two inputs are no longer Low, as discussed above.

TABLE 21-1 A Characteristic Table for the $\overline{RS}$ Flip-Flop[a]

		The State After	
$\overline{R}$	$\overline{S}$	Q	$\overline{Q}$
H	H	Q_0	$\overline{Q}_0$
L	H	L	H
H	L	H	L
L	L	Indeterminant	

Q_0 means no change.
[a]This table emphasizes the "before and after" aspects of the flip-flop.

Table 21-2 emphasizes that the inputs usually are pulses and shows the condition of the flip-flop after the pulses have occurred. This is sometimes called the **characteristic table** for the flip-flop.

TABLE 21-2 Another Characteristic Table for the $\overline{RS}$ Flip-Flop[a]

		The State After	
$\overline{R}$	$\overline{S}$	Q	$\overline{Q}$
H	H	Q_0	$\overline{Q}_0$
⊔	H	L	H
H	⊔	H	L
⊔	⊔	Indeterminant	

[a]This table emphasizes the pulse nature of the inputs. The ⊔ symbol represents a negative going pulse.

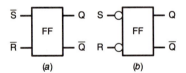

(a)   (b)

FIGURE 21-4 The schematic symbols for the RS flip-flop.

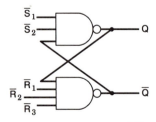

FIGURE 21-5 A multi-input $\overline{RS}$ flip-flop.

The schematic symbols used for the RS flip-flop are shown in Figure 21-4. The version in (b) of the figure has NOTs on the inputs and, hence, responds to positive going pulses on the R and S lines.

There is no reason why there need be only one R and one S line in an RS flip-flop. The circuit shown in Figure 21-5 is an RS flip-flop with two $\overline{S}$ lines and three $\overline{R}$ lines. By carrying out an analysis patterned after the one above, it can be shown that only one of the $\overline{S}$ inputs needs go low to cause the flip-flop to set. Thus the $\overline{S}$ lines and the $\overline{R}$ lines are ORed together by the NAND. (This is another result of DeMorgan's laws.) There is no limit on how many inputs can be ORed together this way.

For historical reasons, there is a certain amount of obscure terminology used with respect to flip-flops and related circuits. The flip-flop is often called a **bistable multivibrator**, since it has two stable states. In contrast, a **monostable multivibrator** has only one stable state; if it is put into its unstable state, it will return to its stable state after a fixed amount of time. Thus a monostable multivibrator can be used as a pulse generator. An **astable multivibrator** has no stable states. As a result, it is an oscillator; it simply keeps changing back and forth. The astable and mono-stable multivibrators are discussed in Chapter 23.

AN ALTERNATIVE RS FLIP-FLOP CONFIGURATION

Flip-flops can also be constructed from NORs. The circuit in Figure 21-6 is also an RS flip-flop. This version is somewhat easier to understand than the one presented before. You should be able to carry out an analysis similar to the one above and demonstrate the following facts:

1. In quiescent state for the flip-flop in Figure 21-6, both the R and S lines are Low.
2. Just as for the previous version of the flip-flop, in the quiescent state, one and only one of the two outputs can be High, the other must be Low.
3. This version of the flip-flop responds to positive going pulses on the input lines. Thus a positive going pulse on the R line results in the flip-flop being set to the Low state, whereas a positive going pulse on the S line results in the flip-flop being set to the High state.

As shown in Figure 21-6, this version of the RS flip-flop is represented by the same schematic symbol as the previous version. This only emphasizes the need to read the data sheets carefully when determining which polarity of signals the inputs of any particular flip-flop require.

While, this version of the RS flip-flop is easier to understand than the one presented first, unfortunately, for technical reasons, it is easier and more efficient to implement flip-flops in integrated circuits with NANDs than with NORs. Hence, most integrated flip-flops are constructed with NANDs, and it is necessary to understand the first version of the RS flip-flop to understand the circuits used in building the more complex flip-flops discussed below.

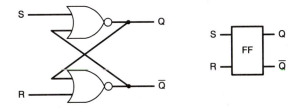

FIGURE 21-6 An RS flip-flop made from NORs and its schematic symbol.

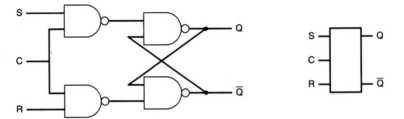

FIGURE 21-7 A clocked RS flip-flop and its schematic symbol. C stands for clock.

CLOCKED RS FLIP-FLOPS

The RS flip-flop discussed above is always sensitive to its inputs. Thus any pulse, including a noise pulse or a glitch on the S line, will result in the flip-flop being set. For some purposes, this is fine. For instance, if a circuit were set up to monitor the opening of a door, a simple RS flip-flop would be what is required. However, in complex circuits, of which calculators and computers are the most familiar but by no means the only examples, there are often times when inputs may not be meaningful. During these times, it is necessary to turn off the inputs of any flip-flops to avoid incorrect settings. As an example, consider what happens when two long binary numbers are presented to the input of a digital adder or multiplier. For some period of time, the outputs are not correct; the length of this time depends on the number of gates through which the signals have to propagate in order to generate the correct answer.

The solution to this kind of problem is to used **clocked** flip-flops, that is, flip-flops whose inputs can be turned on and off depending on the status of some signal called the clock. Sometimes such circuits are called **synchronous** circuits, since they may all operate in synchronism owing to a common clock signal. The circuit in Figure 21-7 is a clocked RS flip-flop. As long as the clock (C) is Low, it does not matter what R and S are. Only when C is High do R and S matter. Because of the NAND in the gating part of the circuit, R and S are sensitive to positive going signals.

In the normal description of the clocked RS flip-flop, it is usually assumed that R and S are levels that will be set up, then a short clock pulse will occur, and as a result the output will change or not change as the case may be. This convention is used in drawing up the characteristic or truth table, Table 21-3. However, for the circuit above, this is not the only mode in which it can operate. During the time when the clock is High, the R and S inputs can change

TABLE 21-3 The Characteristic Table for the Clocked RS Flip-Flop Shown in Figure 21-7

S	R	C	State After Pulse	
			Q	$\overline{Q}$
X	X	L	Unchanged	
L	L	⊓	Unchanged	
H	L	⊓	H	L
L	H	⊓	L	H
L	L	⊓	Indeterminant	

and the output of the flip-flop may change in response to these changes. In this case, the output after the clock pulse will be the status of the flip-flop at the trailing edge of the clock pulse. In a sense, the RS circuit follows the changes while the clock is High.

The basic RS flip-flop can be combined with a bewildering combination of clocking and gating circuits to generate a wide variety of more complex flip-flops. Rather than attempting (and failing) to describe all these combinations, the general types of clocking available, and the general functional types of flip-flops available, will be discussed. A study of any of the data books produced by the integrated circuit manufacturers — see Appendix A — will show that almost all possible combinations of these two characteristics are available.

CLOCKING TYPES

In computers and calculators (and many other circuits), it frequently occurs that the output of a flip-flop is combined in some combinational logic to form part of the input to the same flip-flop. For instance, in a computer or calculator, the number in the accumulator is added to another number and the sum put back into the accumulator. If the accumulator were an array of flip-flops with a simple gating system such as the one shown in Figure 21-8, disaster would result. For while the clock is High, the output of the accumulator will change to reflect the sum; this would become part of the number being added, a new sum would result, and so on.

To prevent this disaster, it is necessary to arrange things so that either the flip-flop is only sensitive to its input at one instant of time, or at most for only a few nanoseconds, or so that the flip-flop decides what to do at one instant and does it at another. These two cases are discussed in slightly more detail below.

Figure 21-9 shows a clock pulse in slightly more detail. Several points along the clock pulse have been given names to facilitate the discussion that follows.

Edge-triggered flip-flops are flip-flops which are built so that the state of the inputs is sampled only at the instant the clock crosses the threshold from Low to High (or vice versa). Although in general it is impossible to create a circuit that responds to only a true instant in time, it is possible to create a circuit that is sensitive to its inputs during so short a time window that any changes in the output of the flip-flop cannot possibly propagate back to the input to confuse the situation. This is because real circuits take time before their outputs change, and signals take time to propagate along wires (6 inches per nanosecond is a reasonable rule of thumb).

There are many possible ways of making an edge-triggered circuit; one very simple scheme is shown in Figure 21-10.

As the timing diagram in Figure 21-10 indicates, the output of the AND does not go True until one propagation delay (t_p) after the input A goes True. At the same time as the output of the AND changes, the output of the NOT goes False. One further propagation delay later, this causes the output of the AND to go False again. Thus the signal called Internal Clock, or Y, in the figure will be True for only one propagation delay every time there is a positive going transition on the line called External Clock, or A.

If the scheme shown in Figure 21-10 is used to generate a gating signal within an integrated circuit flip-flop, the flip-flop will only be sensitive to the

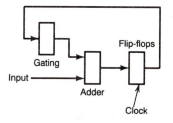

FIGURE 21-8 A situation in which the output of a flip-flop forms part of the input to the same flip-flop.

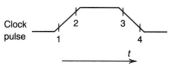

FIGURE 21-9 A clock pulse with all critical times indicated.

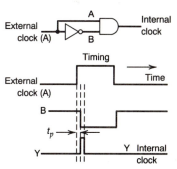

FIGURE 21-10 One way of generating a very short clock pulse.

inputs for a time equal to the propagation delay for one gate; this can easily be as short as a few nanoseconds. It is not possible to obtain a similar result by simply using a very short clock pulse in the overall circuit because the capacitance and inductance of the wiring of a circuit set a limit to the minimum-length pulses that can be used. However, since the distances and capacitance inside an integrated circuit are small, much shorter pulses can be generated and used than can be used externally. Obviously, if for some reason one gate delay results in too short a clock pulse, then several gates can be put in series to generate clock pulses which are one, two, three, and so on, gate delays long.

The master/slave concept provides another way to generate a more complex clocking type. The basic idea is to have two flip-flops connected together by some rather complex gating so that the flip-flop samples the input at one time and changes its output at another. In the master/slave configuration, the master flip-flop samples the inputs and the slave drives the outputs. The gating is arranged so that the following sequence of events happens—refer to Figure 21-9:

1. At time 1, the master and the slave are decoupled. Now, the master can change without affecting the slave.
2. During the time from 2 to 3, the inputs to the master are enabled. During this period, the master monitors the status of the input lines. If the input lines happen to change during this time, the operation of the flip-flop may become much harder to understand. (A simple laboratory exercise with the J-K master/slave flip-flop permits all possibilities to be investigated).
3. At time 3, the inputs are disconnected from the master. Thus the master will no longer change, even if the inputs change.
4. At time 4, the state of the master is transferred to the slave, and the outputs of the flip-flop change. This occurs safely after the time when the inputs are being sampled.

This operation is illustrated in the simplified timing diagram given in Figure 21-11.

Of course, true edge-triggered circuits can also be built in a master/slave configuration. In fact, the possible combinations are almost endless. The moral is that when actually using a flip-flop, read the data sheets very carefully. Pay particular attention to the timing diagrams and the discussion of the clocking and any set-up times required. In general, the schematic symbols alone do not give all the necessary information. As an example, the data sheet for one particular flip-flop is included in Appendix B.

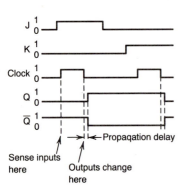

FIGURE 21-11 The timing relationships for a simple master/slave flip-flop.

FLIP-FLOP TYPES

In the preceeding section we discussed the various types of clocking that could be present in a flip-flop. In this section we discuss the functionality of the various types of flip-flops. To pick and use an actual flip-flop, both pieces of information will be needed.

RS Flip-Flops

RS flip-flops have been extensively discussed above. At present, there are integrated circuit implementations of almost any possible combination of features. These include simple $\overline{RS}$ flip-flops, RS flip-flops, and clocked RS flip-flops of several types.

D-Type Flip-Flops

A type-D, or **data**, flip-flop has a single input line. After a clock pulse, its output reflects the state of the Data line before (or during) the clock pulse. Type-D flip-flops are sometimes called **latches** because they latch (or freeze) the state of some line at the instant of the clock pulse. A partial characteristic table and a schematic symbol are given in Figure 21-12. (Omitted from the table are lines dealing with the state of the preset and clear lines—see below.) Typically, type-D flip-flops have additional inputs labeled **preset**, or **set** and **clear**. These are lines that can be used to directly set or reset the flip-flop independently of the clock and data line. (This is a case where using multiple-input NANDs to form the basic RS flip-flop is useful.) As would be expected, type-D flip-flops with both polarities of set and reset lines are available. The inevitable warning: Read the data sheets carefully.

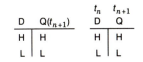

D	$Q(t_{n+1})$		t_n D	t_{n+1} Q
H	H		H	H
L	L		L	L

FIGURE 21-12 A type-D flip-flop and two ways of presenting its characteristic table.

Type-T Flip-Flop

The type-T, or **toggle**, flip-flop also has only one input. Its operation is slightly more complex. If T is Low, then nothing happens when a clock pulse occurs. If T is High, then the output complements on each clock pulse. The schematic symbol and part of the truth table are given in Figure 21-13. The usual warnings apply about reading the data sheet apply.

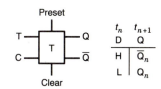

t_n D	t_{n+1} Q
H	$\overline{Q}_n$
L	Q_n

FIGURE 21-13 A type-T flip-flop and part of its characteristic table.

The J-K Flip-Flop

The J-K flip-flop is the most complex and the most powerful of the standard flip-flops. Part of the truth table and a schematic symbol are given in Figure 21-14.

Another approach to understanding the J-K flip-flop is the excitation table, Table 21-4. This table tells what must be done at the inputs of the flip-flop for a given initial state to obtain the desired output.

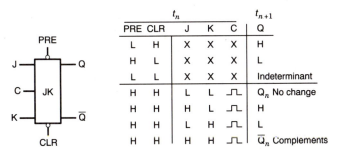

		t_n			t_{n+1}
PRE	CLR	J	K	C	Q
L	H	X	X	X	H
H	L	X	X	X	L
L	L	X	X	X	Indeterminant
H	H	L	L	⎍	Q_n No change
H	H	H	L	⎍	H
H	H	L	H	⎍	L
H	H	H	H	⎍	$\overline{Q}_n$ Complements

FIGURE 21-14 A J-K flip-flop and one view of its characteristic table. Remember that X means "don't care."

TABLE 21-4 The Excitation Table for a J-K Flip-Flop

| | | Must be | |
| | | J | K |
Initial Q_n	Want Q_{n+1}		
L	L	L	X
L	H	H	X
H	L	X	H
H	H	X	L

X means Don't care.

Note that with both the J and K inputs held High, the J-K flip-flop is a type-T flip-flop. Furthermore, with $K = \bar{J}$, the J-K flip-flop becomes a type-D flip-flop. For this reason, there are many variations on integrated J-Ks and not so many for Ts and Ds, since these can be constructed from J-Ks. The J-K flip-flop forms the basis for all counters, shift registers, and the like (see the Chapter 24).

FLIP-FLOP PARAMETERS

Here is a final summary: To pick a flip-flop for some particular purpose, the following must be specified:

1. The logic family being used—see Chapter 22.
2. The type of flip-flop needed.
3. The clocking type needed. This includes polarity, master/slave, and edge-triggered.
4. The signs of the data inputs needed.
5. How many auxiliary inputs are needed: preset, clear, and so on.
6. Timing constraints such as setup times, delays, how fast the flip-flop can toggle, and so on.

The single sample data sheet for a flip-flop in Appendix B illustrates these for a specific flip-flop.

THE MASTER/SLAVE J-K FLIP-FLOP

For reference, the circuit diagram for one possible implementation of a master/slave J-K flip-flop is given in Figure 21-15. An understanding of the operation of this circuit demonstrates a fairly good mastery of flip-flops and logic. It is quite complex, and an analysis will take more than a few minutes, so don't get discouraged too easily. In working through this circuit, remember that the critical elements are the feedback from the two outputs to the input gating circuits. The following rule results:

The master flip-flop can be changed to disagree with the slave, but it cannot then be changed back to agree with the slave.

This rule is one way of summarizing the operation of this flip-flop.

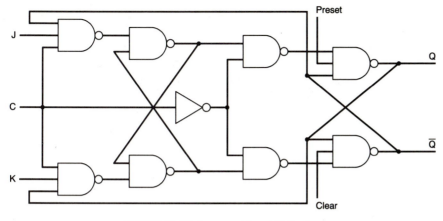

FIGURE 21-15 A master/slave J-K flip-flop.

PROBLEMS

21-1 Analyze the RS flip-flop in Figure 21-5 and show that only one of the $\overline{S}$ inputs needs to go Low to set the flip-flop, and that this is "ORing" the $\overline{R}$ and $\overline{S}$ inputs.

21-2 What happens if both the preset and clear lines in the master/slave J-K flip-flop in Figure 21-15 are held Low at the same time?

21-3 Design the simplest clocked type-D flip-flop having a positive going clock and data input, and a preset and clear line. What polarity signal is needed for the preset and clear lines?

ANSWERS TO ODD NUMBERED PROBLEMS

21-3

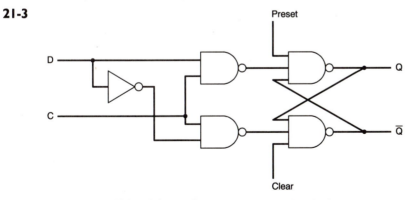

FIGURE 21-16 One solution to Problem 21-3.

Chapter 22

DIGITAL LOGIC PRACTICAL DETAILS

OBJECTIVES

1. A knowledge of a wealth of practical details and terminology concerning digital logic. These include the following:

2. Some knowledge of the history and current status of integrated circuit technology and some of the implications of this history.

3. A knowledge of possible alternative ways to implement simple gates.

4. Familiarity with the three main families of digital logic.

5. A knowledge of the important parameters used to describe logic families.

6. An acquaintance with the details of the TTL and CMOS specifications.

7. A familiarity with open collector and tri-state outputs.

INTRODUCTION

Up to this point, the discussion of digital logic has been free of practical details, with the exception of a few comments about propagation delays. However, practical considerations and technical details are important in building real circuits. So, even though much more could be discussed without inserting any technical details, the reason for many features in present-day circuits are easier to understand when the student has an understanding of the practical, technical details of logical gates, their construction, and their use. This chapter is a digression into the real world of practical details.

A number of topics are discussed in this chapter—all of them are related to the problem of building real, operating circuits. Anyone who actually wants to design and build circuits using digital logic will eventually have to master most of the material in this chapter. On the first reading, some of the material presented in this chapter may seem very complex; however, it is not as bad as it may first appear. Most of this material is relatively simple, and the more complicated parts can be summarized by simple rules.

INTEGRATED CIRCUITS

Simple logical gates can be implemented by using mechanical switches. The earliest digital computers, which were built around the end of World War II and into the late 1940s, were built with relays and vacuum tube circuits to sense the state of the relays and drive other relays. These computers were huge, building-sized machines, used vast amounts of electric power and air conditioning, were highly unreliable, since vacuum tubes burned out frequently, and were slow and not very powerful.

The transistor, invented in 1946 at Bell Laboratories, eventually provided the means that allowed computers to shrink in size and power consumption while growing in reliability and computational power. By the early 1960s, computers constructed with diodes and transistors were commercially available. These machines were regarded as being extremely powerful at the time. These early computers were built from circuit modules that were in turn built from discrete components, that is, resistors, capacitors, diodes, and transistors. In some cases, the circuit modules were designed to have functions similar to those of the basic logical gates. Thus, even then, the design of a computer involved connecting together many simple functional modules to make a total circuit that was more powerful and useful.

The **integrated circuit**, or **IC**, was invented by Jack Kirby at Texas Instruments in 1958. Integrated circuit technology made possible the construction of circuits containing many transistors and resistors in truly microscopic sizes. As a result, by the early 1970s, computers that were made from integrated circuits were commercially available. These computers were much smaller (some could even be picked up by one strong person), used much less power, and were far more powerful than the earliest computers. Integrated circuits are almost universally called **ICs** , or **chips**. The name "chip refers to the fact that the actual IC looks like a tiny chip off a silicon crystal. A short, simple description of the actual process of making an IC is given in Chapter 46. For now, the details of the circuit inside the IC may be ignored.

Since the time of the introduction of the integrated circuit, progress has continued, with the development of different families of logic circuits that can be easily implemented as integrated circuits and through the development of better techniques for cramming more and more circuitry into one integrated circuit. As a result, today, one small, cheap video game purchased from Radio Shack contains a computer more powerful than the earliest commercial computers.

The earliest digital ICs contained only a few independent gates. Because of the advances in understanding the basic semiconductor physics, in the manufacturing technology, and in the design capability, the number of gates that can be put into one IC has grown substantially. Inasmuch as this increase in circuit complex has taken place in a number of steps, names have been given to the various circuit densities. These names are

SSI	Small-Scale Integration	Roughly 1 to 10 gates in a package
MSI	Medium-Scale Integration	10 to 100 gates in a package
LSI	Large-Scale Integration	1000 to 10,000 gates
VLSI	Very-Large-Scale Integration	100,000 or more gates

These names are in general use as a way of describing the level of complexity in a particular integrated circuit. Soon another name will be needed for the next generation in the density race—doubtlessly it will the ULSI. (Ultra-large-Scale Integration).

The term **circuit density** is sometimes used for the number of gates or functions that can be built into one IC. When the current maximum possible density is plotted versus time, it is found that the maximum density has grown exponentially with time. Beginning in 1960 and running up to almost the present, the maximum number of components per IC has increased by roughly a factor of 1000 every 10 years. In recent years, the rate of increase may have decreased slightly, although this may only be a temporary fluctuation in the graph. It is this growth in complexity per IC that has been the most important driving force in the electronics explosion in recent years. At present, single ICs that can store 1,000,000 bits of information or microprocessors having over 300,000 gates are being manufactured, and even denser circuits are being developed.

Integrated circuits are generally very small. The actual silicon containing the circuit, the chip itself, for a typical SSI IC might be 1 mm thick and 2 mm square. Texas Instruments describes a typical SSI chip as about the size of the "m" on a dime. The largest current integrated circuits, the latest microprocessors, are somewhat larger than 1 cm square. The chips themselves are much too small and too delicate for humans to handle, to plug them into circuits, or to otherwise use them. Thus they have to be packaged in some manner.

Very early in the history of ICs, the manufacturers standardized on the packaging of ICs and, as a result, on almost all other modern electronic gear. The basic decision was that connections on circuit boards would be made on a grid having a 0.1-inch spacing. (The Europeans make all sorts of things with 0.254-cm spacings as a result.) ICs were to be packaged in **Dual In-Line Packages**, or **DIPs**, which had two rows of pins (or legs), the pins having a spacing of 0.1 inch in the row and the spacing between the two rows being 0.3 inch. Figure 19-10 is an example. The earliest DIPs had 14 or 16 pins; there are now DIPs available with 8, 14, 16, 20, 24, 40, and 80 pins, to mention only those the author himself has used.

Because almost all ICs have the same pin spacing, it is practical to make universal circuit boards. These boards have rows of holes on 0.1-inch spacing and power and ground lines already provided on the board. Most integrated circuits can be mounted on one of these boards and, hence, almost any kind of circuit can be built on one standard circuit board. These permit great simplification in the laboratory and manufacturing environment. Furthermore, the common use of these boards with their 0.1-inch hole spacing has generated a strong demand for resistors, capacitors, switches, connectors, and all other electrical components having leads on 0.1-inch spacing. At present, almost any component can be mounted on these boards, providing even more simplification in the electronics laboratory.

The most recent microprocessors cannot get enough pins around their perimeter using the 0.1-inch spacing standards. As a result, some very-high-density equipment is now starting to use connections on 0.05-inch centers, and some ICs have more than two rows of pins; in fact, one of the most recent microprocessors, the 80386, looks somewhat like a miniature bed of nails.

The ability to pack more and more logic into essentially the same-sized IC has had a very profound influence on the design of integrated circuits and circuits using ICs. There are two main driving forces in designing new circuits.

First, components (transistors and resistors) and interconnections cost less to make inside an IC than on a circuit board. Thus reducing the component and interconnection counts for actual circuits is one main design goal. This means putting as much as possible of a complex circuit into one chip. There is no real practical limitation here; almost any circuit that anyone might consider wiring up with SSI components can be made on a single IC. But will there be a market for it?

Second, ICs are cheap because they are manufactured in very large numbers in semiautomated processes. The manufacturer must be able to sell enough of any one type of ICs to justify the mass production and to pay for the design and development phase of the circuit. A manufacturer cannot make a very complex IC for a small specialty market. The manufacturer wants to make circuits for which the market will be at least thousands and preferably millions of units.

The result of this is a drive to design more powerful and more flexible circuits, for instance, the universal counter. Additional features, options, and the like can be added at little cost, as long as they do not require many additional external connections. Thus the manufacturer may market one IC with enough features and options built into it that it can be used in many different specialty circuits.

From the user's point of view, this means that it is possible to buy cheap circuits that have great power and complexity. The result is often very elegant, simple solutions to common problems, solutions that would be too complex to use if they were not completely implemented on one IC. It also means that the data sheets for even simple circuits, such as counters, become very large and complex, and the user must spend considerable time reading about the produce before using it.

The author has recently used a single IC, the Intersil ICM 7226, which can be used as a 10-MHz counter, a timer, a frequency meter, and a ratio calculator. This single IC contained everything needed except an external crystal to set the time scale, a switch to determine what function was being used, and the external seven-segment displays to display the result. It also came with a 20-page data sheet. However, it represented the easiest and cheapest way to implement a multidigit BCD counter.

ALTERNATE IMPLEMENTATIONS OF GATES

A quick look in any logic data book or at the short selection table in Appendix B will show that almost any simple logic function needed can be found in an integrated package. However, when actually building a circuit, the most important measure of complexity and cost is how many integrated circuits are needed. The general rule is to use as few circuits as possible. It almost always will be true that not every gate in every package will be used; yet, it almost always seems that some particular gate needed is not available in the packages already mounted on the board. Rather than adding another IC package to the circuit board, it may be possible to combine several unused gates to form the function desired. A few of the many possible combinations are shown in Figure 22-1. Some of these substitutions should be obvious; others can be verified by using truth tables or Boolean algebra. The circuits shown are only the most common substitutions. Obviously, trying to avoid having to

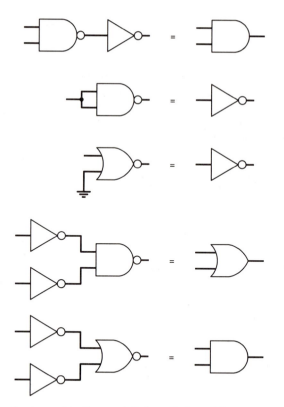

FIGURE 22-1 Several gate combinations and their equivalents. These are sometimes useful in building circuits.

add another integrated package to a circuit is very important when making changes in existing circuits.

FAMILIES OF LOGIC

Integrated logic circuits come in large families. The circuits within one family may have many common features, such as the same High and Low voltages, the same power supply requirements, the same convention for power and ground connections on the packages, and compatibility between the inputs and outputs. Within one family, inputs and outputs may be freely interconnected without much concern as long as the general rules for that particular family are followed. Interconnecting units of different families is more difficult. In some instances, the interconnections between families can be made with little difficulty, in other instances, interface circuits are needed.

The names of the various logic families refer to some feature of the gate circuits, to the manufacturing process, or to the material used. However, except for the historical interest, the names are simply names. At present there are three main families of logic in common use. These are described below.

TTL

The name stands for **transistor-transistor logic** and refers to the details of the circuit that makes up the gates. This family of logic was one of the first ones to gain widespread acceptance and use. It runs at intermediate speeds and uses moderate amounts of power. Ordinary garden-variety TTL flip-flops can operate up to speeds of 15 MHz (see the discussion that follows for more details). Its great virtues include durability (hard to burn out), ease of interconnection, and its ability to drive large multi-input and multioutput buses—see below.

Although it has been repeatedly stated in the literature that TTL is obsolete and will soon be totally replaced by new products, the sales of LSI, MSI, and SSI TTL units continue. A combination of features makes TTL a good product for an introduction to logic in the laboratory. As with all the logic families, TTL has now been divided into a number of subfamilies. These are briefly discussed in the section on details about TTL, below.

ECL

The name stands for **emitter-coupled logic**, again a detail of the gate circuit. ECL is primarily used for very-high-speed operations. It uses a relatively large amount of power and requires transmission line interconnections. As a result, the use of ECL requires a great deal of attention to cooling and circuit layout details. ECL flip-flops can operate at speeds greater than 250 MHz. The fastest portions of the more powerful modern computers are built from ECL logic.

ECL can be used successfully by nonprofessionals, but its use requires attention to all the details that are discussed in the data sheets.

CMOS

This name stands for **complementary metal oxide semiconductor**, which is the basic sequence of material used in the fabrication of the transistors in the logic circuits. CMOS ICs use the least amount of power of any current digital ICs and are generally the slowest of any of the current families. Typical CMOS flip-flops can run at speeds of 10 MHz, although more advanced subfamilies can approach the speeds of the normal TTL flip-flops. However, the power consumption of the CMOS circuits increases as the speed increases.

CMOS has several advantages. It has a larger noise immunity than the other logic families—see below. Because of its slower speed, it does not respond to very fast noise pulses or circuit glitches. Finally, because of lower power consumption, it is possible to pack more gates into one IC without running into heat dissipation problems. Thus CMOS is the preferred technology for VLSI circuits.

CMOS has the disadvantage that the inputs of the circuits can be damaged by static electricity discharge. Thus working on CMOS circuits in cool dry rooms requires special precautions to prevent the buildup of static electricity. New CMOS computers often have wrist grounding straps built into them so that the service people will be grounded to the computer before they touch any of the circuitry.

With the development of newer subfamilies of these logic lines, the distinction between the lines is being blurred. CMOS circuits that operate at normal TTL speeds are appearing; at the same time, very fast varieties of TTL are challenging ECL.

Even more important is the appearance of hybrid families of logic. Thus many current microprocessor ICs use CMOS technology internally but modified TTL-compatible inputs and outputs. Thus a typical microprocessor implementation involves one microprocessor and a number of TTL registers and buffers connected to the inputs and outputs to terminate and drive the lines connected to the outside world. (These are sometimes called the TTL glue needed to make the circuit.) Likewise, the use of ECL internal circuitry coupled with the easier to use TTL external circuitry is another common hybrid.

There are a number of other families of digital logic; some are older and going out of use; others are just being developed. Some of the families often mentioned in older textbooks include **RTL, resistor-transistor logic**, and **DTL, diode-transistor logic**. These are logic types that were common when discrete components were used to build the logic modules but they now have only historical interest. Other newer families based on newer materials, such as gallium arsenide or Josephson junctions, are being used in some experimental circuits.

DIGITAL LOGIC PARAMETERS AND RULES

Within each major family of logic, there is a consistent set of specifications and rules for the use of the circuits. In this section, the most important of these specifications are introduced and discussed in general terms. These discussions include the meaning of the specifications and typical values, but the values are for illustrative purposes only.

Every major manufacturer publishes catalogs and data books for their logic families. Before making extensive use of any logic family, this information is "must" reading. In some circumstances, the data books are available free from the manufacturers; in others, they must be purchased. The major electronic parts distributors stock and sell the books, and some of the most important data books are available at the bookstores serving major universities. Appendix A contains a list of some of the manufacturers and data books.

The major specifications for the use of the circuits are discussed below.

Power Supply Voltages

Each major family is designed to work with either a specific power supply voltage or else a restricted range of voltages. The voltages do not have to be particularly well regulated, but they typically cannot vary by more than 10% without serious consequences. The power supply lines should be well filtered because noise on the power leads can be directly coupled to the output lines. Each family has its own specific recommendations about **decoupling capacitors**, which should be connected to the power leads near the actual circuits.

For the three major families of logic, the nominal power supply voltages and the names usually are

TTL	$V_{cc} = 5.0$ V and ground.
ECL	$V_{ee} = -5.25$ V (-5.0 works quite well) and ground.
CMOS	$V_{DD} = 5.0$ V and ground; however, some CMOS units can be used with V_{DD} as high as 15 V. This has both advantages and disadvantages. See the data sheets.

Logic Voltage Levels

Besides the nominal High, V_H, and Low, V_L, output levels, several other levels are often specified. These include the guaranteed worst case levels, and frequently a decision level or a threshold level V_{th}, which is the dividing line between a High and a Low. These levels depend on the power supply voltages; thus, if the power voltage level drops, all the output logic levels will drop somewhat. A complete specification of the output levels, therefore, has to include limits on the power voltage levels. Furthermore, the details of the guaranteed levels depend on which subfamily of the logic is being used.

For the three main logic families, the nominal values are

ECL	$V_H = -0.75$ V	$V_L = -1.5$ V
CMOS	$V_H \approx V_{DD}$	$V_L = 0$
	(V_{DD} is the power supply voltage)	
TTL	$V_H = 3.2$ V	$V_L = 0.2$ V

Some of these other levels and some of the guaranteed values for the basic TTL family will also be given here as examples only—refer to the data book for any specific subfamily being used. They are

TTL The nominal decision level V_{th} is 1.2 to 1.3 V

V_L is guaranteed to be less than 0.4 V if all loading rules are obeyed.

V_H is guaranteed to be more than 2.4 V.

At an input, any voltage below 0.8 V is guaranteed to be treated as a Low, and any voltage above 2.0 V is guaranteed to be treated as a High.

Noise Immunity

There are many different approaches to defining what is meant by a **noise immunity**, or **noise margin**, specification. This specification is intended to be a measure of how large a noise pulse on an input line can be and yet not cause the logic to malfunction. The best and the most conservative definition is based on worst cases.

Thus for the TTL case above, the noise immunity is 0.4 V, since this is the worst-case separation from the output level to the input decision level. The actual noise margin might be larger, but this is the guaranteed noise margin. (This worst case is the lowest permitted value for a High output, 2.4 V, compared with the highest possible value for the decision point, 2.0 V.)

In general, ECL has a smaller noise margin than TTL. However, the exact noise margin specification depends on the methods of terminating the transmission lines used to interconnect the various units. CMOS typically has a larger noise margin than TTL.

The noise margin is often incorrectly specified as simply the difference between the nominal V_H and V_L levels. Although this permits easy superficial comparisons between logic families, it does not reflect the actual noise margins.

Output Driving Capabilities

The outputs of logic gates must be capable of supplying or sinking current. (**Sinking current** simply means the current flows into the output rather than out of it.) Specifications for the current handling capabilities of the gate outputs while in each logic state are needed to specify the full capabilities of the gates.

In general, as long as only one family of logic is being used, these specifications are not of any great importance; but they are very important when considering interfacing one family to another. Simply as an example, the regular TTL series of gates are capable of supplying up to 400 μA at the High voltage and sinking up to 16 mA at the Low state. CMOS outputs are generally not capable of providing much current, while ECL outputs switch substantial current.

Input Current Requirements

A gate input requires current flowing one way or the other to hold it at a particular logic level. This current is specified in the general specifications for the logic family. As with the specification of the output characteristics, this particular specification is not very important to the typical user of the logic family.

For the regular TTL series, a gate requires less than 40 μA to hold it High and less than 1.6 mA to hold it Low. The current required at an ECL input is dependent on the termination used at the input. For CMOS, virtually no dc current is required to hold an input at either logic level; however, the inputs have a moderately large capacitance, so a reasonable amount of charge must be supplied to change an input from High to Low, or vice versa.

Fanout

The number of inputs of logic gates that can be connected to one output and yet still operate is a number that is of great usefulness to the average user of a logic family. This number is the **fanout** for that particular family. Clearly, the fanout is determined by the specifications for the output driving capabilities and the input loading specifications.

As can be seen from the sample numbers given above, TTL units have a guaranteed fanout of 10. The ECL fanout is determined by the termination scheme being used. For CMOS, essentially an infinite fanout is possible, but the speed of operation decreases with increasing fanout because of the parallel capacitance for inputs in parallel.

Speed

The **propagation delay** of a typical gate is a characteristic of the logic family. The specifications for a family give the typical propagation delay as well as more detailed information about how this delay varies for transitions from High to Low and from Low to High, and how it varies as a function of load, as a function of power supply voltage, as a function of temperature, and other possible variables. Another parameter that is sometimes of interest is how fast the output actually changes once it starts to change. This is called the **transition time** for the gate. Very fast transition times increase the cross talk between adjacent wires and make for a noisy environment. As a result, transition times should not be very short compared with the total propagation delay. Thus normal TTL has a propagation delay of about 9 nsec and transition times of about 6 nsec. (Transition time is related to slew rate for an op-amp; it really is a measure of how fast the output transistors can switch states.)

Since actual integrated circuits may contain different numbers of gates between the input and the output, information about the propagation delays for each type of input to the output is generally given in the specifications for the individual circuits.

Unless a particular application calls for pushing a logic family to its extreme speed limits, most of these delay specifications are not very useful. All that really matters in most cases is the typical propagation delay and how fast a normal J-K flip-flop can toggle. For the basic families, these are shown in Table 22-1.

TABLE 22-1 Nominal Speeds of the Basic Logic Families

Family	Gate Delay	Toggle Rate	
TTL	9 nsec	15 MHz	Normal TTL
CMOS	9 nsec	30 MHz	High-speed CMOS
ECL	2 nsec	125 MHz	10,000 series

Interfacing to Other Families

Unfortunately, it is often necessary to use integrated circuits from two or more different families in a circuit. Thus the problem of interconnecting the various logic families always arises. Most of the manufacturers of the currently popular logic families also make integrated interface circuits. Circuits are available in the ECL family that function as interfaces to and from TTL. (Note that these circuits must have two power supplies, $+$ and -5 V.) There are special high-voltage TTL circuits that can handle the 7 or even 15 V which some CMOS circuits use.

At present, the data sheets for all the major logic families contain sections on interfacing to other families. These include typical circuits. The easiest way to deal with the interface problem is simply to follow the suggestions in the data sheets.

DETAILS ABOUT TTL

To more fully illustrate some of the types of information available for the various logic families, somewhat more detail about the TTL family will be given in this section and about CMOS in the next section. More information is given about TTL than CMOS for several reasons: because of the ubiquitousness of TTL circuits, because the properties are easy to measure in the laboratory (milliamperes and microamperes are much easier to measure than nanoamperes), and because much of the terminology used in the field comes from TTL.

The data given here are for the standard TTL line; the various TTL subfamilies are briefly summarized in the last part of this section. The data pre-

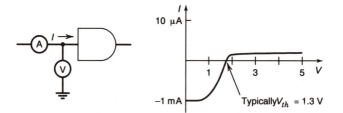

FIGURE 22-2 A typical voltage–current relationship for a TTL input.

sented here should be taken only as an illustration; read the data sheets carefully for the particular logic family you are using. Finally, a data sheet for one particular TTL (LS subfamily) gate and the CMOS equivalent is given in Appendix B.

Inputs

An open TTL input acts as if it were High. To hold an input Low, current must flow out of the input (must be pulled out of the input and somehow sunk to ground.) As a result, TTL is sometimes called **current sinking logic**. Figure 22-2 illustrates the measurement scheme and the voltage–current relationship for a typical TTL input. Note the asymmetrical scales on this graph. A negative current means that the current is flowing out of the input.

There are two quantities that are guaranteed in the specifications:

1. To hold an input high (>2.4 V) requires less than 40 μA supplied to the input.
2. To hold an input low (<0.4 V) requires less than 1.6 mA to be sunk from the input.

As can be seen from the graph, at room temperature, the typical input does far better than these guaranteed values.

One caution is in order here. While an open TTL input acts as if it were High, it is quite susceptible to noise. It is not good practice to count on an input being permanently high if it is left open. Good practice requires connecting these inputs to the power supply through a 1 kΩ resistor. Several inputs to be held High can be connected to one resistor.

Transfer Characteristic

The transfer characteristic for a gate is a plot of its output versus its input. The measurement scheme and a typical plot are shown in Figure 22-3. As can

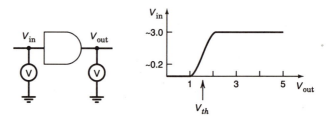

FIGURE 22-3 The transfer characteristic for a typical TTL gate.

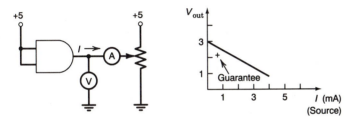

FIGURE 22-4 Voltage–current relationship for a typical TTL output that is sourcing current.

be seen from this plot, the output switches rather rapidly from Low to High when the input crosses some threshold voltage, which typically is about 1.3 V.

Output Characteristics

There are two different conditions for which the output characteristics are specified. These are when the output is trying to be High and when it is trying to be Low. Figure 22-4 shows the voltage–current relationship for an output that is trying to be a High, or is sourcing current. The guarantee for the family is that any output can source more than 400 μA with the output held above 2.4 V. As the figure shows, the typical output does somewhat better than this.

Figure 22-5 shows the voltage–current relationship for an output that is trying to be a Low, or is sinking current. The guarantee for this condition is that a gate can sink more than 16 mA of current with the output voltage remaining below 0.4 V. As the graph shows, the typical unit does far better than this at room temperature.

Fanout

By comparing the various guaranteed values, it is clear that the manufacturer guarantees that any gate can drive at least 10 inputs and remain completely within all operating specifications. An examination of the typical plots shows that a typical gate can successfully drive many more inputs than just 10. However, using this fact is not good practice. The various plots degrade rapidly as a function of temperature, and there is always a possibility of getting a unit that is far from being typical. Conservative design practice means not exceeding the fanout limit of 10.

There are special gates called **buffers**, which have extra-heavy-duty outputs with guaranteed fanouts of 15, 20, or 30, depending on the particular unit. In addition, two gates can be placed in parallel to get twice the fanout. Thus there is no real excuse other than either laziness or greed for exceeding the fanout limitations of TTL.

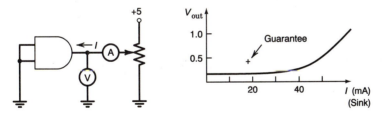

FIGURE 22-5 Voltage–current relationship for a typical TTL output that is sinking current.

Subfamilies of TTL

All of the preceding discussions have been about normal TTL circuits. However, during the 15 or so years that the family has been in heavy use, several variations on the basic gates have been devised. These have either increased the operating speed or decreased the power consumption or, more recently, both. Essentially all the SSI and MSI TTL gates have part numbers of the form 74xx; thus, the standard quad-NAND is the 7400. The other subfamilies are identified by a letter in the part number, such as 74H00 or 74LS00. Table 22-2 summarizes the various TTL subfamilies.

TABLE 22-2 Nominal Characteristics of the Various TTL Subfamilies

Family	Designation	Propagation Delay	Flip-Flop Maximum Frequency	Power Consumption
Normal	74xx	9 nsec	15 MHz	10 mW
High-speed	74Hxx	6 nsec	35 MHz	22 mW
Low-power	74Lxx	33 nsec	2.5 MHz	1 mW
Schottky	74Sxx	3 nsec	75 MHz	19 mW
Low-powered Schottky	74LSxx	9 nsec	25 MHz	2 mW

As Table 22-2 makes clear, at present the LS series is the best choice for general-purpose use. Finally, this table does not exhaust the varieties of TTL available. New subfamilies keep appearing as the various manufacturers attempt to capture the market by introducing new technological innovations.

DETAILS ABOUT CMOS

A typical CMOS input draws remarkably little current as long as it is not changing. A typical value for an input is 0.00001 μA at room temperature. The guaranteed worst case is 0.1 μA at room temperature and 1 μA over the entire operating temperature range.

A CMOS output acts like a direct connection to either ground or V_{DD} through a resistor. Thus as long as no current is flowing, the High voltage is essentially V_{DD} and the Low voltage is 0 V (ground); the current limits depend on the particular CMOS series being used. For the Motorola High Speed CMOS series, an output is guaranteed to be able to source or sink up to 4 mA without the output voltage exceeding the guaranteed values. In this way, as long as only levels are considered, CMOS fanout is almost unlimited, about 4000, but operating speed considerations limit this greatly (see the discussion that follows).

Finally, a CMOS gate as a whole requires very little power as long as it is not changing its output state. Again, a typical gate draws 0.0005 μA, and the guaranteed worst case is 0.5 μA per gate over the normal temperature range. There is an important caution here, however: Unused CMOS inputs that are not tied to either V_{DD} (through a resistor) or to ground can drift to an intermediate voltage at which the gate starts drawing large currents and may even self-destruct. Thus unused inputs must be tied to the appropriate logic levels even for gates that are not being used.

When no logic levels are changing, CMOS circuits draw very little power. However, the gate inputs and outputs and the wiring between gates all have capacitance. When an output switches from one logical level to the other, this capacitance must be charged or discharged as the case may be. This charging and discharging results in increased power consumption while the process takes place. Furthermore, the time required for an output to switch from High to Low, or vice versa, is determined by the RC time constant, where R is the effective source resistance of the gate and C is the total capacitance of the output with all the wiring and inputs tied to it. As a result, the propagation delay, the transition or switching time, and the power consumption all depend on the fanout and the nature of the wiring being used. For most units, typical values for the propagation delay and the transition times are given, assuming that a total capacitance of 50 pF is attached to the output. However, the data sheets typically give formulas to calculate the power consumption, the propagation delay, and the transition time as a function of capacitance and frequency.

The data books also give formulas to calculate the power consumption and propagation delays as a function of the operating frequency and the total capacitance involved. These expressions illustrate two important points. For the power consumption, a sample formula is

$$P_D = (C_L + C_{PD})V_{cc}^2 f + V_{CC}I_{CC} \qquad (22\text{-}1)$$

where

P_D = power dissipated, in μW

C_L = total load capacitance present at the output, in pF

C_{PD} = a measure of internal capacitance, called power dissipation capacitance, given in pF and is 20 pF per gate for this particular device

V_{CC} = power supply voltage, in V

f = operating frequency, in MHz

I_{CC} = the quiescent current, 2 μA for this device

In a similar manner, the propagation delay for any value of load capacitance other than 50 pF can be calculated. The equation for this is

$$t_{PT} = \frac{t_p + 0.5V_{CC}(C_L - 50 \text{ pF})}{I_{OS}} \qquad (22\text{-}2)$$

where

t_{PT} = total propagation delay

t_p = the propagation delay for a 50-pF load

C_L = the actual load capacitance

I_{OS} = the maximum short-circuit current, about 18 mA at room temperature.

The power consumed by a CMOS circuit rises as the operating frequency rises and at the highest operating speeds approaches the power consumption of LSTTL. The expression for the propagation delay indicates why the very large fanouts possible with CMOS are rarely used—they reduce the maximum operating speeds too much. Finally, remember that the specific values and formulas given here apply only to one particular type of circuit. It is always necessary to read the data book carefully before using any logic family.

SPECIAL OUTPUT TYPES

In designing complex instruments or computers, the situation where one gate should drive a particular line at one time and another gate should drive it at another time occurs commonly. This is somewhat analogous to a telephone party line. Although there are ways to solve this problem with normal gates, they are all expensive in terms of wire and gates. What is needed is the ability to tie the outputs of several gates to one line, as shown in Figure 22-6. This configuration is called a **wired**, or **dotted**, OR (or, by some people, a wired or dotted AND).

The intent of the configuration shown in Figure 22-6 is that at some time the wire, which is generally called a **bus** (since it can carry many different pieces of information), is sometimes to be driven by one gate and sometimes by another. Considerble care has to be taken to ensure that at no instant of time more than one gate drives the bus.

The configuration shown in Figure 22-6 cannot be implemented with ordinary gates of the types that have been discussed thus far. The output of these gates are always either High or Low; thus, the signal on the data bus will be the result of some number of gates trying to push the bus High while others are trying to pull it Low. As a result, a wired OR cannot be implemented with normal gates.

There are two common solutions to this problem. The first one introduced was the open collector gate; more recently, the tristate gate was introduced. These are each discussed below. The discussion is all in terms of TTL, since it was the TTL versions of these gates that gave the terminology to the field; but CMOS gates of these types also exist.

The output of a normal TTL gate consists of two transistors connected in series between ground and V_{CC}. Because of the shape of the schematic for this configuration, it is called a **totem pole** output. This kind of output can either sink current to ground or source current from the V_{CC} (5 V). A second feature of this output configuration is that the output is basically foolproof—it can

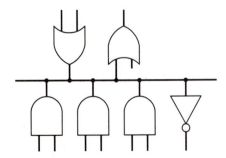

FIGURE 22-6 A wired OR configuration.

be shorted to ground, to the 5 V supply, or to anything else in a TTL circuit and not be damaged.

Some gates have only one transistor in the output; this is called the open collector output configuration. This output can sink current to ground, but it can supply no voltage to the bus. Thus an open collector can pull a line to ground but it cannot pull the line up to 5 V. Some external component will have to be used to pull the line High. A typical implementation of a number of different signals gated onto one bus using open collector output gates is shown in Figure 22-7. The resistor, about 1 kΩ, to 5 V pulls the line positive. In the figure, three different signals, S_1, S_2, and S_3, can each be separately gated onto the data bus by the separate gating signals G_1, G_2, and G_3. Although this particular example may not look very useful, when we realize that the number of gates on the data bus can be changed at will and that the gates do not have to be anywhere near each other, this becomes a very useful tool.

But there is one very important fact about data buses implemented with open collector gates. These **data buses inherently use negative logic**. Consider the bus shown in Figure 22-7:

If there are no outputs tied to the bus, the bus is High.

If there are multiple gates tied to the bus, but no gate signals are High, the bus is High.

If one particular gate signal is High, say, G_2, and S_2 is Low, the bus is still High.

If one particular gate and the associated signal are both High, say, G_2 and S_2, then the data bus is Low.

Thus the inactive state on the bus is High, and the active state is Low. Only a Low on the bus can carry any information, or the signal on the bus is assumed False unless the bus goes Low. This corresponds to negative logic. It does not cause any real problems, because inverting a signal is easy, but it does cause endless confusion in designing and debugging circuits.

Tristate outputs are a more recent and more elegant solution to the data bus problem. Although they have many advantages, their main advantage may well be that they do not require the use of negative logic and, hence, they prevent many design mistakes. The nature of the tristate gate is that, when the gate is selected, it acts like a normal gate, its output is either High or Low,

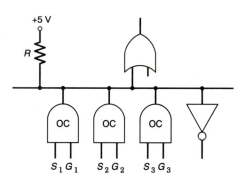

FIGURE 22-7 A Wired OR implemented with open collector gates.

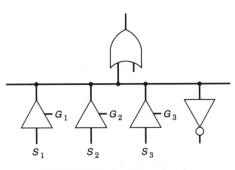

FIGURE 22-8 A tristate data bus.

and it can supply current to the bus or sink current as the case may be. However, when the gate is not selected it is essentially disconnected from the circuit, and it neither supplies nor sinks current; it functionally is simply a very large passive impedance connected to the line. An implementation of a data bus using tristate gates is shown in Figure 22-8, which also illustrates at least one schematic convention for tristate outputs.

Although the gate inputs on tristate logic are typical TTL signals, they are not combined with the data input, but merely turn the output gates off and on—that is, from their high-impedance passive state to their active state.

One of the main disadvantages of the open collector configuration is that the parallel capacitance of all the collectors attached to the bus can substantially slow down the possible speed of the bus. Tristate outputs do not present as much capacitive loading in their high-impedance (off) state. In new designs, using the tristate implementation rather than the open collector implementation, is highly recommended. Many of the VLSI units, such as memories and microprocessors, are compatible only with the tristate implementation of buses.

Chapter 23

CLOCKS, TIMERS, AND ONE-SHOTS

OBJECTIVES

1. Be familiar with the nature of and the uses of one-shots, timers, and crystal-controlled oscillators.

2. Be able to calculate the timing components for the 74121 one-shot.

3. Be able to calculate the timing components for the 555 used as an astable multivibrator.

INTRODUCTION

The design of an electronic circuit begins with a specification of the function to be performed by the circuit, Then, the circuit to carry out that function is designed. If the original function is particularly complex, an intermediate step of dividing the primary function into functional subunits may be useful. It is very important to recognize when the desired function (or subfunction) is a well-known function, has a name, and is available in an easy-to-use integrated package. Much of the ability to do this comes from experience; however, even experienced designers spend time reading the current electronics literature watching for new functional units and new ideas. A beginner has to build up a library of circuits, names, and functions in his or her mind to be able to think about problems. Aimlessly thumbing through data books, while often successful, is not very efficient. This and subsequent chapters in this section as well as many other chapters in this book are intended to provide some of this background. In this chapter, several circuits and related terminology are introduced. The intent is to provide the tools and ideas to think with, not to provide a mastery of, the circuits. Use of any of these circuits will still require the data books but, at least, you should know the name of the circuit.

This chapter contains an introduction to the three main types of digital circuits that bring time into the digital logic world. These are the circuits that either generate single pulses or series of clock pulses. The names of the circuits are the one-shot, the timer, and the crystal-controlled oscillator. For each of these circuits, the general principles and uses are briefly discussed, and one particular implementation of the circuit is illustrated.

ONE-SHOTS

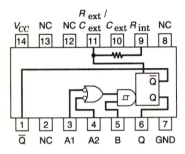

FIGURE 23-1 The pinning and function table for a 74121.

One-shots, or **monostable multivibrators,** are circuits that generate fixed-length output pulses after receiving an appropriate trigger signal. Although the details of the circuits vary widely, the length of the output is generally determined by the charging of a capacitor. A **trigger,** or **start,** signal sets the output ON and starts the charging of a capacitor. When the voltage on the capacitor reaches some predetermined value, the output is turned OFF, the capacitor is returned to its initial condition, and the circuit is ready to be triggered again. By changing the value of the capacitor or the charging current, the length of the output pulse can be varied.

A basic TTL one-shot is the 74121. Figure 23-1 shows the most important parts of the information provided in the data sheets for this circuit. There are a number of conventions being used in the function table, Table 23-1. First, the output side of the table indicates that the $\overline{Q}$ output is always the complement of the Q output, that the quiescent (inactive) state of the one-shot has the Q output Low, and that the active state results in a positive going pulse from the Q output. In the input part of the table, an X means "don't care," that is, it doesn't matter what the state is, an upward arrow means a positive transition, and a downward arrow means a negative transition.

TABLE 23-1 Function Table

Inputs			Outputs	
A1	A2	B	Q	$\overline{Q}$
L	X	H	L	H
X	L	H	L	H
X	X	L	L	L
H	H	X	L	H
H	↓	H	⊓	⊔
↓	H	H	⊓	⊔
↓	↓	H	⊓	⊔
L	X	↑	⊓	⊔
X	L	↑	⊓	⊔

Thus the function table says that if the B input is High and if one or the other of the two A inputs is High and if there is a downward transition of the other A input, the output will be triggered. Likewise, if one or the other of the two A inputs is Low (both may be Low) and if there is an upward transition on the B input, then the output will be triggered.

As the schematic diagram indicates, the B input goes to a Schmitt trigger. The data sheet for the 74121 indicates that the hysteresis for the B input is about 200 mV. This means that once the B trigger input signal has crossed the threshold level in the upward direction, it must go down by 200 mV before it can trigger the one-shot again; refer to Chapter 34 for a more extensive discussion of Schmitt triggers and hysteresis. Because of this hysteresis, very slowly rising signals can be used with the B input without danger of retriggering. The A inputs do not have Schmitt trigger inputs, so only rapidly changing inputs—that is, TTL signals—should be used on these inputs. Because of these different capabilities, the A inputs are often called edge-triggered inputs because they best respond to changing TTL signals; the B input is called a level-triggering input because it can be used with slowly changing signals or levels.

Once the output has started, its length is independent of further transitions of the inputs and depends only on the values of the external R and C. The output pulse may be between 40 nsec and 28 sec, depending on the external timing components. If no external timing components are used, that is, if R_{int} (pin 9) is connected to V_{CC} (pin 14) and no external capacitor is used, the output pulse will typically be 30 to 35 nsec (the necessary capacity is provided by the parasitic internal capacity).

For normal operation, the width of the output pulse is given by the expression

$$t_w(\text{out}) = 0.7 C_T R_T \qquad (23\text{-}1)$$

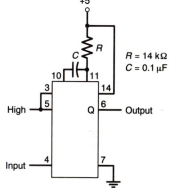

FIGURE 23-2 A 1-msec one-shot circuit.

If R_T is measured in kilohms and C_T in picofarads, then t is in nanoseconds. C_T may be in the range of 10 pF to 10 μF, whereas R_T may be in the range of 2 to 30 kΩ.

A typical use of the 74121 to generate a 1-msec pulse triggered by a negative going pulse on the A2 input is illustrated in Figure 23-2. The 74121 has been used as the illustration in this section. However, the 74123 is the most commonly used one-shot. This particular IC has two independent one-shots with fewer inputs for each section in one package.

There are other one-shots available that differ in one way or another. One common variation concerns **retriggering**. A **retriggerable** one-shot starts the timing cycle over if another trigger occurs while the output is already High. Thus a rapid series of triggers can hold the output continuously High from the first trigger to the preset delay after the last trigger. A **non-retriggerable** one-shot produces only one fixed-length pulse after a trigger, no matter how many triggers occur while the output is High. Because of this variety of one-shots, look at the data books carefully to make sure that there is not another unit that is more appropriate for your particular application. For reference, keep in mind that the 74121 is non-retriggerable whereas the 74123 is retriggerable.

Although it is entirely possible to use one-shots without any idea of how they work internally, it is sometimes useful to have an idea of how one could be built from simpler components. A simplified one-shot circuit is shown in Figure 23-3. The operation of this simple one-shot is as follows. The output of an RS flip-flop is both the output of the one-shot and the control of a switch. While the output is Low, the top plate of the capacitor is held at ground. When a input occurs, the flip-flop flips, the output goes High, and the switch changes so that the capacitor is charged by the current flowing from the voltage source through the resistor R. When the voltage on the capacitor reaches some predetermined fraction of the power supply voltage V_{CC} (a fraction determined by the two resistors in the voltage divider), the comparator changes state, causing the flip-flop to flop back, the output to go Low, the capacitor to be grounded again, and the circuit to be back in the state in which it began. The length of the output pulse depends on the values of R and C which can be changed without changing any other feature of the circuit.

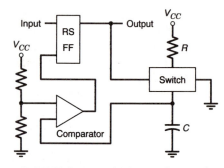

FIGURE 23-3 A simple design of a one-shot.

TIMER CIRCUITS

Integrated timer circuits represent one of the interesting developments in IC design. The circuits consist of a number of high-quality functional blocks that are combined in one IC but that are interconnected externally. Thus one IC can be used for many different functions. As an example, the 555 timer described below can be used as a monostable multivibrator (one-shot), as an astable multivibrator (an oscillator), as a linear voltage ramp generator, as a missing pulse detector, and as a pulse width modulator. This circuit has so much versatility that a whole book on its use has been written—by Balen (see Appendix A for the complete reference).

The 555 timer is the most common timer circuit in use. The functional block diagram of the 555 circuit is given in Figure 23-4. For reference, the data sheets for the 555 are included in Appendix B. As the block diagram indicates, the 555 contains all the components needed to implement the simple one-shot of Figure 23-3. However, since the components are not interconnected, they may be connected together in other ways to produce all of the other functions mentioned above. To make the module even more functional, the 555 may be used with V_{CC} between +5 and +18 V, and the output can source or sink up to 200 mA.

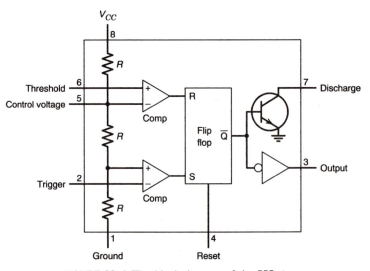

FIGURE 23-4 The block diagram of the 555 timer.

A configuration in which the 555 operates as an astable multivibrator is shown in Figure 23-5. In this configuration, the unit is connected so that it triggers itself and, as a result, the output oscillates continually. The external capacitor C alternately changes to $\frac{2}{3}V_{CC}$ and then discharges to $\frac{1}{3}V_{CC}$ (the values of $\frac{2}{3}$ and $\frac{1}{3}$ are set by the voltage dividers feeding the comparators). The external capacitor charges through resistors R_A and R_B. The charging time (output High) is given by

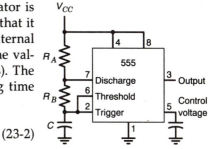

FIGURE 23-5 A 555 used as an astable multivibrator.

$$t_1 = 0.695(R_A + R_B)C \qquad (23\text{-}2)$$

The discharging time (output Low) is given by

$$t_2 = 0.695R_B C \qquad (23\text{-}3)$$

The total period T is given by the sum of these two expressions:

$$T = t_1 + t_2 = 0.695(R_A + 2R_B)C \qquad (23\text{-}4)$$

The **duty cycle,** or DC, or the ratio of the time when the output is Low to the total period, is

$$DC = \frac{R_B}{R_A + 2R_B} \qquad (23\text{-}5)$$

There is actually much confusion in the literature about the exact definition of duty cycle. Here, it is defined as the time off/total time ratio. In other places, it is defined as the time on/total time ratio; and in still other places, it is defined as time off/time on ratio, or the inverse of this. In all cases, the intent of the definition is clear, but any time you find yourself dealing with duty cycle numbers, make sure that everyone involved means the same thing by these numbers.

With the appropriate choices of external timing components, the period of the oscillation can range from microseconds to hours. Note that with this particular set of definitions, the duty cycle cannot be more than 50% and the off time must be less than the on time. If an oscillator having a shorter on time than off time is needed, simply invert the output of the 555 with a normal TTL inverter. By using a 555 astable multivibrator operating at twice the desired frequency and a JK flip-flop as a divide by two circuit, a symmetrical clock pulse can be obtained. Finally, remember this is just one of the many possible uses of the 555 timer IC. The data sheet shows many more, as do some of the books listed in Appendix A.

CRYSTAL-CONTROLLED OSCILLATORS

A number of materials exhibit the property known as **piezolelectricity**, a property that involves mechanical strain and electric fields. When a piezoelectric crystal is subjected to a strain (squeezed or twisted), an electric field develops in the crystal and, hence, a voltage appears between the faces of the crystal. Conversely, when a voltage is applied to the faces of the crystal, the crystal changes shape slightly. Although quartz is the most commonly used piezoelectric material, many other materials exhibit the property and are in use in various applications. Besides the use discussed below, piezoelectric materials are used in phonograph pickups (the old crystal cartridges), in some new high-quality speaker systems, in strain gages, and in pressure transducers, as well as in many other applications.

A quartz crystal has a mechanical resonant frequencies at which it will oscillate if it is excited. If a tiny quartz crystal is carefully cut, this resonant frequency can be in the megahertz region. The resonant frequency depends only on the physical properties of the crystal (these include size, temperature, and external pressure). So, as long as the crystal is protected from external changes, the frequency will be constant. A carefully cut, packaged, and thermally controlled crystal can be used as an ultrastable frequency reference—radio stations formerly kept their frequencies constant to ±10 Hz out of 1 MHz with crystal-controlled oscillators. With advances in technology and increased ingenuity, crystals that are remarkably insensitive to thermal and mechanical variations are now available.

Crystals have become fairly common in modern life, even if a knowledge of what they are and what they do has not. People happily buy crystals for their citizen's band radio, for their police scanners, and for their ship-to-shore radios. As you may have guessed, it is a tiny quartz crystal that gives the name "quartz" to modern clocks and watches. The operating frequency for these crystals is 32.768 kHz, that is 2^{15} counts per second.

Computers, communications circuits, timers, frequency meters, and many other instruments need to have accurate, stable clock pulses available. A quartz crystal, together with a simple circuit to generate digital pulses, fits this need perfectly. As would be expected, all of the circuitry needed to make a complete crystal-controlled digital oscillator is available in one IC. A number of these crystal-controlled digital oscillator circuits are available in the IC lines that have been manufactured as support chips for the microprocessor chips.

As an example of the simplicity of these circuits, one implementation of a crystal-controlled oscillator is shown in Figure 23-6. In this particular IC, there are actually two independent units. These oscillators can be used either as crystal-controlled oscillators or as voltage-controlled oscillators (refer to Chapter 37). Then **en** input is an enable input, that is, an off/on switch where a Low means on. The **cont** and **R** inputs are respectively frequency control and frequency range inputs for use when the circuit is used as a voltage-controlled oscillator.

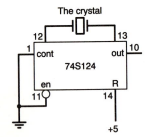

FIGURE 23-6 A crystal-controlled digital oscillator.

DISCUSSION

Three different circuits have been discussed above, yet they all can be used to make digital oscillators (Problem 23-3 asks how to make an astable multivibrator out of the 74121). This is a very normal situation in electronic design—there almost always will be more than one way to achieve the desired function. In most instances, the hardest part of any design is picking out the best way. In this particular situation, the 74121 can be used to make a very fast TTL-compatible oscillator whose frequency is easily changed by changing one or two components. The 555 can be used to make an oscillator having a much longer period; again, its period can easily be changed and it can be made TTL-compatible or used with other logic families. The 74124 can be used as an ultrastable TTL-compatible oscillator. Which is best for a particular use depends on the design goals of that particular use. Finally, it is often the case that there is no obvious best choice. Then, it comes down to which is easier or which you are more comfortable about using.

PROBLEMS

These problems can also be used as simple laboratory exercises.

23-1 Pick timing components for a 74121 that will result in pulses
(a) 100 nsec long. (b) 500 nsec long. (c) 100 μsec long.

23-2 Design a circuit, using several 74121 circuits, that is triggered by a negative going edge, produces a 100-nsec-long output pulse on one output line, and, when this pulse is finished, produces a 100-μsec pulse on a second output line.

23-3 Using the results of Problem 2, create an astable multivibrator circuit from two 74121s that has an on time of 100 nsec and an off time of 100 μsec. When this circuit is first turned on, it may start oscillating all by itself, or it may come on in a stable off mode. A single push button switch can be used to start the circuit oscillating. Include this in your design.

23-4 Pick timing components for 555 astable multivibrators that will result in (a) an on time of 100 msec and an off time of 5 msec. (b) an off time of 10 μsec and a duty cycle of 40%.

ANSWERS TO ODD NUMBERED PROBLEMS

23-1 (a) $C = 50$ pF and $R = 2.86$ kΩ. (b) $C = 100$ pF and $R = 7.14$ kΩ.
(c) $C = 0.01$ μF and $R = 14.3$ kΩ.

23-3

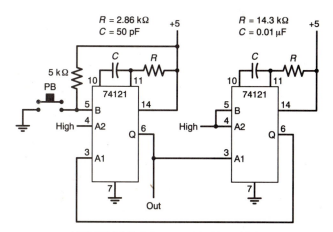

FIGURE 23-7 Solution to Problem 23-3.

Chapter 24

REGISTERS, SHIFT REGISTERS, AND COUNTERS

OBJECTIVES

1. Be familiar with storage registers, shift registers, and counters.

2. Understand some of the possible variations in counter design.

INTRODUCTION

The purpose of this chapter is to introduce the most common and most important circuits based on the flip-flop. These are the storage register, the shift register, and the counter. For each of these circuits, the basic functions are discussed in general terms, and then one or two possible implementations or variations on the circuit are demonstrated. This is only an introduction to the very large number of these circuits that are available. The actual selection of one of these circuits to be used in a particular application will require reading the manufacturer's data books. Many other applications will be found in the **TTL Cookbook** and the **CMOS Cookbook**—see Appendix A for detailed references.

A REVIEW OF THE J-K FLIP-FLOP

Since most of the circuits discussed in this chapter are based on the J-K or the closely related type-D flip-flop, this chapter begins with a review of the most important features of the J-K flip-flop. The activation table, Table 24-1, applies to the J-K flip-flop shown in Figure 24-1. This table shows the state of the output of the flip-flop after the situation shown in the left-hand side of the table occurred. The subscription $n + 1$ refers to the condition after, whereas the subscript n refers to the condition of the flip-flop before the situation shown in the left-hand side of the table occurs.

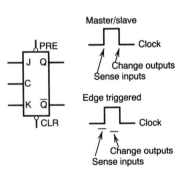

FIGURE 24-1 Symbol and timing details for J-K flip-flops.

TABLE 24-1 Activation Table for a J-K Flip-Flop

Pre	Clr	J	K	C	Q_{n+1}
L	H	X	X	X	H = 1
H	L	X	X	X	L = 0
L	L	X	X	X	Indeterminant
H	H	L	L	⊓	Q_n (no change)
H	H	H	L	⊓	H
H	H	L	H	⊓	L
H	H	H	H	⊓	$\overline{Q}_n$ (complements)

Perhaps the most important feature of the J-K flip-flop is the timing relationships between the inputs and the changes of the output. The inputs are sensed at one instant of time, whereas the outputs change at a later instant of time. This is obviously true with the master/slave flip-flop discussed in Chapter 22, but it is also true with the edge-triggered flip-flop. These timing relationships are also illustrated in Figure 24-1.

In the rest of this chapter, the state of a flip-flop will be described by saying that it holds a 0 or a 1. Saying a flip-flop holds a 0 implies that the Q output is Low and the $\overline{Q}$ output is High; similarly, a flip-flop that holds a 1 has its Q High and its $\overline{Q}$ Low.

STORAGE REGISTERS

A single digital signal can be used to carry or to transmit a single piece, or a single **bit**, of information. Conventions exist for representing numbers, letters, and other information as collections of multiple bits of information—see Chapters 25 to 27, and 54 for further information about these conventions. Although the need to use and store such information is most evident in computers and calculators, it is also true in many other circuits.

A single flip-flop can be used to store or remember a single bit of information. To store numbers, letters, or other more complex types of information, it is necessary to store multiple bits of information. The simplest form of storage is a **register**, or a **latch**, a row of type-D flip-flops with a common clock and reset lines. These are normally packaged in groups of four or eight bits. One example is the 74LS379 shown in Figure 24-2, with its function table in Table 24-2. This is a four-bit latch or register with a common clock, parallel input, and both true and complementary outputs.

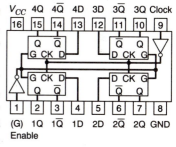

FIGURE 24-2 The pinning and function table for a 74LS379.

TABLE 24-2 Function Table

Inputs			Output
G	Clock	Data	Q
H	X	X	Q_0 (unchanged)
L	↑	H	H
L	↑	L	L
X	L	X	Q_0 (unchanged)

This particular unit is an edge-triggered flip-flop with a clock and an additional enable line. When the enable is High, the clock is ignored; that is, the output will not change. When the enable is Low, an upward transition on the clock line will cause the input to be sensed and then, perhaps, the outputs to change. Since there are only four independent units in this package, there are enough pins so that both the Q and the $\overline{Q}$ output from each unit can be brought out.

Obviously, there are many possible variations of this circuit, depending on how many outputs per flip-flop (Q or both Q and $\overline{Q}$) and how many inputs there are. Since the primary limitation on the amount of storage in one IC is the number of pins required, a great deal of ingenuity has been devoted to getting more storage per pin. By the use of tristate outputs together with some internal and external gates, it is possible to use the same pin as either the input or the output of the flip-flop. This permits 16 bits of storage in a 24-pin IC. See the data books for the many variations on this type of circuit.

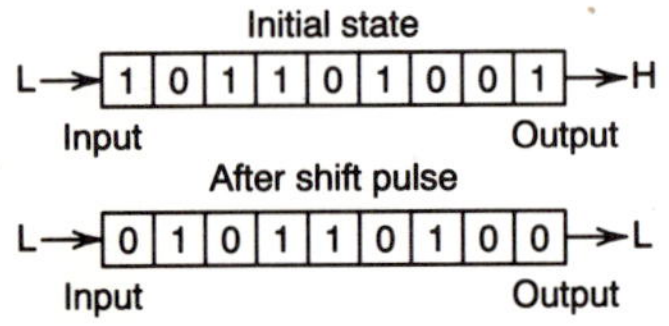

Initial state

L →| 1 | 0 | 1 | 1 | 0 | 1 | 0 | 0 | 1 |→ H

Input   Output

After shift pulse

L →| 0 | 1 | 0 | 1 | 1 | 0 | 1 | 0 | 0 |→ L

Input   Output

FIGURE 24-3 A schematic shift register.

SHIFT REGISTERS

The purpose of a **shift register** is to hold a string of bits and to shift the string one place to the left (or right) on command. An example of this situation is shown in Figure 24-3. The initial state of all the flip-flops, the status of the input, and the status of the output are all indicated in the first part of the figure; in the second part, the state of the flip-flops and the status of the input and output are indicated after the shift command has occurred. In this case, the shift command has caused the pattern to shift one place to the right.

It is easy to implement a shift register with a set of J-K flip-flops. The circuit in Figure 24-4 does this. In this particular implementation, a clock (shift) pulse will cause the pattern to shift to the right because the Q and $\overline{Q}$ of each stage are connected to the J and K inputs of the next stage to the right. For instance, if flip-flop B contains a 0, then its Q is Low and, hence, the J of flip-flop C is Low; likewise the $\overline{Q}$ of B is High and the K of C is High. From the activation table, Table 24-1, it is easy to see that the next clock pulse will cause flip-flop C to contain a 0—that is, the 0 has been shifted from B to C. In a similar manner, the contents of each of the flip-flops will be shifted right one bit when a clock pulse occurs. The contents of the A flip-flop after a shift pulse will depend on the status of the input J and K lines just before the clock pulse.

Many different versions of the basic shift register are available as prepackaged ICs—see a TTL or CMOS data book, for examples. A few of the possible-variations will simply be listed here. With some extra gating, the shift register

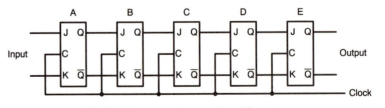

FIGURE 24-4 A realization of a shift register.

can be arranged so that it can shift either left or right depending on a second signal; in this way, bidirectional shift registers can be produced. A second variation concerns the ability to load all the bits of the shift register at once, which is called **parallel input, PI,** or to load them via a **serial input, SI,** as the example above does. Another variation depends on the ability to read all the bits at once or read only the last one, which are called **parallel output, PO,** or **serial output, SO,** respectively. Obviously, combinations are possible, and most have been manufactured, namely PIPO, PISO, and so on. The 74LS299 is a typical example of a modern shift register. This is an eight-bit universal shift/storage register. It has multiplexed tristate inputs/outputs so that it can be used in either a PI or a PO mode; it can load, store, shift left, and shift right. The length of the shift register is another variable. Very-long-shift registers (256 bits or more) having serial inputs and outputs were used as solid state memories before high-density integrated memories were available.

Shift registers have many uses. Examples include sequentially turning on one bit at a time in a cyclic process (see Chapter 36 on the analog-to-digital converter for such an example). Another example is controlling an array of lights, as a memory or as a delayed memory — delayed memory answers questions such as: "What was the state of this bit eight cycles ago?" A common use of shift registers is for **parallel-to-serial conversion**, or vice versa. An example of the need for this occurs in computer communications. Within a computer, data are transmitted in **parallel** via an 8-, 16-, 32-, or even 64-bit-wide bus (see Chapter 27), while communications between computers over a telephone line must take place one bit at a time or **serially**. Shift registers can be used for this conversion. A pair of registers, one a PISO and the other a SIPO, can be combined to form a series communications link between two computers. Obviously, the 74LS299 mentioned above can serve the purpose.

One hybrid circuit that has many of the features of a shift register but that is quite different is an **FIFO, first in, first out**, register. This circuit is arranged like a shift register, but the first bit shifted in is immediately available at the output. FIFO registers can be implemented with combinations of counters, gating logic, and random access memory.

BINARY COUNTERS

If the circuit shown in Figure 24-5 starts with all the flip-flops containing 0 and a sequence of clock pulses occurs, the following sequence of states of the flip-flops will occur.

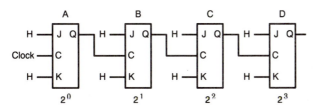

FIGURE 24-5 A four-bit ripple counter. The value associated with each bit of the counter is indicated below the flip-flop.

Status of Bits A B C D	Number of Pulses Received
0 0 0 0	0 (Initial state)
1 0 0 0	1
0 1 0 0	2
1 1 0 0	3
0 0 1 0	4
1 0 1 0	5
0 1 1 0	6
1 1 1 0	7
0 0 0 1	8
.	.
.	.

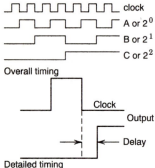

FIGURE 24-6 Timing for a ripple counter.

Obviously, this is a binary counter, with the bits reversed from left to right; that is, the least significant bit is the leftmost bit (A), and the most significant is the rightmost bit (D). Clearly, this counter counts up to 15 and then starts over at 0 again. See Chapter 54 for an explanation of binary numbers.

This particular counter above is called a **ripple counter**, or **ripple scaler**, because the transitions ripple through the circuit. Figure 24-6 shows both an overall timing diagram and a detailed timing diagram for this counter. To see why this is called a ripple counter, consider the following sequence of events. (This discussion is written for a counter based on master/slave flip-flops; a similar one can be constructed for edge-triggered flip-flops, with the same results.) The sequence to be described starts with all the bits of the counter containing a 1 (all the Qs High) and the clock High. Then, the clock goes Low; a short time later (the delay time), the Q output of the flip-flop A goes Low. However, this output is the clock input to flip-flop B; hence, a short time later (another delay time) the output of B will go Low. This second transition occurs **two** delay times after the trailing edge of the external clock pulse. In a similar manner, the third flip-flop will change after three delay times, and so on. Thus, the transitions caused by the external clock pulse can be seen rippling through the counter in a dominolike ripple—hence, the name of the counter.

Because of this rippling through of transitions, a ripple counter takes on wrong values for short periods of time after a clock input. For instance, if the counter contained a 3, and an input occurred, the actual sequence of events observed would be that the counter would go from 3 to 2 for one delay time, then to 0 for another delay time, and finally to 4. These incorrect values may possibly cause difficulties. Since the delay times for modern ICs are very short (typical values are 20 nsec for normal TTL flip-flops), even the time required for a transition to ripple through 10 or 20 stages of flip-flops is reasonably short. If a counter is simply counting events and will be read when the events stop, there is no problem. But if the counter is continually being read by fast digital circuits, then problems can occur. For instance, say, some process was being started each time a counter passed through the zero state. This state can easily be detected by NORing together all the Q outputs. When the counter reads zero, the NOR output will be High. This zero detector will correctly identify the zero state, but it will have short **glitches** when the counter is going from state 3 to state 4, and so forth. If the electronics can respond to pulses one delay time long, which can easily happen with modern high-speed logic circuits, then incorrect results will occur.

To avoid glitches or incorrect intermediate states, it must be ensured that all the flip-flops in the counter get their clock pulse at the same time and, hence, that all the flip-flops change at the same time. This is called a **synchronous counter**. The circuit shown in Figure 24-7 does this. The only problem with this arrangement is that the gates get very wide as the number of bits in the counter gets large. Units are available that can be cascaded together, thus providing a hybrid ripple/synchronous counter. In general, synchronous counters can operate faster than can ripple counters.

As would be expected, there is a bewildering array of possibilities for each of the basic binary counters. With a little extra gating, counters can be produced that can count up or count down, depending on an external control line. Provisions can be made for carry in/carry out or borrow in/borrow out for cascading count up/down modules to make even longer counters. Some counters can be preset to a particular number. Others can only be reset to 0 or perhaps set to all 1s. A careful study of the data books is in order before using any counter. The 74193 is a good example for an introduction to the complexities. It is a four-bit synchronous binary counter which can be used to count up or count down; it also has carry and borrow outputs for cascading more than one stage.

VARIATIONS ON BINARY COUNTERS

A **modulo N** counter is a counter that has N distinct states. Thus a modulo 3 counter is one which would count 0, 1, 2, 0, 1, 2, and so on. In these terms, a n-bit binary counter is a modulo 2^n counter. However, the terminology is generally reserved for counters that count up to some number which is not a power of 2 and then resets; thus, modulo 5, 7, and 10 counters are typical.

There are many ways to implement modulo counters. One way is to preset the counter to $n - 1$, count down to 0, and preset again. Another way is to set up the gating between the various stages of the counter to ensure that the correct sequence of numbers will result. In either procedure, considerable care must be devoted to worrying about glitches and illegal count sequences.

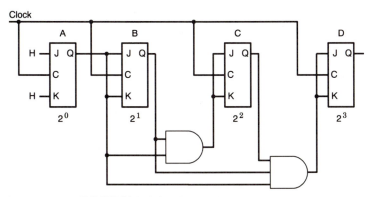

FIGURE 24-7 A four-bit synchronous counter.

Older and more formal texts on digital electronic design are filled with examples and information about how to do this (see Appendix A for references). Some of the modern universal counter ICs come close to totally eliminating the need to design these counters.

The current breed of humans seem to prefer numbers expressed in decimals. Although binary counters are fast and efficient, the first reaction of most users is to convert the result into a decimal number. Although a four-bit binary counter can count from 0 to 15, it can also be set up as a modulo 10 counter so that it counts in the sequence 0, 1, 2, 3, 4, 5, 6, 7, 8, 9, 0, 1, and so on. If this is done in the usual manner, the resulting numbers are the first 10 of the binary sequence from 0 to 15. This is called **binary coded decimal,** or **BCD** (see Chapter 54).

Because this is such a common situation, a wide variety of BCD counters are available. In fact, counter ICs are frequently issued in pairs, both with identical pinning, but one of which operates as a binary counter and the other of which is a BCD counter. For instance, the 74192 is the BCD version of the 74193 discussed above. Even more complex units are available that can be switched from binary to BCD, depending on the status of an external control line. Again, a determined search of the data books is needed to understand the richness of the choices available.

Chapter 25

MORE COMPLEX COMBINATIONAL LOGIC

OBJECTIVES

1. Gain a familiarity with the following complex functional logic units: decoders, bus drivers, multiplexers, demultiplexers, adders, ALUs, parity generators.

INTRODUCTION

The purpose of this chapter is to introduce a number of the more complex types of combinational logic circuits available in integrated form. These circuits are all solutions to problems that occur sufficiently often in digital logic design to make it worthwhile for manufacturers to make integrated versions of the circuits. The circuits include decoders, multiplexers and demultiplexers, adders, multipliers, ALUs and parity generators.

Again, let me repeat that the purpose of this chapter is to provide a vocabulary of terms and circuits to be used when one of these circuits is needed. However, a careful study of the manufacturer's data sheet is necessary before actually using one of these circuits.

The sample circuits mentioned in this chapter are all TTL circuits simply because, at present, TTL data books seem to be more common than data books for the other logic families. However, all the circuits discussed also exist in the other major logic families. Finally, there is no particular order for the circuits in this chapter.

DECODERS

A **decoder** circuit takes information presented in one form and decodes or converts it into another form. As might be expected, there are many different kinds of decoders or converters. There are decoders that will convert binary numbers into BCD, or BCD numbers into binary (the 74184 and 74185 pair). Likewise, there are decoders that will convert BCD numbers into one of the

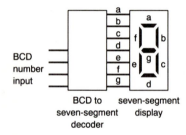

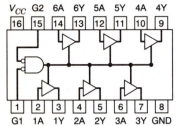

FIGURE 25-1 The use of a BCD to seven-segment decoder.

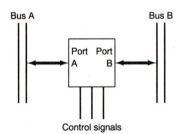

FIGURE 25-2 The 74365 hex bus driver.

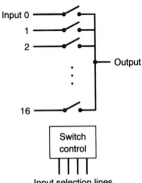

FIGURE 25-3 A conceptual bidirectional bus controller.

FIGURE 25-4 The basic digital multiplexer.

other codes that have been used to code decimal numbers, for example, BCD to Gray Code.

A common decoder circuit takes a BCD digit and converts it into suitable driving signals for seven-segment LED displays. A typical use of this circuit is shown in Figure 25-1. The logic for one segment of this decoder is worked out in Problem 20-2. The 7447 is one example of this kind of decoder.

There are many variations on this circuit to allow for different kinds of displays, to deal with leading zero suppression, to control the decimal point, and to control multiple digits of display with only one decoder and some auxiliary circuitry including a shift register. The data books contain detailed application notes for these units.

BUS DRIVERS

As described in Chapter 22, it is often necessary to gate data onto a data bus. In a computer or similar system, the data buses often are 8, 16, 32, or 64 bits wide. Thus large numbers of bits of data have to be gated onto the bus at the same time. Although this can be done with a lot of separately gated SSI tristate buffers, it is much more convenient to package the buffers together so that the gating is done by a common signal. The resulting circuit is called a **bus driver**. The 74365, shown in Figure 25-2, is such a unit. It contains six tristate buffers controlled by a common gate circuit.

In computer systems, there are often several independent data bus systems. At some times, data from one of these buses must be gated onto the other bus; at other times, the data flow is the opposite direction; and at still other times, the buses are to be independent. Although this can be done with a number of bus driver circuits, like the one described above, it is more convenient to combine all of these together in one package. A conceptual schematic of such a device is shown in Figure 25-3.

The control signals in Figure 25-3 determine the operation of the circuit; the options that can be exercised are to gate data from bus A onto bus B, from bus B to bus A, or to disconnect the two buses—disable both sets of tristate outputs. Since the only real limitation on the complexity of the circuit is set by the number of external pins and since this bidirectional controller only needs one or two more pins than a single bus driver, a considerable savings in IC packages and external interconnections is possible. One such **bus controller** is the 74S428, which is intended to be used with the 8080A microprocessor IC. Bus controllers specifically intended to be used with each of the major microprocessor families exist.

MULTIPLEXERS/DEMULTIPLEXERS

A digital **multiplexer** functions as an electronic multiposition switch. A simple conceptual multiplexer is shown in Figure 25-4. This unit has 16 digital input signals, four **input selection lines**, or **address lines**, and one output. The switch control logic treats the input address lines as a binary number or address and closes the switch that is addressed. Thus if the input address lines are a binary representation of the number 14, the 14th switch will be closed and the output will be the same as input 14. Because N binary bits can represent 2^N separate numbers, but the largest number is only $2^N - 1$, the inputs are always numbered starting with 0 and not 1.

A multiplexer has obvious use in a computerized control system. For example, a computerized energy control system for a building may need to monitor the status of a number of different windows. A simple microswitch at each window can be used to tell if the window is open or closed; if these signals are brought to a multiplexer, then the computer can address each window sequentially and determine if it is open or closed.

The 74150 is almost exactly the same as the circuit shown in Figure 25-4 — refer to the data sheet in Appendix B. This unit has 16 digital inputs numbered from 0 to 15, four select inputs, and one output. In addition, the 74150 has a strobe or enable input that permits the cascading of multiple units to form larger multiplexers, that is, 32 to 1, 64 to 1, or wider. This particular multiplexer is called a 16-line to 1-line data selector. There are many variations on this basic unit: 4 to 1 units, 8 to 1 units, dual 8 to 1 units, as well as variations on the number and types of the enable inputs and variations on the outputs — inverting, noninverting, open collector, tristate, and so on.

As was mentioned in Chapter 20, multiplexers provide a very elegant way to build circuits that reproduce arbitrary truth tables. The circuit shown in Figure 25-5 reproduces the truth table, Table 20-3. This certainly is a very simple circuit, but at first it may not be evident that it really reproduces Table 20-3. A careful comparison of the data sheet for the 74150 in Appendix B and Table 20-3 reveals several differences that must be taken into account when using this circuit.

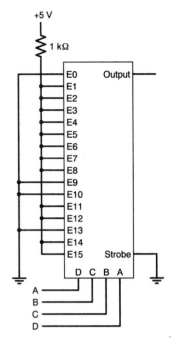

FIGURE 25-5 A multiplexer circuit that reproduces truth table, Table 20-3.

First, in Table 20-3, input variable D varies most rapidly (changes every line), whereas input variable A changes least rapidly. The data sheet for the 74150 shows that the A select line is the most rapidly varying line and that the D line the least rapidly varying line. This is only a naming problem. By connecting the input signal corresponding to A in Table 20-3 to the D select input of the 74150, connecting input signal B to the C select input of the 74150 and so on, as indicated in Figure 25-5, this problem is eliminated.

The next problem is that the lines in Table 20-3 are numbered from 1 to 16 and from the top down, whereas the data inputs of the 74150 are numbered from 0 to 15 and in an order that corresponds from the bottom up in Table 20-3. Again, this is simply a problem in naming. Associating input E0 of the 74150 with line 16 of the truth table, input E1 with line 15, and so on, and finally input E15 with line 1 of the table solves this problem

Finally, the output of the 74150 is the NOT of the selected input. This again is a problem that is easy to fix. Table 20-3 has four lines for which the output of the circuit is to be true; these are lines 3, 6, 7, and 16. Following the equivalence set out above, this means that if inputs E13, E10, E9, and E0 are all connected to a Low (False) and all the other lines are connected to High (True), the output of the circuit will reproduce Table 20-3.

Thus all of the complexities in using a multiplexer for reproducing an arbitrary truth table are "paper and pencil" type complexities; they have to do with deciding which signal is to be connected to which input of the IC. Once this is done, the circuit itself is very simple.

The **demultiplexer** is the inverse of the multiplexer. The conceptual design for the multiplexer in Figure 25-4 could equally serve as a design for the demultiplexer if the inputs were labeled outputs and the output were replaced by a High or a Low. In a typical demultiplexer, the input address lines or select lines specify which output is to be active; all of the other output lines are left in an inactive state.

The 74154 is an example of a demultiplexer; it has four select inputs and 16 outputs. Which output is active depends on the number represented by the four select inputs. Thus if the four inputs represent the binary number 7, then output 7 will be the one that is selected. For the 74154, a selected output is Low whereas an unselected output is High. The 74154 is called a 4-line to 16-line demultiplexer, or decoder. In general, multiplexers and demultiplexers are manufactured in pairs; thus, the 74154 is the inverse unit for the 74150. As a result, all the comments above about enable lines and other variations also apply to demultiplexers.

The following example is presented as an indication of how these units can be combined to create larger, more capable systems. One 4-line to 16-line decoder can be combined with sixteen 16-line to 1-line multiplexers to create a 256-line to 1-line multiplexer. The basic outline of the circuit is indicated in Figure 25-6.

It takes eight address lines to select 256 lines, so there are eight input address lines. Four of these lines go to all the 16-to-1 multiplexers in parallel; the other four go to the 4-line to 16-line decoder. The 16 output lines of this decoder are then the enable lines for the multiplexers. Thus four of the address lines select which multiplexer is active, and the other four determine which input is selected in that multiplexer. In this design, it is assumed that the multiplexers have tristate outputs so that they can be connected together in a wired-OR configuration. Demultiplexers to generate multiple enable lines are also used in building large memory systems.

The circuits described above are digital multiplexers, that is, the signal are either Highs or Lows. However, analog multiplexers and demultiplexers also exist. These are quite similar; for instance, a 16-to-1 analog multiplexer would function just as the digital one, except that the data input to the 16 inputs would be analog and the data output would also be analog. The four select lines would still be digital.

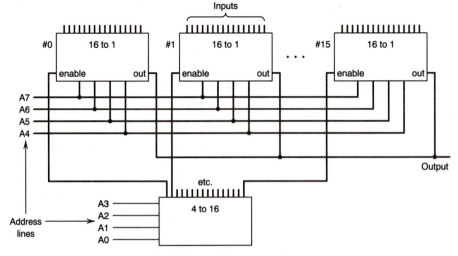

FIGURE 25-6 A 256-to-1 multiplexer.

ADDERS

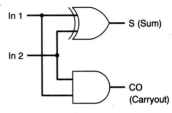

FIGURE 25-7 A half adder.

Two binary bits, when added together, can give rise to only three different answers—0, 1, and 1, with a carry to the next most significant. The truth table, Table 25-1, shows these possibilities for two bits called In 1 and In 2. A quick inspection of this table shows that the sum of the two bits is the exclusive or, XOR, of the two bits, while the carry out is the AND of the two bits. Thus a circuit that can add two bits can be constructed as shown in Figure 25-7. Since this circuit has no provisions for a carry in, it is called a **half adder**.

TABLE 25-1 Truth Table for a Half Adder

In 1	In 2	Sum	Carry Out
0	0	0	0
0	1	1	0
1	0	1	0
1	1	0	1

An adder that can sum two bits plus a carry in to produce a sum and a carry out is a **full adder**. A full adder can be built by adding the two bits to produce an intermediate sum and then adding the sum to the carry in to produce the final sum. The carry out is simply the OR of the two carries; thus, a full adder is built from two half adders and an OR. This is shown in Figure 25-8. Obviously, this can easily be implemented in one IC. Likewise, ICs that will add wider binary numbers can be implemented. One such IC is the 74283, which adds two four-bit binary numbers plus a carry in to get a four-bit sum and a carry out. See the data sheets for more information.

ARITHMETIC UNITS

Just as ICs exist that will do a multibit sums, so also do ICs exist that will do other operations. For instance, the 74284/74285 combination is a circuit which multiplies two four-bit binary nuumbers to get an eight-bit binary product. Likewise, the 7485 is an IC that compares the magnitude of two four-bit binary numbers and generates an output that indicates equality or which number is bigger.

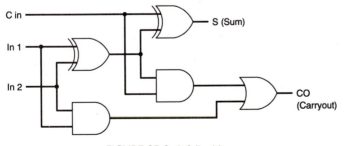

FIGURE 25-8 A full adder.

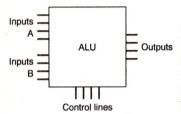

FIGURE 25-9 An outline of an ALU.

In a computer, it is sometimes necessary to add two numbers, sometimes multiply them, sometimes compare them, at other times OR them or AND or XOR them. In this chapter and in previous ones, units that can do each of these functions have been discussed. The next obvious step is to create a single IC that can do any of these functions. A block outline for such a circuit is shown in Figure 25-9. This circuit is shown as having two four-bit inputs and a four-bit output. The output is the function of the inputs that is specified by the control lines. Thus the output might be A + B, A − B, A XOR B, and so on. This type of unit is called an **arithmetic logic unit**, or **ALU**.

The 74181 Arithmetic Logic Unit/Function Generator is a TTL ALU. The unit can add, subtract, or compare the numbers, add or subtract one, as well as form all the logic combinations of the two inputs. Obviously, the use of such a unit is moderately complex. The fact that there are 12 pages of data sheets for this unit is some indication of its complexity.

Of course, the ALU does not represent the limit of complexity for integrated circuits. The tendency toward more and more capability and flexibility in one IC continues, with the development of the microprocessor, a complete computer in one IC, being one extreme of this trend.

PARITY GENERATORS

The final circuit discussed here, the parity generator, is a formerly obscure circuit that has become quite common because of increased use in computer-related circuits. In modern computers, information is transmitted in packets that are some multiple of eight bits, a unit called a **byte**. Unfortunately, errors occur in information transmission and storage. The use of a parity bit is the most common way of attempting to detect when an error has occurred.

The basic idea is to pick a convention, either **even parity** or **odd parity**; then, a ninth bit, the parity bit, is added to each byte. The parity bit is set to ensure that the number of bits in the nine-bit combination of the byte plus parity bit is even or odd, respectively. This is illustrated below:

Data (in Binary)	Parity Bit Odd Parity
00100100	1
01101101	0

Data (in Binary)	Parity Bit Even Parity
00100100	0
01101101	1

The parity bit is transmitted with the data, stored with the data, and generally carried along through the computer.

If a single bit of the nine-bit byte plus parity bit packet changes because of an error, then the parity will no longer be correct (verify this). At any time, the reliability of the data can be checked by recalculating the parity and comparing it to the existing parity bit. In this way, most errors can be detected. Parity checking schemes are used in most memory systems and in essentially all communications schemes as a reliability check.

More complex schemes exist to both detect errors and to make possible the correction of most errors. These are called **error correcting codes** and are an important topic in computer science.

After the extended introduction above, little need to be mentioned about the actual circuits. The 74180 is a typical unit; it has eight data inputs and can be used to generate or check either even or odd parity bits

Chapter 26

MEMORIES

OBJECTIVES

1. Become familiar with the terminology used to describe memory systems and memory chips.

INTRODUCTION

The purpose of electronic memories is the same as that of human memory—to retain information so that it can be used at a later time. Although human memory is able to store a wide variety of items, such as names, languages, images of scenes, smells, digital electronic memory is limited by being able to remember only sequences of zeros and ones or Highs and Lows. Much of the effort invested in computer science at the present time is devoted toward finding efficient ways to store and retrieve information about languages, images, and perhaps even smells in digital memories.

It is hard to remain unaware of the importance of electronic memories. Discussions of computer systems involve extensive considerations of memory size, type, speed, organization, and management. Even discussions of international trade and international politics eventually get down to the basic elements not only of oil, cars, shoes, steel, wheat, but also memory ICs. This is heady company.

This chapter is an introduction to the ideas and language used to describe memories, with particular emphasis on integrated digital memories—the famous memory chips. Essentially no details are provided, but rather the chapter is in the form of an expanded glossary to give the background needed to read and comprehend the literature about memories and computers.

HISTORICAL PERSPECTIVE

The development of electronic memories or electronic storage elements has been an integral part of the development of computers. As a result, the language used to describe memories is the language of computer architecture and computer science; the converse is also true. In fact, early computers were called "stored program devices," or "stored program calculating machines," indicating that, at least in the beginning, one of the most important features of a computer was the memory.

A wide variety of memories or storage elements have been used within the relatively short computer era. The earliest computers used relays and punched paper tapes as memories. Other early elements included acoustic delay lines (an electromechanical analog of a shift register) and magnetic drums. Since then, punched cards, magnetic tapes, magnetic disks of many types, and magnetic cores have been used. In the last decade, integrated solid state memories have come to dominate the computer memory market. This chapter mainly concentrates on the terminology required to understand solid state memory.

MEMORY CLASSIFICATIONS

There are a number of basic classifications that are used in describing memory systems, involving mainly twofold distinctions. These are discussed in this section.

Perhaps the most important classification is between random access memory and sequential access memory. In **random access memory**, or **RAM**, any element may be accessed at any time with essentially the same ease and delay as any other element. There is no required order in which information must be stored in, or read from, the memory.

In **sequential access memory** the information is read out of the memory in the same order in which it was stored. For instance, a shift register can serve as a sequential access memory. Magnetic tapes are another example of sequential access memories; the tape is processed from beginning to end and the data are always encountered in the same sequence. Disk units are a hybrid of sorts; large blocks of information can be accessed in a random mode, while within the large blocks, the information can only be accessed in sequential mode.

In modern computer systems, a distinction is generally made between high-speed storage (or memory) and mass storage. The distinction is based on the speed and ease of retrieval of the information stored in the memory and, at least historically, on the size of the memory. **High-speed memory** is just what the name implies; it is memory that can be accessed most rapidly, generally as rapidly as the processor can operate. In a modern computer, this means that any element of the high-speed memory can be read or written in a time that is less than 1 μsec and that may be as short as 100 nsec. In general, high-speed memory is some sort of RAM, but this was not always true in early computers. In the past, high-speed memory was generally very expensive and was the limiting feature for most computer uses. Even at present, memory limitations are important, as any PC advertisement shows.

Mass storage, on the other hand, was some sort of memory that was slower, might be harder to access, and was generally much larger than the high-speed memory. In the last decades, most mass storage has been based on magnetic drums, disks, and tapes. However, new mass memory devices based on video disk, video tape, or compact audio disk technology are appearing for computer systems. Frequently, mass storage systems are sequential access systems, meaning that the information must be read out in the same order in which it was written. Mass storage systems are generally quite slow by computer standards; access times for high-speed disks can be 8 msec or longer, and delays of up to hundreds of seconds are possible with massive tape systems. Finally, while the sizes of high-speed memories have been growing rapidly, so have the sizes of mass storage devices. It is fairly normal for a computer system to have a mass storage system with a total capacity at least several orders of magnitude larger than that of the high-speed memory system of the computer.

Another distinction that is made among memory systems is the distinction between permanently mounted and dismountable memories. The distinction should be fairly clear. **Permanently mounted memory** is attached to a particular computer and cannot be removed without taking the computer apart. **Dismountable memory** media, on the other hand, can be removed from the computer, stored elsewhere, and perhaps even be put on another computer to transfer information from one computer to another. This distinction applies mainly to mass storage media. The internal hard disk on a PC is permanently mounted, whereas the floppy diskettes and magnetic tapes are obviously dismountable. However, the advent of very large ROMs (see below) carrying system programs or even the microcode that defines the computer has made it possible to build machines whose personalities can be changed by removing one ROM and plugging in another ROM. This is most obvious in the calculator market where various program packs can be purchased and plugged into the calculator. However, such changes are also feasible in large computer systems where new systems or upgrades in the details of the computer operations are distributed and installed by exchanging ROMs.

A fourth distinction is between volatile and nonvolatile memories. This refers to what happens to the contents of the memories when the power is removed. **Volatile memories** forget when the power is turned off, whereas **nonvolatile memories** remember their contents when the power is turned off. Clearly, magnetic tapes and disks are nonvolatile. Most IC RAM systems are volatile whereas core memory systems are nonvolatile. Most mass storage systems are nonvolatile. In situations in which it is important to maintain the contents of all the memory during power failures, such as for banks, emergency equipment, and military equipment, something must be done to ensure the nonvolatility of the memory. The choices are auxiliary power supplies and/or core memories.

MEMORY ARCHITECTURE

This section deals with the basic architecture of integrated random access memory. In almost all modern computers, information is organized into units of eight bits called **bytes**. Memory is organized in terms of bytes; each byte has an address that uniquely identifies it. Thus memory may be thought of as a long row of numbered bytes. The computer writes a piece of information into memory by specifying the address and the data, and the memory system stores the data into the correct byte. Similarly, the computer reads a piece of information from the memory by specifying the address and then accepting the data returned by the memory system.

It should be noted that all references to memory are from the point of view of the computer. Thus the computer **writes** data to memory when it wants the data stored in the memory, and it **reads** the data when it wants to retrieve data that was stored earlier. In some computer systems, bytes are organized into larger units; for instance, four bytes might be called a **word**. In these systems, there will be a convention for converting references to words into references to bytes in the memory system.

Although the overall memory of the computer is organized as a long numbered row of bytes, the organization of the individual memory IC or chip is not necessarily the same. The individual RAM memory chips may be bit oriented or byte oriented. Thus a chip having the same total storage capabilities might be organized as 128k × 1 bit—that is, as 128 thousand individually addressable bits; or it might be organized as a 16k × 8bits—that is, as 16 thousand individually addressable bytes.

To make a 128k-byte memory system out of either of these two arrangements would take eight of these chips in either case, but the arrangement would be different. However, if the memory were to include a parity bit, it could be made with nine of the 129k × 1 bit chips, but it would require either a different chip or a more complex arrangement with the 16k × 8 bits chips. Thus the organization as a single long row of bits allows more flexibility. As a result, at present, the densest memory chips (densest in terms of the most information stored in a single package) are organized as a single long row of numbered bits.

Of course, once a module that functions as a 128k-byte memory unit has been assembled, a number of these modules can be cascaded with additional address decoders/demultiplexers to make even bigger memories in a manner very similar to that shown in Figure 25-5 for cascading multiplexers.

Finally, it should be pointed out that all statements of memory size are based on the powers of 2. In all references to k in memory size, such as 128k RAM, the k refers to 1024, or 2^{10}. Thus a 128k RAM actually has 128 × 1024, or 131,072, bits; and a 1-megabit memory really has 1024 × 1024, or 1,048,576, bits.

A conceptual design of an integrated memory IC is shown in Figure 26-1. The functions of the various parts of this memory are described as follows.

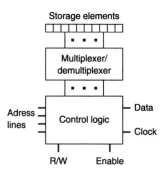

FIGURE 26-1 A block diagram of a memory chip.

The **storage elements** actually store the data in the memory. Each element has a unique number or address. Current ICs may contain a million or more of these elements.

The **address lines** are the input lines that carry the binary number or address that specifies which of the storage elements is to be accessed. If there are n address lines, then the memory chip may contain 2^n elements. This means that a 256k-bit memory must have 18 address lines and a megabit memory chip must have 20 address lines.

The **R/W line** is a control input to the memory chip that determines whether the data are to be stored in the addressed element or are to be read out of the element.

The **data line** carries either the input data to be written into the memory, or is the data being read from the addressed element. The meaning of the data line obviously depends on the state of the R/W line.

The **clock** is used to synchronize operations within the memory chip with the external operations. For instance, on a write cycle, the clock tells the memory chip that the data at the data input are valid and may be stored in the internal storage element.

There are one or more **enable** lines that collectively determine if this particular memory IC should take part in the current operation. This permits the cascading of memory chips to make even larger systems, as discussed above.

The **multiplexer/demultiplexer** contains the circuits that fan out the clock, the R/W signal, and the data line within the array of storage elements and that select which individual storage element should respond. Although this process is conceptually similar to that discussed in the previous chapter, a little reflection will show that decoding 18 address lines into 262,000 lines that are then the gating signals for 262,000 flip-flops is a hopeless task. Somehow the multiplexing and demultiplexing must be done in stages. There are many different ways this can be done, resulting in terminology such as 2-dimensinal arrays, $2\frac{1}{2}$-dimensional arrays, and so on. The details of this are far beyond the scope of an introductory text.

The evolution of new techniques and the shrinking of the circuits for doing multiplexing and demultiplexing has been one of the main reasons for the continual increase in memory densities. (The IC memories started with 16 bits only a few years ago and are now up to a million bits.)

The current generations of integrated solid state memories use a number of different techniques for actually storing the information within the IC. These range from using actual flip-flops to arrays of charge storing capacitors. The technique used is determined by the particular design goals for the particular memory type; thus flip-flops are needed for RAMs, capacitors are used for DRAMs, and simple connections are used for ROMs (see below for definitions of this terminology). The shrinking of the basic storage cell and the reduction in power consumption per bit is the other main reason for the continual increase in memory densities.

AN EXAMPLE

To make all of this more concrete, a very small simple memory system is shown in Figure 26-2. This is a 512 × 2 bit memory system built from 128 × 1 bit memory chips, indicated as M in the figure, and one 2-line to 4-line demultiplexer. To address 512 elements requires nine address lines, indicated as A_0 to A_8 in the diagram. For simplicity, A_0 through A_6 are drawn as one heavy line to keep the diagram from being too complex. The memory ICs are in two columns corresponding to the two bits of data; the two data lines are indicated as D_0 and D_1.

As shown, the seven least significant bits of the address lines are fanned out to all the memory ICs, as are the R/W line and the clock line. The two data lines are each fanned out to all the memory ICs in the appropriate column; thus, D_0 is connected to all the ICs in the bit 0 column. The two most significant address lines go to the input side of the 2-line to 4-line demultiplexer. Depending on the number specified by these lines (0 to 3), the appropriate output is active. Each output is used to select or enable all the memory ICs in one row of the memory. For example, if address lines A_7 and A_8 specify number 1, only the two memory ICs in row 1 are enabled.

The memory ICs have a bidirectional input/output data port; the output portions are tristate. Thus, the enable lines ensure that only one memory IC in each column will be active, and the R/W line determines if the active IC will be placing the data on the line into the memory or driving the line with the status of the addressed element. In either case, the actions will be synchronized by the clock signal.

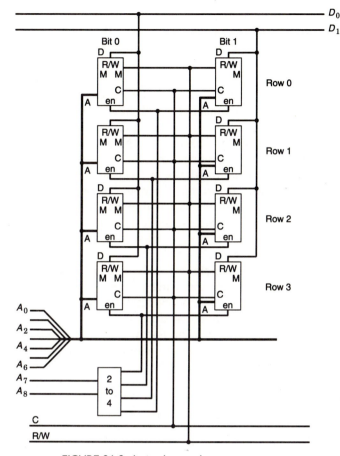

FIGURE 26-2 A simple sample memory system.

It should be quite evident how to make the memory bigger. To make it wider, just add more columns. Thus adding six more columns would produce a memory able to store 512 bytes of data. Furthermore, adding more rows will make the memory larger in the sense of having a bigger maximum address. Using larger memory ICs, such as 128k × 1 ICs, would convert this into a 512k × 2 bit memory with only the addition of more address lines. Finally, this whole circuit could be regarded as just one bank of memory and could be selected by using an enable line on the demultiplexer. Thus this very simple example shows essentially all of the pieces needed in designing a memory system. Further details would be beyond the scope of a general introduction to the topic. The data sheets for memory ICs contain sample circuits for memory systems that include all the details.

IC MEMORY TYPES

Memory comes in many different types. Each type has its own particular advantages and uses. Each type is identified by some combination of code words. In this section, a number of these words will be defined.

RAM, which has already been defined, stands for **random access memory**. This means that any element in the memory can be accessed as easily and as quickly as any other element. Further data can be both written into and read out of this kind of memory.

ROM stands for **read only memory**. This is a form of random access memory that can only be read; that is, the contents of the memory cells cannot be changed by the user. The contents of the cells have been set during the manufacturing process and never change, barring circuit failure. A great deal of information can be put in one ROM, since the actual storage elements are simply connections or missing connections. Because ROMs are made like any other IC, they are only made in batches of thousands or more. Hence, ROMs provide an excellent way to provide the operating systems needed to run computers, calculators, computerized washing machines, or any other device that will be built in large numbers and that will not change. Obviously, ROMs are worthless for developing an operating system when everything changes every day.

PROM stands for **programmable read only memory**. This is a form of ROM for which the contents of the memory can be written by the user by using special hardware. It provides a way for a developer to write a new operating system, say, for a computer-controlled washing machine, burn it into a PROM, and then test it out in a machine. The hardware needed to burn in, or program, PROMs is not very complex and is available as a device connected to a PC. Once a PROM has been programmed, it cannot be changed; hence, if a change has to be made, the current PROM is thrown out and a new one is programmed. However, since one PROM type can be used in many different applications by simply programming it in different codes, PROMs are suitable for development work and for systems where only a few copies will be needed (a few means less than many thousands).

EPROM stands for **erasable programmable read only memory**. This is a form of PROM that can be erased, usually by exposing the IC to ultraviolet light for one-half hour or so. Once the EPROM has been erased, it can be reprogrammed. Obviously, the advent of EPROMs more or less ended the use of PROMs, since reuse is more economical than throwing out.

EEPROM stands for **electrically erasable programmable read only memory**. This is the end of this line of development. Under normal use, these units act like ROM; however, there are provisions for electrically erasing the contents of the ROM and writing new contents into the cells while the unit is in the circuit. For instance, a developer can repeatedly change the washing machine control program by using a development system, but when the IC is installed in a production washing machine where the lines needed to erase and reprogram the cells are not connected, the contents of the memory are safe from change.

Static memory is based on flip-flops or some similar circuit. Once the cell has been set, it will always remain set until either something else is deliberately written into it or the power is turned off. Static RAMs are sometimes called **SRAMs**.

Dynamic memory is based on charge stored on a capacitor or some other process that decays away in time. A dynamic memory system must have some provisions for reading and rewriting each cell before the information stored in the cell can totally decay away. This process is called **refreshing**. Dynamic memory ICs contain most of the circuitry needed to automatically refresh; however, the refreshing process has to be interlocked with the normal reading and writing cycles. All microprocessors contain the circuitry designed to implement the refresh cycles when the processor is not accessing the memory. Naturally a dynamic RAM is often called a **DRAM**.

MEMORY SPECIFICATIONS

There are a number of different items that have to be specified to completely specify a memory. Obviously, one item is type, that is, SRAM. A second parameter is the digital family that is compatible with the memory. Today, most memories are CMOS memories having limited capatibility with LS TTL circuits. Obviously, it is necessary to specify the size and arrangement of the memory. Finally, memories come in various speed ranges. One particular memory IC may come in versions that can complete a read cycle in 250 nsec, in 120 nsec, or in 90 nsec. Obviously, the faster versions will cost significantly more than the slower ones.

Chapter 27

A SIMPLE CONCEPTUAL DIGITAL COMPUTER

OBJECTIVES

1. Become familiar with the basic design and operation of a modern computer.

INTRODUCTION

This chapter is a very brief introduction to computers. It has two purposes: first, to show at least one context in which many of the circuits discussed in the previous chapters are used; and, second, as an introduction to some of the language, concepts, and techniques used in the literature to describe specific computer systems. This discussion is about a simple, generic computer system; no specialization and essentially no refinements are discussed. References to specific computers and the reference books describing them are given in Appendix A. The discussion applies equally well to all sizes of computers—microcomputers, minicomputers, as well as so-called main frame computers. As with most of this book, some terms must be used in the discussion before they can be fully defined. As a result, several readings of this material will be required to fully understand it.

COMPUTERS AND PROGRAMMING LANGUAGES

At some level, computers are exceedingly stupid. They must be told in excruciating detail what to do on a step-by-step basis. The lists of these instructions constitute a program. The programs, together with any data they are to process, are stored in the computer memory system. Computer memories are organized by bytes and by larger units called words. One word contains one or more detailed instructions as well as either the data or the address of the data to be processed. Of course, digital memory stores only 0s and 1s; the computer interprets these strings of binary data as instructions. Thus programs consist of long lists of binary numbers, some of which are the instructions to

232

the computer and some of which are data needed by the program. Programs in this form are said to be in **machine language**, since this is the actual form the computer needs and processes. Useful programs range in length from a few thousand bytes to many millions of bytes of instructions. Because it is fairly traditional to list programs one statement or instruction to a line, it is fairly common to refer to the length of a program in terms of **lines of code**. In machine language, each line of code generally corresponds to one instruction for the computer.

Obviously, humans do not write long programs in machine languages. Most programs are written in some **higher level language**, such as Ada, C, Basic, COBOL, FORTRAN, or Pascal. Before a program written in one of these higher level languages can be executed on a computer, another program, called either an **interpreter** or a **compiler** depending on its detailed function, translates the program from the higher level language into a machine language program that can actually be run on the computer. One line of a program written in a higher level language generally translates into many lines of machine language code.

An intermediate step in this translation process is often a program written in an assembly language. An **assembly language** is a symbolic language that corresponds on a line-by-line basis with the machine language. However, the assembly language uses mnemonic names for instructions rather than binary numbers, and generally symbolic names for address of the data. Thus instructions have names such as "add," "load," and "store"; internal registers in the computer have symbolic names like "R1," "R2," "PC"; and addresses in memory can be given symbolic names. Another translation program, called an **assembler**, converts the assembly language program into the actual machine language program executed by the computer.

Although writing a program in assembly language is still much more tedious and time consuming than using a higher level language such as Basic, many programs are written in assembly language. All the short little sections of programs used in this chapter will be written in a simple generic assembly language.

GENERAL STRUCTURE

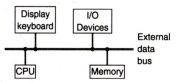

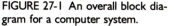

Figure 27-1 shows an overall block diagram of a computer. As shown, there are three main sections: the **central processing unit**, or **CPU**; the **memory**; and the **input/output**, or **I/O**. These sections are all connected to an external bus that carries address, data, and control information. The CPU is the heart of the computer; it is where all computations are made, where all comparison or decisions are made, and where essentially all control of the computer operation originates. The memory is where both the program and any data are stored. The general nature of memory systems was discussed in Chapter 26. The I/O section of the computer provides the pathways for transferring information and programs into and out of the computer. This includes keyboards and display screens as well as other routes, including telecommunications, printers, plotters, and magnetic tapes. Mass storage devices such as disk units, floppy diskettes, video tapes, and compact discs are usually considered as I/O devices.

FIGURE 27-1 An overall block diagram for a computer system.

Although most modern memory systems are organized by bytes (with or without a parity bit included), all but the smallest computer systems generally deal with data and instructions in larger or wider units than just a byte. Generally, this larger unit is called a **word**. Computer systems are commonly identified by their word size, for example, as a 16-bit or a 32-bit computer. In most computer systems, each reference to memory results in one word of data being transferred to or from memory. Thus for a 32-bit computer, each reference to memory will result in 32 bits, or four bytes, of data being written to or read from memory.

Each computer system will have some convention about how word addresses convert into byte addresses and how the bytes are organized within the word. For instance, for a hypothetical 32-bit system, a word might be addressed by specifying the first and most significant byte. Thus reading address 16,000 would result in the word consisting of the bytes with addresses 16,000, 16,001, 16,002, and 16,003 being read, with the first being the most significant part of the word and the last, the least significant.

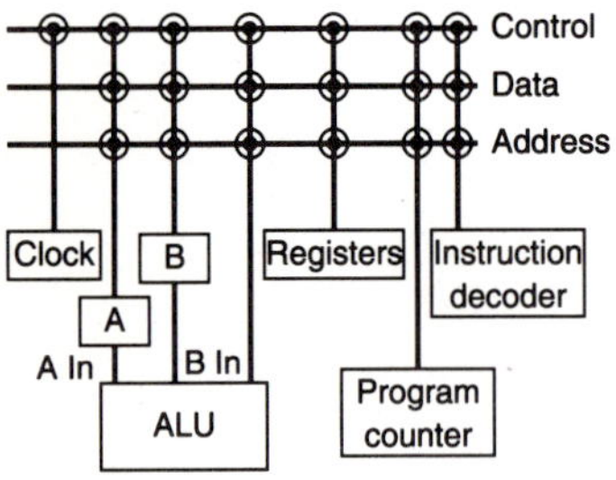

Fig. 27-02 • W-254
With enclosing open circles

FIGURE 27-2 A block diagram of a computer CPU.

THE CPU

Figure 27-2 shows a block diagram of a typical simple CPU. It consists of a number of functional units all arranged around an internal bus system. This bus system is indicated here in somewhat more detail than in the overall block diagram. The internal bus includes an address bus, a data bus, and a control bus. Their functions are discussed in the paragraphs that follow.

Each of the functional units is shown as being connected to all sections of the internal bus; however, these connections are shown inside circles to indicate that they are not just passive connections but active connections. At any instant of time, a particular portion of the CPU may be reading information from a part of the bus, writing information out onto the bus, or simply be disconnected from the bus. (These connections are the typical bus connections discussed previously.) In fact, as shown below, the operation of the computer is determined by controlling these connections.

All operations of the computer will involve coordinated operations by different portions of the computer. For instance, at some instant of time, the computer memory may be putting the contents of some location in memory onto the data bus. At the same time, after allowing for the required signal propagation delays, the information is read into an internal register in the CPU. Clearly, timing signals need to be provided to synchronize these operations. The **clock** is typically a crystal-controlled digital oscillator that generates one or more timing signals which are connected to almost all portions of the computer to allow synchronization of the operation. In general, all operations involving the data buses will proceed according to the following rule. Any circuit that is placing a signal on a data bus should gate the signal onto the bus at the leading edge of the clock pulse and keep the signal gated onto the bus throughout the clock pulse. Any circuit reading a signal from a data bus should strobe (read) the signal at the trailing edge of the clock pulse. In this way, adequate time is allowed for the propagation of the data through the system and for any noise, ringing, and cross talk caused by the changing of many signals to die away before the data are read. Of course, it is the job of the computer designer to ensure that the clock speed, the logic type, the lead length, the shielding, and other features are all such that when this rule is followed, correct, reliable operations are obtained.

The **arithmetic logic unit**, or **ALU**, is where all the actual computations are done, that is, all the additions, subtractions, multiplications, logic operations, shifts, and the like. Much of the growth in the power of computers has been due to the increasing capabilities of the ALU sections of the computers. Early computers could add, subtract, and do simple logic operations. Multiplication of integers required the use of a multiple-step program (software) that carried out the shift and add process, which corresponds to binary multiplication (see Chapter 54). Now essentially all computers have hardware that can do integer multiplication and division. This has resulted in significant gains in the speed of computer operation. Likewise, early computers treated numbers in **floating point notation** by means of software. Later, more powerful computers had hardware that could carry out arithmetic operations on floating point numbers much more rapidly than could the software procedures. Again, this has caused a considerable increase in the speed and power of the computers.

Some of the circuitry involved in an ALU was discussed in Chapter 25. Little more can be said here without getting deeply involved in the details of computer design. However, it should be pointed out that a typical ALU has two data inputs and one data output besides a number of control inputs. If the CPU has only one data bus, then there need to be latches (storage registers) on the inputs. These are indicated in Figure 27-2 as latch A and latch B. Their use and some other options are discussed below.

The **registers** are essentially local memory. They are arrays of type-D flip-flops (latches) which can be used to store information from the buses or to provide information to the buses. Their advantage is that they can be accessed much faster than information can be written to, or read from, the external memory system. Thus if an intermediate result is generated in the midst of some calculation, the program will run faster if this result is temporarily stored in a local register rather than being stored in external memory and then read back in later when it is needed.

Again, much of the increase in the power of modern computers has been due to the increased number of and use of these general purpose, uncommitted registers. Early computers had very few registers; a minimum number might be the two registers connected to the ALU, the program counters (see below), and one other register. Modern computers have many more of these registers; aside from the 3 special purpose commited registers mentioned above, typical systems have somewhere between 16 to 128 general purpose registers. Their use in a sample calculation will be mentioned below.

The **program counter**, or **pc**, is really just another register with a very special purpose. It contains the address of the memory location that contains the next instruction to be fetched from the memory system and brought into the CPU for execution. Its use is described in the discussion that follows.

The **instruction decoding logic** is the portion of the CPU where the actual machine instructions are decoded and converted into the appropriate internal operations sequence required to carry out the desired operation. This portion of the CPU generates most of the signals on the control bus. The signals, together with timing information from the clock, determine what each piece of the computer system is doing at any particular instant. The correct sequence of operations ensures that the program desired is carried out. The operation of this section and much of the rest of the computer can best be illustrated by considering the detailed operations that take place in executing a very simple program.

A SAMPLE PROGRAM

A program consists of a list of instructions and associated data items. A typical instruction contains enough information to specify some particular operation and to address or contain the data needed for the operation. For instance, a simple instruction sequence that corresponds to the operation of

$$A = B + C$$

might be as follows:

Instruction 1. Read the data from the memory location corresponding to location B and put it into register 1 in the CPU.
Instruction 2. Add the word in the memory location corresponding to C to the contents of register 1, put the results in register 2.
Instruction 3. Store (write) the contents of register 2 into the memory location corresponding to A.

Of course as discussed above, instructions are really just strings of bits—0s and 1s. It is the function of the instruction decoder section of the CPU to translate the binary information into the appropriate sequence of operations needed to carry out the instructions just described. Of course, data also consist of a string of bits. The only real difference between data and instructions is the interpretation of the bits. In fact, at times it makes sense to modify an instruction by treating it as data and performing arithmetic operations on it. At other times, errors in programs result in data being interpreted as instructions with essentially random results.

For an illustration, the list of instructions above might be written in a very simple assembly language as follows.

load B, R1	(load B into register 1, R1)
add C, R1, R2	(add C to R1, result into R2)
store R2, A	(store R2 into A)

The portion of each line in parentheses are comments explaining the mnemonic instructions.

At the end of each instruction, the next instruction to be executed is fetched from the memory and put into the instruction decoder. This is generally called an **instruction fetch cycle**. A very detailed description of the operation of the computer during the three instructions above begins with the instruction fetch cycle that brings in the first of these instructions.

Instruction Fetch Cycle

The instruction decoding section puts the appropriate signals onto the internal and external control lines to ensure that the following events happen:

1. To tell all parts of the computer that the CPU has taken control of the address lines, to indicate to the memory section that the next operation involves a read from the memory at the address indicated by the address lines, and to gate the contents of the program counter onto the address lines. This will be called starting a read cycle in the following.
2. In response to these control signals, the memory system gates the contents of the appropriate location onto the data lines.
3. After waiting the appropriate amount of time to allow the memory to respond, the instruction decoder strobes the information from the data lines into the instruction decoder logic.
4. Since, in general, the next instruction to be executed will be in the next memory locations, the last step of the instruction fetch operation is to increment the program counter so that it contains the address of the next sequential instruction.
5. Next the interpretation and execution of the next instruction can begin.

First Instruction

The first instruction is an instruction to load an internal register from memory. The steps involved are as follows:

1. First, another memory fetch cycle is required. It proceeds almost exactly as above. The control logic starts another read cycle; the address specified this time comes from the address portion of the instruction.
2. The memory responds to this read cycle exactly as it did before.
3. After the appropriate delay, the instruction decoder causes the contents of the data lines to be strobed into the general register R1.
4. Since this is the end of the first instruction, another instruction fetch cycle is needed. The program counter has been incremented so that it now addresses the location containing the next instruction; thus another instruction fetch proceeds exactly as before.

Second Instruction

The second instruction involves fetching another piece of data from the memory, adding it to the contents of a general register, and storing the results in another general register. This will involve several extra steps:

1. First, another memory read cycle is needed to read the data in the word addressed by C into the CPU. This proceeds exactly as above, except that at the last step, the instruction decoder section causes the contents of the data lines to be strobed into register A attached to the ALU.
2. Next, the instruction decoder causes the contents of register R1 to be gated onto the internal data bus and, after allowing a suitable delay for signal propagation, causes the contents of the data bus to be strobed into register B, which is also attached to the ALU.
3. At this time, the two inputs are present at the input of the ALU. The instruction decoder logic now causes the output of the ALU to be gated on the internal data lines and again, after a suitable delay, causes the data lines to be strobed into the general register R2. (This multiple use

of the data bus is the reason latches are needed on the A and B inputs of the ALU.)

4. This is the end of the second instruction so, another instruction fetch cycle is needed.

Third Instruction

This is a simple memory write. It proceeds much like a read cycle with the obvious differences:

1. The instruction decoder logic sets the appropriate control lines to indicate that the CPU is taking control of the external bus, that the next operation is a memory operation, and that data are to be written into the memory location addressed.
2. At the same time, the decoder logic gates the address from the instruction onto the address lines and the contents of register R2 onto the data lines.
3. The memory section responds to these control signals by writing the contents of the data lines into the appropriate memory location in an operation synchronized by the clock.
4. This is the end of this instruction, so another instruction fetch is begun. Since this is also the end of this short program example, the discussion ends here.

COMMENTS ON THE SAMPLE PROGRAM

What follows is a number of comments about these instructions—they are in no particular order.

Generally, the internal bus is more complicated and extensive than the external bus. For microcomputers—computers on a single IC—the number of pins on the package sets a limit on how many signals can be brought out of the microcomputer onto the external bus. Furthermore, some of the signals on the internal bus would have no meaning on the external bus and, likewise, some of the external signals would have no meaning on the internal bus. The most important lines, such as the data lines, the address lines, and the most important of the control lines, are the same both in the internal and the external buses. However, there may be some way of isolating the internal from the external lines to allow different operations to take place at the same time.

Many computer instructions do not require use of the external bus, especially if there are many general-purpose internal registers in the processer. Also, there are often times when external devices may need to take control of the external bus. As an example, once a modern disk controller has been set into operation by the appropriate instructions from the CPU, it is able to read material directly from the disk and put it into memory without any further instructions from the CPU. This allows the computer to appear to do two things at once and, hence, have more apparent power. For this to be possible, there need to be gates on the address, data, and some control lines so that the internal and external buses can be separated at some times and can be connected at other times. Also, some mechanism is needed so that only one device is attempting to control the external bus at a time. This mechanism is generally called **bus arbitration**.

The **width**, or number of bits in each portion of the buses, has much to do with the power of the computer. Obviously, if a computer deals with 32-bit words, then it will be much easier to read and write data to and from memory if the data bus is 32 bits wide. In an effort to keep the external pin count down, some early microprocessors had wider internal buses than external buses, but the drive to more and more powerful computers has forced external buses to be as wide as the normal word width in the computer system.

The width of the address bus determines the largest memory address that can be used, and this in turn sets a limit on how large simple programs can be. An eight bit-wide address bus can only address 256 words of memory; a 16-bit bus can address 64k words (really 65,536 words), 24 bits can address 16,777,216 words, and 32 bits can address about 4×10^9 words (really 4,294,967,296 words). The recent growth of computer power has resulted in an ever-increasing width of the address buses and an ever-increasing total memory space available for use in the computer systems.

In the sample program above, it is clear that the existence of two special-purpose registers on the input of the ALU causes some awkwardness in programming. Another option would be to have two additional internal data buses, the A bus and the B bus, which are connected to the A and B inputs of the ALU. Then, all operations could have more or less that same form. The sample program above could be recast to fetch B and put it in R1, fetch C and put it in R2, gate R1 onto bus A, gate R2 onto bus B, and the output of the ALU onto the data bus and R3. In this way, no register would be treated any differently than any other. All numbers fetched from memory would be available in internal registers for possible future use, and the complier, interpreter could make more efficient use of a large number of internal general registers. Clearly, this is a place where the designer must make some decisions at the very start of the design process for a computer. As the ability to put more and more gates and buses onto one IC has developed, the internal structure of the CPU has gotten more complicated so that, in general, the computers can operate faster.

In general, it is nice to be able to contain one or more instructions in one computer word. Thus, in a 32-bit computer, it would be nice to ensure that most instructions require less than 32 bits. One possible way of doing this is to allocate eight bits to being an **instruction code** and the rest of the word to being either addressing information, data, or additional instruction information—this would be determined by the first eight bits. With this choice, the computer would have at most 256 different basic instructions (the maximum number that can be addressed by eight bits).

It would be very inefficient if every instruction that addressed a memory location had to have a full 32-bit address on a computer in which the word size was 32 bits and the address bus width was also 32 bits. As a result, most computers have a number of ways of generating the actual address for a memory reference. The most common methods are to form the sum of the address portion of the instruction with one or two registers called **index registers** or **base registers** to generate the actual address put onto the address bus. In this case, there is usually a very-high-speed adder unit that actually generates this sum that is the address. This adder is independent of the ALU, and its operation is automatic. There are many different variations on the ways of doing address in machine language programs. One of the most complex and important sections of any data book that describes the details of computer system has to do with the number of addressing options, how they operate, and for what they are most useful. Much of the increase in the power of computers has been due to the growing diversity of address types and the growing sophistication of their use in programs.

In general, program instructions are executed in sequential order unless an instruction explicitly calls for a change in the order of execution. Such an instruction might be in the form: "If register 1 contains a number greater than the number in register 2, then execute the instruction in location xxxx, otherwise execute the next sequential instruction." When this instruction is executed, if the test is true, then the address xxxx is placed in the program counter and then the next instruction is fetched—the sequential operation is broken. If the test is not true, nothing is done to the program counter and the sequential operation continues.

As part of every instruction fetch, the program counter is incremented so that it contains the address of the next sequential instruction. While this could be done by means of the ALU, it clearly makes more sense to create the program counter in such a way that it can be incremented by 1 (or 2 or 4, depending on the word width and the addressing conventions) without the use of any additional hardware. Clearly, if the program counter is a counter that can be parallel loaded, a single count pulse at the appropriate input will cause the counter to increment. This is the source of the name "program counter."

Further details here would be an error. For more information, consult the manual for any specific computers—see Appendix A for specific references.

INTERFACING

Although very few people design and build computer systems, many want to connect a laboratory instrument to a computer. This involves building an interface. At some level, building an interface requires a detailed understanding of the external bus of the computer, what each signal does, and what conventions there are in the use of this bus. To make the design of interfaces easier, the manufacturers of most of the major microprocessors also make **peripheral interface adaptors**, or **PIA**s, which are single ICs that connect to the external bus and do much of the logic required to interface most types of devices to the computers. These PIAs are still quite complex devices; whereas the data book for a microprocessor might be 500 pages long, the data sheets (book?) for a PIA may only be a few pages long. Nonetheless, they are very useful and to study them is well worth the effort. Appendix A contains some references.

Interfacing is sufficiently common that add-on cards are available for many PC systems that can be used to interface most devices to the PC. Typically, these cards include a PIA and some additional circuitry to make their use easier. In general, the use of one of these cards to interface a simple laboratory instrument requires no more knowledge than is presented in this book. It does, however, require extensive study of the data sheets, instruction manuals, and a fanatical attention to detail.

A sample interface of an A-to-D converter to a computer system is worked out in some detail in Chapter 39.

PART FOUR

ANALOG CIRCUITS

This part of the book is an introduction to analog electronics. The use of the name analog is fairly new; it is used to distinguish this class of electronic circuits from digital electronics. In digital electronics, the voltages and currents take on only discrete values, and any information carried by the signals is contained in the meaning assigned to each value. In contrast, in analog electronics the voltages and currents can take on all values within some operating limits, and information can be carried by the size of the voltages or currents or both. Not too many years ago, essentially all of electronics was analog electronics.

Analog circuits are made up of amplifiers, attenuators, and filters. They are described in terms of gain, phase shift, frequency response, stability. At present, most analog circuits are built around integrated operational amplifiers. The first two chapters of this section (Chapters 28 and 29) are an introduction to operational amplifiers and the most common simple circuits made from them. Chapter 30 is a historical interlude, an introduction to the analog computer, the first major use of operational amplifiers. Chapter 31 discusses a number of practical details and much of the terminology about operational amplifiers, whereas Chapter 33 presents a brief introduction to formal amplifier theory—circuit analysis techniques and terminology specific to amplifier circuits. Finally, Chapters 32 and 34 examine several of the more important, more complex circuits based on operational amplifiers.

Analog circuits are sometimes called **linear** circuits because for many of the circuits there is a linear relationship between the inputs and the outputs of the circuits. The data books produced by the manufacturers for these circuits will sometimes refer to analog circuits or linear circuits.

PREREQUISITES

For all the chapters in this section, there is a general set of prerequisites. These are
1. A familiarity with basic circuit analysis—Chapter 3.
2. A familiarity with decibels—Chapter 53.
3. Some terminology related to AC circuit analysis is used throughout—Chapter 8.

4. Occasional references are made to the Thevenin equivalent impedance, but in general the ability to use the theorem is not required—Chapter 6.
5. Chapters 28 and 29 are prerequisites for the following chapters in this section.

   A list of additional prerequisites follows.

CHAPTER 28 OPERATIONAL AMPLIFIER BASICS

No prerequisites.

CHAPTER 29 BASIC OPERATIONAL AMPLIFIER CIRCUITS

1. Be familiar with the basics of op-amp circuits and the analysis of these circuits—Chapter 28.
2. Some of the terminology used in this chapter is more fully defined in Chapter 31.

CHAPTER 30 ANALOG COMPUTERS

1. A familiarity with elementary calculus.

CHAPTER 31 OPERATIONAL AMPLIFIER PARAMETERS

No prerequisites.

CHAPTER 32 THREE SPECIAL-PURPOSE AMPLIFIERS

1. A familiarity with the basic operational amplifier circuits and their limitations—Chapters 28, 29, and 31.

CHAPTER 33 FORMAL AMPLIFIER THEORY

1. A knowledge of the terminology and specifications for op-amps—Chapter 31.

CHAPTER 34 COMPARATORS AND SCHMITT TRIGGERS

1. A familiarity with simple logic circuit conventions—Chapter 19.

Chapter 28

OPERATIONAL AMPLIFIER BASICS

OBJECTIVES

1. Become familiar with the terminology used to describe operational amplifier circuits.

2. Become familiar with the normal infinite gain analysis scheme for op-amp circuits.

3. Have an understanding for the errors made in the infinite gain approximation.

4. Become familiar with the concept of ground or common.

5. Know what feedback is.

INTRODUCTION

This chapter begins by describing the basic concepts of an amplifier circuit. This includes introducing the concepts of feedback and ground as well as defining a number of terms. Next, the integrated circuit operational amplifier is introduced and its main properties are briefly discussed. Finally, the gain of the basic inverting operational amplifier circuit is calculated, and the methods for analyzing ideal operational amplifier circuits are discussed.

AMPLIFIER BASIC CONCEPTS

Amplifiers are the most important building blocks used in analog electronics. They are active devices usually having two input terminals and two output terminals (see Figure 28-1). The general goal of most descriptions of an amplifier is to separate out its effects—the relationship between the output and the input—from the internal details of the amplifier. Thus the amplifier is to be treated as a black box described by a relatively small number of parameters. The output of the amplifier is some function of the input. In the simplest cases, the output is a linear function of the input, that is, a constant times the input. The amplifier may respond differently to ac inputs than to dc inputs; for instance, it may respond only to one or the other. The amplifier

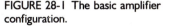

FIGURE 28-1 The basic amplifier configuration.

243

response may be a strong function of the frequency of the input signal. In general, only a little information about the amplifier is needed. The most important properties of an amplifier are its gain, its input impedance, its output impedance, and its frequency response. These are described in more detail below.

The **voltage gain** of an amplifier is defined as

$$G_v = \frac{v_{\text{output}}}{v_{\text{input}}} \tag{28-1}$$

where both v_{output} and v_{input} refer only to signals and do not include any biases or offsets. Implicit in this definition is the assumption that the amplifier is linear, at least, over some range of input voltages; this means that doubling the input voltage will double the output voltage.

In general, the gain is a complex number, because the input and output may differ in phase as well as in magnitude. For most circumstances, only the magnitude of the gain is important; however, when considering amplifier stability, this is not true. The **current gain**, or **power gain**, of an amplifier can be defined in a similar manner. Gains are usually measured in decibels, (see Chapter 53).

The **input impedance** is the impedance seen looking into the input of the amplifier. It describes the loading effect the amplifier will have on a signal source.

The **output impedance** of the amplifier is the impedance seen looking back into the amplifier; this is the Thevenin equivalent output impedance of the amplifier. It affects the ability of the amplifier to provide output current.

The **frequency response** of the amplifier describes performance of the amplifier as a function of the input frequency. All of the parameters describing the amplifier may change as the input frequency changes. In the simplest case, the input and output impedances are purely resistive, and only the gain changes as a function of frequency.

Of course, there are many other parameters needed to specify fully the external details of an amplifier, but the four above are the most important. A much more extensive list of parameters is given in Chapter 31, where operational amplifier specifications are discussed.

The general amplifier discussed thus far has two input and two output leads. Not all amplifiers have this many; as a result, there is a certain amount of terminology used to describe the basic amplifier configuration. An amplifier that has two inputs and responds to the difference between the two inputs is called a **difference amplifier** or sometimes is said to have a **differential input**. An amplifier that has two outputs that are 180° out of phase with each other is said to have a **differential output** or, sometimes, to have **double-ended** outputs. Some amplifiers have one of the output lines connected to the common point of the circuit. These are said to have **single-ended** outputs. Likewise, some amplifiers have one of the input lines connected to the circuit common; these are said to have single-ended inputs.

Amplifiers need not have gains greater than 1; they can make signals smaller. In this case, they are often called **attenuators**, or **active attenuators**, to distinguish them from simple voltage dividers. Amplifier circuits having very strongly shaped frequency responses can be built. These are called **filters**, or **active filters**.

Often, a portion of the output of the amplifier is connected back to the input; this is called **feedback**. This situation is shown in a schematic way in Figure 28-2. If the feedback makes the change in the output caused by a given input change smaller than it would have been if there were no feedback, then it is **negative feedback**. If the feedback makes the change in the output greater than it would have been without the feedback, then it is **positive feedback**. A more formal approach to feedback is outlined in Chapter 33. For now, these two intuitive definitions will suffice.

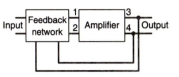

FIGURE 28-2 A feedback network and an amplifier.

Finally, a few words must be said about the topic of ground. Ground is probably the most confusing and complex topic in all of electronics. Most circuits have one point (or wire) that is common to many parts of the circuit. This point is called **common**, or **ground**, or sometimes **earth** since, in principle, this point could be connected to the earth and the operation of the circuit would be unchanged. The existence of the ground in a circuit is what makes it possible to clip one lead of the voltmeter or the oscilloscope to the circuit ground and to poke around with the other lead making meaningful measurements. (This fact is also related to the conservative nature of the electrostatic field.)

Yet the very simplicity of the concept of ground often leads to difficulties in making measurements. A high-fidelity amplifier may have a perfectly well-defined ground point; an oscilloscope used to study the operation of the amplifier may also have a well-defined ground point, and the **two grounds may not necessarily be the same.** There can well be an ambituity when measuring a signal with respect to ground—which ground?

Furthermore, connecting the two instruments together may cause the operation of one or the other to change dramatically because of the interaction between the two grounds. People who measure small signals over large distances spend much of their time worrying about grounds, ground loops, and similar related problems. From a beginner's point of view, the grounding of interconnected circuits often appears to be a black art.

Because there is an implicit ground in the power system to which the instruments are connected, poorly designed or poorly built circuits can have grounds that differ by 110 V. People have been electrocuted when they touched the chassis (ground) of some electronics while in contact with the earth, say, a water pipe. Grounds can be a real problem. In any case, the schematic symbols for common, or ground, are shown in Figure 28-3. Although some writers use these symbols to represent different things, such as circuit board common, chassis connection, and earth connection, there is no universally accepted convention for the use of these symbols.

FIGURE 28-3 Schematic ground symbols.

OPERATIONAL AMPLIFIERS

Amplifiers are built for a wide range of purposes ranging from detecting the incredibly weak signals from radio astronomy telescopes or satellite broadcasts to giants producing many horsepower of output to power the speakers of a rock band, a football stadium, or a political demonstration. Today, however, except for some very special purpose input or output stages, most amplifiers are built around integrated operational amplifiers.

Operational amplifiers are high-gain, differential input amplifiers. The name **operational amplifier**, which is universally shortened to **op-amp**, comes

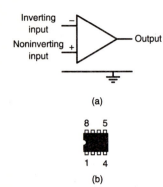

(a)

(b)

FIGURE 28-4 A. The schematic symbol for an operational amplifier. B. A full size sketch of a 741 from above.

from the fact that they can be used to perform mathematical operations such as addition, subtraction, integration—see the next chapter. Op-amps were first developed during World War II when they formed the basis for analog computers that were used to solve very difficult differential equations involved in engineering design and military strategy.

The original op-amps were large, complex vacuum tube circuits, which cost thousands of dollars, that did not have particularly good specifications. Today, integrated circuit op-amps having a wide range of often spectacularly good specifications are available. The most common op-amp, the 741, is available in an eight-pin dip configuration (dual in-line package) for less than 25 cents in large quantities and for less than $1 a piece from Radio Shack.

The schematic symbol for an operational amplifier is given in Figure 28-4. For general information, a roughly full-sized sketch of a 741 in the eight-pin dip package is also included.

This amplifier has a differential input and a single-ended output. The output voltage is measured with respect to the circuit common, or circuit ground. The input to the amplifier is the difference between the two input leads. These leads are usually identified by plus and minus symbols and are called the **inverting** (−) and the **noninverting** (+) inputs. Because the meaning of these two different inputs is often confusing, they will be explained three different ways.

First, if the noninverting input is more positive than the inverting input, the output of the amplifier is positive. If the inverting input is more positive than the noninverting input, the output is negative. Using V_+ as the voltage of the noninverting input with respect to ground and V_- as the voltage of the inverting input, this can be written as

$$\text{If } V_+ - V_- > 0 \qquad \text{then} \qquad V_{\text{output}} > 0$$
$$\text{If } V_- - V_+ > 0 \qquad \text{then} \qquad V_{\text{output}} < 0$$

Second, if the noninverting input becomes more positive while the inverting input is unchanged, the output will become more positive. Likewise, if the noninverting input becomes more negative, the output will become more negative. If the noninverting input is unchanged and the inverting input becomes more positive, the output will become more negative.

Finally, another way of saying the same thing is: The output will be in phase with a signal applied to the noninverting input, and it will be 180° out of phase with a signal applied to the inverting input.

Several important items are not shown in the schematic symbol in Figure 28-4. The op-amp is an active device; it must have a power source to function. In general, op-amps need two or more power supply voltages (one of which may be ground). To simplify the schematic diagrams, these are rarely shown, but they must be provided in actual circuits or else nothing works. (Forgetting to connect up the power leads is a very common mistake in wiring simple prototype circuits or laboratory exercises.) There may be other leads that can be used for some purpose in controlling the op-amp. For instance, the 741 has two leads that may be used to reduce the effects of the input offset voltage (see below). For illustrative purposes, the pin connections for the 741 in the eight-pin dip package are given as follows:

Pin Number	Function
1	Offset null adjustment
2	Inverting input
3	Noninverting input
4	Negative power (−15 V max)
5	Offset null adjustment
6	Output
7	Positive power (+15 V max)
8	No connection

A fairly complete specification of the properties of an amplifier takes a large number of parameters. An extensive list of these parameters is given in Chapter 31. Here, the seven most important properties will be briefly discussed. Included in this discussion are typical values of these parameters for the 741, what range of values are available in common op-amps, and what the value would be for an ideal op-amp. Note that, although the 741 is not a particularly good operational amplifier by modern standards, it is still capable of serving many functions. Because of its low cost and common use, it is a very good device for use in introductory laboratories. For reference throughout this section, a data sheet for the 741 is included in Appendix B.

The gain of an operational amplifier without a feedback loop is called the **open loop gain** and is usually given the symbol A. The output of the amplifier is A times the difference in the input voltages, that is,

$$V_{\text{output}} = A \times (V_+ - V_-) \qquad (28\text{-}2)$$

Note that this expression contains all the sign conventions about the inverting and noninverting inputs that were given above.

An ideal op-amp would have a very large gain, as close to infinite as possible. However, considerations of stability of the amplifier make it difficult to produce amplifiers having voltage gains much greater than 100 db (the output is 100,000 times the input!) The 741 has a nominal gain of 103 db, and typical values for standard op-amps range from 60 to 120 db. Op-amps intended for very-high-frequency use or for other special purposes may have smaller voltage gains.

The **input impedance**, or **input resistance**, is the load seen by a signal source connected between the two input terminals. For an ideal op-amp, this would be infinite. For the 741, the input resistance is nominally 2 MΩ; op-amps having values in excess of 10^6 MΩ exist. However, since the nature of the circuit in which the op-amp is used can make the apparent input impedance appear much larger or much smaller than this value, this parameter is less important than the input bias current.

The **input bias current** is the current that flows in the input lines. In general, for the amplifier to operate, a current must flow in the input lines; the direction of this current depends on the details of the op-amp and the input circuit. The current flows through whatever is connected to the input of the op-amp. Its importance is determined by its effects in this portion of the circuit. Ideally, this circuit should be zero. For a 741, the nominal input current is typically 0.1 μA, and op-amps with input bias currents as small as 0.01 pA (10^{-14} A) are available.

The **output impedance** is the impedance seen looking back into the op-amp from the output. This is the Thevenin equivalent impedance of the op-amp used in a circuit without feedback. For the 741, this impedance is nominally 75 Ω; for high-power op-amps, it is a small fraction of an ohm, while for very low-power op-amps, it can be several thousand ohms. Ideally, it would be zero. However, again the circuit in which the op-amp is used can greatly change the apparent value of the output impedance. As a result, the important parameters become the maximum output current and the maximum output voltage swing for the op-amp. For the 741, the maximum output current is about 20 mA, and the output voltage can swing to within about 2 V of the power supply values.

The **input offset voltage** is an imaginary input voltage that would account for any output voltage that occurs with no input voltage present. Op-amps are not perfect; even with no input voltage present, there will be some output voltage. This is generally specified in terms of an equivalent **input offset voltage** (the voltage needed at the input to cause the output to be zero). Ideally, this voltage would be zero; however, for the 741, it is typically 2 mV. Because this is a large voltage compared with possible inputs to op-amp circuits, provisions are made for adjusting the operation of the op-amp so that this voltage is zero (see the data sheet for the particular op-amp being used). As a result, the really important parameter becomes the stability in this offset voltage; that is, how much does the voltage change with changes in temperature or time? For a 741, this is about 3 μV/ °C; stabilized op-amps with drifts as low as 0.01 μV/ °C are available.

The **frequency response** of a typical op-amp used in an open loop situation is not particularly good. The nominal frequency response for the 741 is shown in Figure 28-6 below. Typically, op-amps have a flat frequency response from dc to some nominal value (about 100 Hz for a 741) and then a response that falls off at a constant rate. (For internally compensated amplifiers, this value should be 6 db per octave; see Chapter 33.) The value at which the gain falls to unity is often specified. This is roughly 1 MHz for the 741; op-amps with unity gains at 100 MHz or more are available.

The **slew rate** is a parameter that is related to the frequency response. Because of the nature of the op-amps output and compensation, no matter how fast its input changes, its output cannot change any faster than some limiting value called the **slew rate**, or **output slewing rate**. This rate is usually measured in volts per microsecond. For a 741, the value is nominally 0.5 V/μsec. Ideally, this value would be infinite, and op-amps with slew rates exceeding 100 V/μsec exit.

There are many other parameters describing the operation of an op-amp, but these are the most important ones. A more complete list with more precise definitions and additional discussion will be found in Chapter 31.

ANALYSIS OF THE BASIC INVERTING AMPLIFIER CIRCUIT

Probably the most common op-amp circuit is the configuration shown in Figure 28-5. This is an inverting amplifier. For an ideal op-amp, the output is given by

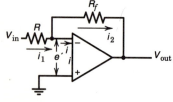

FIGURE 28-5 An inverting amplifier with the currents and voltages needed for analysis indicated and given names.

$$V_{out} = -\frac{R_f}{R}V_{in} \qquad (28\text{-}3)$$

The minus sign in the expression means that the amplifier is inverting. This means that if the input goes positive, the output goes negative. A very important feature of this circuit is that the gain depends only on the two external resistors, not on the gain of the op-amp itself. This is discussed in more detail in the following paragraphs.

The expression for the gain of this amplifier circuit can be derived in two ways. First, an analysis of the circuit of Figure 28-5 is carried out for a real op-amp; at the end of the derivation, the parameters of the real op-amp are replaced by those of an ideal op-amp. The second way is to set up rules and procedures for the analysis of ideal circuits by making any approximations at the very beginning.

To assist in the analysis of the circuit, a number of currents and one intermediate voltage have been indicated in the Figure 28-5. In this configuration, the noninverting input is connected to ground (0 V) and the input voltage is connected to the inverting input via the resistor R. The voltage at the inverting input is indicated as e'. R_f connected between the output and the inverting input provides negative feedback.

First, KCL applied to the inverting input gives

$$i_1 = i_i + i_2 \qquad (28\text{-}4)$$

But

$$i_1 = \frac{V_{in} - e'}{R} \qquad (28\text{-}5)$$

and (watching the signs)

$$i_2 = \frac{e' - V_{out}}{R_f} \qquad (28\text{-}6)$$

The definition of the open loop gain gives

$$V_{out} = -Ae'$$

or

$$e' = -\frac{V_{out}}{A} \qquad (28\text{-}7)$$

Combining these four equations gives

$$\frac{V_{in}}{R} + \frac{V_{out}}{AR} = i_i - \frac{V_{out}}{AR_f} - \frac{V_{out}}{R_f}$$

$$V_{out} = \frac{i_i}{1/R_f + 1/AR + 1/AR_f} - \frac{V_{in}/R}{1/R_f + 1/AR + 1/AR_f} \qquad (28\text{-}8)$$

This result is neither very nice nor very informative, but it simplifies with just two assumptions about the ideal amplifier. First, the ideal operational amplifier is assumed to have infinite input impedance and zero input bias current. This means that $i_i = 0$ and that

$$i_1 = i_2 \tag{29-9}$$

This assumption causes the whole first term in Equation 29-8 to go away. Thus

$$V_{out} = -\frac{V_{in}/R}{1/R_f + 1/AR + 1/AR_f}$$

$$= -\frac{R_f}{R} V_{in} \frac{AR}{AR + R_f + R} \tag{28-10}$$

Second, the ideal amplifier is assumed to have an infinite gain. This means that e' is zero or that the terms above involving $1/A$, such as $1/AR$ and $1/AR_f$, can be dropped. This yields the result promised at the beginning:

$$V_{out} = -\frac{R_f}{R} V_{in} \tag{28-3}$$

Thus the gain of this circuit (consisting of the op-amp and the two resistors) is

$$G_v = -\frac{R_f}{R} \tag{28-11}$$

Note that this is the gain of the entire circuit; it is sometimes called the **closed loop gain** because of the closed feedback loop in the circuit. This is to be contrasted with the open loop gain A, which applies to the op-amp alone without any feedback. Again, the closed loop gain for a circuit involving an ideal op-amp depends only on the external components, not on the op-amp itself.

Equation 28-3 can be obtained more quickly by making the two assumptions or approximations or both about the op-amp at the very beginning. In fact, the normal method of analyzing an op-amp circuit is to do this; then later, if the effects of the imperfections are of interest, they can be taken into account. These two assumptions are discussed below.

First, no current flows into or out of the inputs to the op-amp. This is equivalent to assuming that the input bias current is zero and the input impedance is infinite. In the diagram of Figure 28-5, this means that $i_i = 0$ and $i_1 = i_2$. Because the analysis of op-amp circuits almost always involves applying KCL to the inverting input (summing the currents at this input), the inverting input is often called the **summing point** in the circuit.

Second, the gain of the op-amp is infinite. For the circuit above, this means that $e' = 0$. This is sometimes called the **virtual ground** approximation, or **infinite gain** approximation, since the operation of the op-amp and the feedback loop is to make the inverting input appear to be at ground. (For other configurations, see the next chapter.) This approximation means that the voltages at the inverting and the noninverting inputs are the same.

With these two assumptions, the analysis of the circuit proceeds in two trivial steps. KVL applied to the summing point gives

$$i_1 = i_2 \tag{28-12}$$

Since e' is 0, the expressions for the currents are very simple and Equation 29-12 becomes

$$\frac{V_{in}}{R} = \frac{-V_{out}}{R_f} \tag{28-13}$$

and Equation 28-3 follows immediately.

As a verification that this circuit involves negative feedback, consider the following sequence of events. An external signal causes the input to go slightly positive, which means that the inverting input goes slightly positive. Owing to the action of the amplifier, the output goes slightly negative. This change in the output is connected back to the inverting input by the feedback resistor and decreases the effect of the initial change; this is negative feedback. Thus the op-amp works through the feedback resistor to keep the difference between its input voltages equal to zero.

It is always important to check the accuracy of any assumptions made during the analysis. The problems provide several chances to calculate the actual numerical effect of assuming that the gain of an op-amp is infinite in real circuits. However, here is another approach. If the output of a 741 were 5 V and if it had an open loop gain of 100,000, the actual input voltage e' would be

$$e' = \frac{V_{out}}{A} = \frac{5}{100,000}$$

$$= 50 \ \mu V$$

which is certainly very small.

There are two relatively common situations in which these assumptions break down: (1) when the output cannot change fast enough to keep up with the required changes, that is, when it is trying to change faster than its slew rate; and (2) when the output cannot reach the final voltage needed because it is trying to exceed either its maximum voltage limits. This situation is usually described by saying that the amplifier is **saturated**. If either of these two limits is reached, then the usual analysis of the op-amp circuit no longer holds true, and it is much harder to figure out what is happening. However, since this usually represents a breakdown of the circuit, the normal process is to find out what are the limits on the operation of the circuit and to ensure that they are never reached.

A demonstration of the breakdown of the analysis and, in particular, the breakdown of the virtual ground approximation makes a fairly nice experiment. Set up a simple inverting amplifier and connect a scope between ground and the inverting input. As long as the amplifier is operating normally, no signal (on a millivolt scale) will be seen. However, once the input amplitude is turned up to the point where the output saturates, a signal will be seen at the inverting input. Similarly, if a nominal amplitude is picked and the frequency is raised, dramatic effects are seen at the inverting input once the amplifier becomes slew rate limited.

DISCUSSION OF THE PROPERTIES OF THE INVERTING AMPLIFIER

The gain G of the circuit has been found to be

$$G = \frac{-R_f}{R}$$

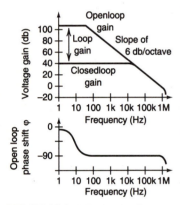

FIGURE 28-6 A Bode plot for a 741.

The input impedance of the overall circuit is R because of the virtual ground approximation. Since the inverting input is at ground, the input signal generator sees R connected to ground as the input.

There is a very important and confusing point here: two different gains and input resistances have been mentioned, and they must be kept clear. There are parameters for the open loop circuit, that is for the op-amp alone, and for the closed loop circuit. Thus the open loop gain of the op-amp is A and the closed loop gain is $-R_f/R$ for the inverting amplifier. Likewise, the open loop input impedance for the 741 is about 2 megohms, whereas the closed loop value is R. Only rarely are op-amps used without negative feedback loops: as a result, the primary parameters of interest are the closed loop values. The open loop values are of interest only to the extent that they determine the closed loop values. This will be discussed in subsequent chapters. But, to conclude the presentation here, always keep in mind which set of parameters are being discussed.

The graph in Figure 28-6 shows the frequency response for a 741 alone and used in a typical inverting amplifier configuration having a gain of 100. Thus the effect of the negative feedback is to greatly improve the apparent frequency response of the circuit. The **loop gain** is the difference between the open loop gain of the amplifier and the gain of the circuit with feedback. Note that the loop gain decreases as the frequency increases. The loop gain is really a measure of the quality of the op-amp assumptions.

The plot of the log of the op-amp gain versus the log of the frequency is sometimes called a **Bode plot**. Sometimes, the phase shift of the op-amp is also included on the Bode plot. The Bode plot is discussed further in Chapter 33.

The effects of other imperfections in the op-amp can be easily be found. The usual procedure is to analyze the ideal circuit, then assume that the various imperfections are independent, and solve for the effect of each one of interest in turn. (Note that this is equivalent to assuming superposition or assuming that the circuit is linear.) Several examples are given in the next chapter, and an equivalent circuit showing many of the imperfections is given in Chapter 31.

PROBLEMS

28-1 For an op-amp having an open loop gain of 100,000 in a circuit having $R = 1$ k and $R_f = 100$ k, calculate the nominal gain and the actual gain.

28-2 For a inverting amplifier having $R = 10$ k and $R = 1$ Meg, calculate the nominal gain, the real gain, and the approximate error in the nominal gain if the op-amp has a open loop gain of 1, 10, 100, 1000, 10^4, 10^5, 10^6, and infinity.

28-3 For an op-amp having an open loop gain of 10^5 used in inverting amplifier circuits having $r = 1$ k and $R_f = 10$ k, 100 k, 1 Meg, and 10 Meg, calculate the nominal gain, the actual gain, and the percentage error in each case.

28-4 Can you derive any rules for estimating errors from these examples?

ANSWERS TO ODD-NUMBERED PROBLEMS

28-1 Nominal gain is -100; actual gain is -99.9
28-3

R_f	Nominal Gain	Actual Gain	% Error
10 k	-10	-9.999	0.01
100 k	-100	-99.9	0.1
1 Meg	-1000	$-990.$	1
10 Meg	$-10,000$	$-9091.$	10

Chapter 29

BASIC OPERATIONAL AMPLIFIER CIRCUITS

OBJECTIVES

1. Be familiar with a wide range of standard op-amp circuits.

2. Be able to identify the function of most common op-amp circuits.

3. Be able to design circuits having the desired output.

4. Be familiar with some of the important problems in the standard op-amp circuits.

5. Be familiar with some of the techniques used to analyze the effects of imperfections in op-amp circuits.

INTRODUCTION

This chapter is a catalog of the most common and useful operational amplifier circuits. There is a very brief discussion of the uses and properties of each circuit. For some circuits, one or more of the properties may be derived as an illustration of the methods used to analyze op-amp circuits. This chapter continues the discussion of the previous chapter.

REVIEW OF IDEAL OP-AMP ANALYSIS

Unless stated otherwise, the following is assumed for all circuits: (1) No current is flowing into or out of the input leads of the op-amp; and (2) the op-amp has infinite gain, so that the two inputs are at the same voltage. The ideal op-amp also has zero offset current and voltage, unlimited frequency response, infinite slew rate, infinite input impedance, zero output impedance, and no limitations on the output voltages and currents, but only the first two properties are important in the derivations. The effects of imperfections in the op-amp will be discussed in some cases.

A CATALOG OF CIRCUITS

The Inverting Amplifier

It was shown in the previous chapter that the output of the **inverting amplifier** in Figure 29-1 is given by

$$V_{out} = -\frac{V_{in} R_f}{R} \qquad (29\text{-}1)$$

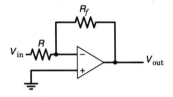

FIGURE 29-1 The inverting amplifier.

and the input impedance of this circuit is

$$R_{in} = R$$

As a special case, if $R = R_f$, then $V_{out} = -V_{in}$ and the output of this amplifier is simply the negative of the input; that is, the amplifier inverts the input.

There are some limitations on the possible range of values for R and R_f. R cannot be too large because the input bias current must flow through it or through R_f without inducing too large a voltage drop, since this voltage drop will be part of the effective input voltage. In addition, R_f cannot be too small or the op-amp's maximum output current limitation may be exceeded—remember, one end of R_f is at a virtual ground.

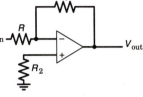

A Better Inverting Amplifier

The circuit in Figure 29-2 is also an inverting amplifier; its output is given by FIGURE 29-2 Another version of the same expression as the previous example. The only difference is the resistor added in the noninverting input lead. If this resistor is chosen correctly, the effects of the input bias current are almost eliminated. To do this, R_2 should be equal to the parallel combination of R_f and R. (In all these expressions, it is assumed that the signal source is a very-low-impedance source. If this is not true, then the signal source impedance must be added to R.)

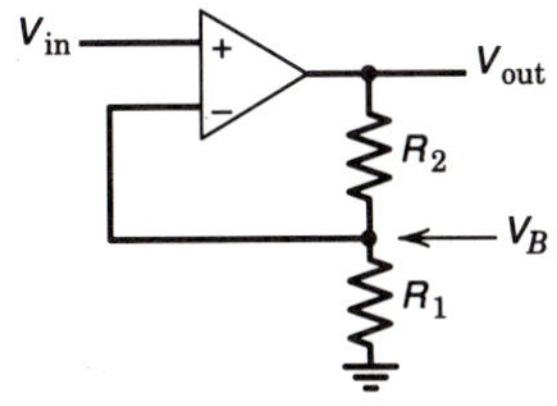

FIGURE 29-2 Another version of the inverting amplifier.

Noninverting Amplifier

The circuit in Figure 29-3 is called a **noninverting amplifier**. In this configuration, the input goes to the noninverting input and a voltage divider returns a fraction of the output voltage to the inverting input. Because this is an ideal operational amplifier, the two inputs must be at the same potential; hence, the voltage indicated as V_B in Figure 29-3 must be equal to V_{in}. So

FIGURE 29-3 A noninverting amplifier.

$$V_{in} = V_B = V_{out} \frac{R_1}{R_1 + R_2}$$

and

$$V_{out} = V_{in}\left(1 + \frac{R_2}{R_1}\right) \qquad (29\text{-}2)$$

This is a general-purpose amplifier that does not change the sign of the signal—hence, the name. The circuit also has a very high input impedance

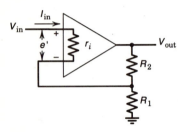

FIGURE 29-4 The circuit used to calculate the input impendance of the noninverting amplifier.

and a low output impedance. Note that this statement refers to the circuit as a whole, that is, to the op-amp and the feedback loop together, as opposed to just the op-amp alone. The following two derivations illustrate the methods for calculating these parameters.

To calculate the input resistance (impedance) for this circuit, the equivalent circuit shown in Figure 29-4 is used. Here, r_i is the input resistance of the op-amp itself. The input resistance of the entire circuit is determined by applying a voltage to the input terminal and measuring the current that flows in it. Thus

$$R_{in} = \frac{V_{in}}{I_{in}}$$

But

$$I_{in} = \frac{e'}{r_i}$$

$$V_{out} = -Ae'$$

and from Equation 29-2,

$$V_{out} = V_{in}\left(1 + \frac{R_2}{R_1}\right)$$

So, working backward,

$$e' = -\frac{V_{out}}{A}$$

$$= -V_{in}\frac{R_1 + R_2}{R_1 A}$$

Ignoring the signs and collecting all this gives

$$I_{in} = \frac{V_{in}(R_1 + R_2)}{r_i R_1 A}$$

and finally

$$R_{in} = \frac{r_i R_1 A}{R_1 + R_2} \tag{29-3}$$

R_{in} becomes very large as A becomes large. Thus, for this noninverting case, the apparent input impedance is so large that any effects of the op-amp circuit on the signal source will be dominated by the bias currents.

There is a minor error in the derivation above. Equation 29-2 was derived by assuming that the gain of the amplifier was infinite; then suddenly, in this second derivation, it was treated as if it were finite. Doing the calculation exactly would yield additional correction terms whose importance decreases as A^2. Thus this derivation got the leading or most important term of the result. This is a fairly normal situation. Evaluating the effects of imperfections on op-amp circuits can become very complex. But since these evaluations are an approximation at best, it is quite reasonable to approximate the derivations to get only the leading terms.

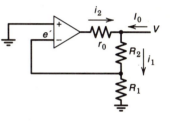
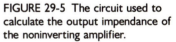

A trick is used to calculate the output impedance. The input is grounded and a voltage is applied to the output. This is shown in Figure 29-5. Here, r_o is the output resistance of the op-amp itself. The derivation is similar to the one above. The apparent output resistance of the circuit is given by

FIGURE 29-5 The circuit used to calculate the output impendance of the noninverting amplifier.

$$R_{out} = \frac{V}{I_o}$$

From the diagram it is easy to verify that

$$I_o = i_1 - i_2$$

$$i_1 = \frac{V}{R_1 + R_2}$$

$$i_2 = \frac{Ae' - V}{r_o}$$

and

$$e' = -\frac{VR_1}{R_1 + R_2}$$

So

$$i_2 = \frac{AVR_1}{r_o(R_1 + R_2)} - \frac{V}{r_o}$$

$$I_o = \frac{V}{R_1 + R_2} + \frac{AVR_1}{r_o(R_1 + R_2)} + \frac{V}{r_o}$$

$$= \frac{Vr_o + AVR_1 + V(R_1 + R_2)}{r_o(R_1 + R_2)}$$

and

$$R_{out} = \frac{r_o(R_1 + R_2)}{r_o + AR_1 + R_1 + R_2} \tag{29-4}$$

which tends to zero as A becomes large.

The Voltage Follower

The limiting case of the noninverting amplifier is the **voltage follower** shown in Figure 29-6. For this circuit,

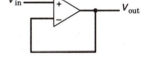

FIGURE 29-6 The voltage follower.

$$V_{out} = V_{in} \tag{29-5}$$

The input impedance becomes

$$R_{\text{in}} = AR_i$$

and the output impedance is

$$R_{\text{out}} = \frac{r_o}{A}$$

Thus this circuit is simply a buffer and an impedance transformer. Note that although this circuit has no voltage gain, it can have very large power gains due to the changes in impedance levels.

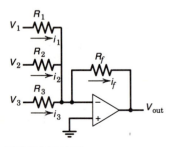

FIGURE 29-7 The summing amplifier.

The Summing Amplifier

The output of the **summing amplifier** circuit shown in Figure 29-7 is given by

$$V_{\text{out}} = -\frac{R_f}{R_1}V_1 - \frac{R_f}{R_2}V_2 - \frac{R_f}{R_3}V_3 \qquad (29\text{-}6)$$

If $R_1 = R_2 = R_3 = R$, this becomes

$$V_{\text{out}} = -\frac{R_f}{R}(V_1 + V_2 + V_3) \qquad (29\text{-}7)$$

Thus the original circuit gives a weighted and inverted sum of the input voltages. The second form is more obviously an adder. This circuit can be extended to any number of inputs. The ultimate limit on the number of terms that can be added is the decrease in the loop gain that occurs as each term is added. Since the input resistors are all in parallel from the summing point to ground, there is some equivalent resistance, R_{eq}, from the summing point to ground. The effective gain the op-amp is being asked to produce is R_f/R_{eq}, which increases as more and more inputs are added. As the effective gain increases, the loop gain for the amplifier decreases and errors in the operation of the circuit increase.

Because the inverting input is a virtual ground, there is no interaction between the various voltage sources. It is this fact that makes this circuit work. A noninverting summing circuit cannot be built with only one op-amp.

Finally, if the derivation of the gain is not obvious, it follows directly from the application of KCL to the summing input of the op-amp. Thus

$$i_1 + i_2 + i_3 = i_f$$

Using the virtual ground approximation, these currents become

$$i_1 = \frac{V_1}{R_1}, \text{ etc.}$$

and the result above follows immediately. Note that this is the circuit that makes obvious why the summing input is given this name.

Difference Amplifier

Figure 29-8 shows a **difference amplifier**. It is no mistake that the resistors in the two inputs have the same name; they must have the same value if this circuit is to work correctly. The output is given by

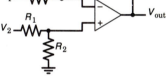

$$V_{\text{out}} = \frac{R_2}{R_1}(V_2 - V_1) \qquad (29\text{-}8)$$

FIGURE 29-8 A difference amplifier.

The derivation of this result is as follows. The voltage at the noninverting input is

$$V_+ = V_2 \frac{R_2}{R_1 + R_2}$$

Since the voltage at the inverting input is the same as the noninverting input, applying KCL to the inverting input gives

$$\frac{V_1 - V_+}{R_1} = \frac{V_+ - V_{\text{out}}}{R_2}$$

Eliminating V_+ and the rearranging gives

$$\frac{V_1}{R_1} - \frac{R_2}{R_1}\frac{V_2}{R_1 + R_2} = \frac{V_2}{R_1 + R_2} - \frac{V_{\text{out}}}{R_2}$$

Solving this for V_{out} gives Equation 29-8.

For this circuit to work well, the resistors need to be carefully matched. Furthermore, because the summing point is not at ground, several inputs cannot be added at the inverting input without change the weighting of V_2.

Current-to-Voltage Converter

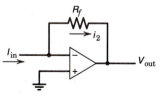

The circuit shown in Figure 29-9 is a **current-to-voltage converter**. The output of this circuit is given by

FIGURE 29-9 Current-to-voltage converter.

$$V_{\text{out}} = -R_f I_{\text{in}} \qquad (29\text{-}9)$$

This follows directly from the usual analysis. The circuit obtains its name from the fact that its input is explicitly a current and its output is a voltage. In some sense, all the circuits above could also be called current-to-voltage converters. The name depends on the point of view and the use of the circuit.

Constant Current Source

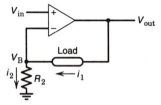

The circuit shown in Figure 29-10 maintains a constant current given by V_{in}/R through the load. Thus it is a **constant current** source that is controlled by V_{in}. The analysis follows directly from the two assumptions about op-amp circuits:

FIGURE 29-10 A constant current source.

$$i_1 = i_2$$

$$V_B = V_{\text{in}}$$

But

$$i_2 = \frac{V_B}{R_2} = \frac{V_{in}}{R_2}$$

and, hence,

$$i_1 = \frac{V_{in}}{R_2} \tag{29-10}$$

no matter what the impedance of the load. Here, as in all these derivations, it is assumed that there are no limits on the output of the op-amp, or at least that any limitations are not exceeded.

The primary problem with this circuit is that the load must float; that is, one end of the load cannot be connected to ground. A number of modifications of this circuit avoid this problem; however, they take us beyond an introduction to op-amps. See the references in Appendix A for more information on this circuit.

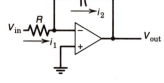

FIGURE 29-11 An integrator.

The Integrator

The circuit in Figure 29-11 functions as an integrator. For a general input, the usual analysis scheme gives

$$i_1 = i_2$$

$$\frac{V_{in}}{R} = -C_f \frac{dV_{out}}{dt}$$

or

$$V_{out} = \frac{1}{RC_f} \int_0^t V_{in} \, dt \tag{29-11}$$

where it is assumed that $V_{out} = 0$ at time $t = 0$.

If the input is an ac voltage, this circuit can also be treated as a simple inverting amplifier or

$$V_{out} = V_{in} \frac{Z_f}{Z}$$

$$= -V_{in} \frac{-j/\omega C}{R}$$

$$= \frac{V_{in} j}{R\omega C} \tag{29-12}$$

as an active low-pass filter. A simple application of calculus demonstrates that these two results are really the same—simply integrate an ac input and compare the result to this second case.

The main problem with this circuit is that the effects of the input offset voltage and the input bias currents are also integrated. Thus even with no input, the output tends to drift off into saturation. Either some form of low-frequency rolloff or an ultrastable op-amp must be used to control this drift.

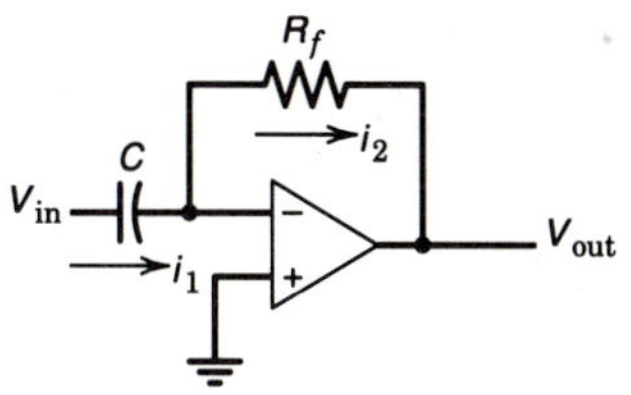

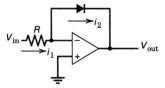

FIGURE 29-12 A differentiator.

The Differentiator

The circuit in Figure 29-12 is a **differentiator**. This follows directly from the usual analysis scheme:

$$i_1 = i_2$$

$$C\frac{dV_{in}}{dt} = -\frac{V_{out}}{R_f}$$

$$V_{out} = -R_f C \frac{dV_{in}}{dt} \tag{29-13}$$

Again, if the input happens to be a pure sine wave, an ac voltage, this can be treated as an inverting amplifier:

$$v_{out} = -v_{in}\frac{Z_f}{Z}$$

$$= jv_{in} R_f \omega C \tag{29-14}$$

Thus this is a high-pass filter. Again, these two results are equivalent.

Nonlinear Amplifiers

If a nonlinear device is used either as the feedback element or in the input, then the output is no longer linearly related to the input. Figure 29-13 shows one such case. In this figure, a diode is used as the feedback element. The voltage–current relationship for a diode is exponential (refer to Chapter 41). Thus

FIGURE 29-13 A logarithmic amplifier.

$$i_2 = B \exp\left(\frac{V_{out}}{\alpha}\right)$$

where B and α are constants. The usual analysis applies:

$$i_1 = i_2$$

$$\frac{V_{in}}{R} = B \exp\left(\frac{V_{out}}{\alpha}\right)$$

and

$$V_{out} = \alpha \log_e \frac{V_{in}}{RB} \tag{29-15}$$

Thus the output of the circuit is related to the logarithm of the input.

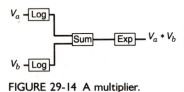

FIGURE 29-14 A multiplier.

If the diode were in the input lead, then the output of the circuit would be proportional to the exponential of the input, or to the antilogarithm of the input. Using these two circuits, the circuit in Figure 29-14 is an analog multiplier. Variations on this circuit can give rise to circuits that can divide, take reciprocals, or take square roots. Many of these combinations are available in commercial packages. But, again, more details about these nonlinear circuits would take us far beyond the scope of an introduction.

More Circuits

This catalog of op-amp circuits could go on almost forever. As an example of another whole class of circuits, there are the **active filters**, that combine op-amps and RC circuits to produce circuits that have carefully controlled frequency responses. However, the mathematics of analyzing the circuits and the complexity of designing a circuit that has the desired response is far beyond the scope of an introduction. Other classes of circuits exist, but the catalog must end at some point.

DISCUSSION AND REFERENCES

The material presented thus far in these two chapters is adequate for a person who wants to be familiar with op-amps and perhaps be able to identify what a circuit is doing. However, to design and build op-amp circuits that really work, more information is needed. Chapter 31 introduces more vocabulary and a number of additional parameters describing the operation of op-amp circuits. In addition, there are many good references for more advanced circuits. Some of these references are given in Appendix A.

PROBLEMS

29-1 Design circuits whose outputs represent (a) $3x - 5y$. (b) $12 - 6(dy/dt)$.

29-2 Design circuits whose outputs represent (a) $7x + 6y - 17z$. (b) $2\int x\,dt - 6y + 3(dz/dt)$.

29-3 Derive an expression for the output of the circuit in Figure 29-15.

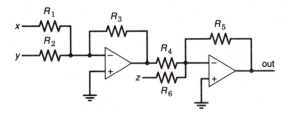

FIGURE 29-15 The circuit for Problem 29-3.

29-4 Derive an expression for the output of the circuit in Figure 29-16.

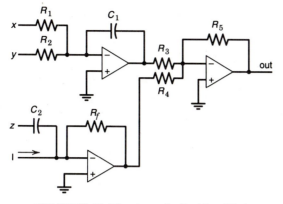

FIGURE 29-16 The circuit for Problem 29-4.

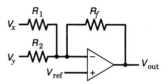

FIGURE 29-17 The circuit for Problem 29-5.

29-5 Derive an expression for the output of the circuit in Figure 29-17.

29-6 Derive an expression for the output of the circuit in Figure 29-18.

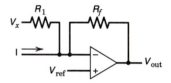

FIGURE 29-18 The circuit for Problem 29-6.

ANSWERS TO ODD NUMBERED PROBLEMS

29-1 (a) A solution is given in Figure 29-19, with R in the range of 1k to 10k Ω. (b) A solution is given in Figure 29-20, with $R = 10$k Ω, $RC = 6$, and $C = 60$ μF.

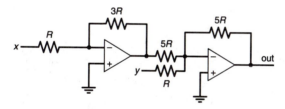

FIGURE 29-19 The solution to Problem 29-1a.

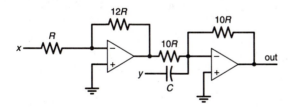

FIGURE 29-20 The solution to Problem 29-1b.

29-3

$$\text{Out} = \frac{R_5}{R_4}\left(\frac{R_3}{R_1}x + \frac{R_3}{R_2}y\right) - \frac{R_5}{R_6}z$$

29-5

$$\text{Out} = V_{\text{ref}}\left(1 + \frac{R_f}{R_x} + \frac{R_f}{R_y}\right) - \frac{R_f}{R_x}V_x - \frac{R_f}{R_y}V_y$$

Chapter 30

ANALOG COMPUTERS

OBJECTIVES

1. Be familiar with the basic ideas and techniques involved in setting up and using an analog computer.

2. Be able to design an analog computer that will solve a set of differential or linear equations.

INTRODUCTION

This chapter is a historical interlude. The analyses of many real situations result in sets of coupled differential equations. Examples of these situations include multistep chemical reactions, such as the metabolizing and clearance of drugs from the human body, the flexing of an airplane wing, and the forecasting of the weather. Prior to the invention of digital and analog computers, most of these sets of equations were insoluble except when simplified by the most drastic and unrealistic approximations.

Digital and analog computers came into use at roughly the same time. But because the techniques for solving sets of differential equations with analog computers were at first more obvious, for many years analog computers were used extensively in the investigation of situations that gave rise to complex sets of differential equations. As a result, the analog computer was often called a **differential analyzer**.

Now that many techniques for solving differential equations on digital computers have been developed and digital computers have increased in power by many orders of magnitude, there no longer are many cases that are *best* treated with an analog computer. In fact, computer scientists sometimes debate if there are any such cases. As a result, the analog computer has become much less common and may totally fade away. However, the techniques used in analog computers are quite interesting by themselves.

A SIMPLE EXAMPLE

The differential equations that describe electronic circuits often have exactly the same form as the differential equations describing other physical problems. As an example, a weight hung on a spring in a system having friction (the damped harmonic oscillator) is described by the same differential equation that describes the charge in an *RCL* circuit. Thus the electrical circuit can be considered an analog to the mechanical circuit, or vice versa.

This particular analogy is not very useful because the differential equation can be solved exactly. However, in a case where the differential equations cannot be solved, it is often possible to build an electronic analog to the situation, measure the behavior of the electrical system, and thus learn about the analogous system in a better, faster, safer, cheaper, or more practical way than by simply observing it. This electronic analog is a form of an analog computer.

The basic idea behind the analog computer is to create an electrical circuit that is described by essentially the same differential equation as the one to be solved. In this way, the voltages and currents in the circuit correspond to the variables of the physical system of interest. The various circuit parameters correspond to the physical parameters in the system. By varying the parameters of the circuit, the behavior of the physical system can be investigated over a wide range of parameters.

A very simple example will illustrate this process (see Figure 30-1). This example will be done backward; that is, we start with the circuit, find the differential equation it satisfies, and then identify a physical system that has the same differential equation.

This circuit consists of two integrators, so

$$V_B = K - \left(\frac{1}{RC}\right) \int V(t)\, dt$$

and

$$V_{out} = K' - \left(\frac{1}{RC}\right) \int V_B\, dt$$

Thus

$$V_{out} = K' - \left(\frac{1}{RC}\right) \int K\, dt + \left(\frac{1}{RC}\right)^2 \int \int V(t)dt^2 \qquad (30\text{-}1)$$

where k and k' are two constants of integration. Taking the second derivative of this equation gives us

$$V(t) = (RC)^2 \frac{d^2 V_{out}}{dt^2}$$

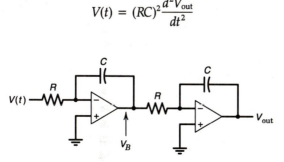

FIGURE 30-1 A simple analog computer.

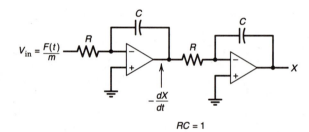
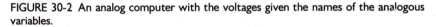

$$V_{in} = \frac{F(t)}{m}$$

$$-\frac{dX}{dt}$$

$$RC = 1$$

FIGURE 30-2 An analog computer with the voltages given the names of the analogous variables.

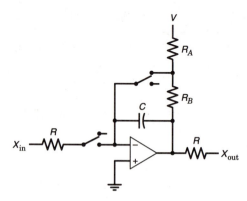

FIGURE 30-3 A way of setting initial conditions.

or

$$\frac{d^2V_{out}}{dt^2} = \frac{1}{(RC)^2} V(t) \qquad (30\text{-}2)$$

This is a very common differential equation. For instance, Newton's second law, which relates the acceleration of an object of mass m subject to a force $F(t)$ gives

$$\frac{d^2x}{dt^2} = \frac{F(t)}{m} \qquad (30\text{-}3)$$

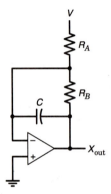

FIGURE 30-4 The setting of initial conditions.

To make the identification of the terms clearer, let $RC = 1$. Then the input to this circuit corresponds to F/m, the output to x, and the intermediate voltage to $-dx/dt$. This is illustrated in Figure 30-2.

But to completely solve the differential equation, the initial conditions must be specified; that is, the values of x and dx/dt at the time $t = 0$ must be set. This can be done with the aid of some switches. Consider the circuit shown Figure 30-3. With both the switches in the down position, this circuit is just the integrator discussed above; the connection from the output to the voltage source through the two resistors has no effect other than to change the output current that the op-amp must supply to generate the appropriate output voltage. However, when both the switches are in the up position, the resulting circuit is that shown in Figure 30-4.

The output of the circuit in Figure 30-4 is

$$X_{\text{out}} = V\frac{R_B}{R_A}$$

or the value of X_{out} can be set to any desired value by adjusting V, R_A, and R_B. If both the switches are initially up and both are moved to the lower position at the same time, then this is an integrator whose output was equal to $-V(R_B/R_A)$ at the instant the switch was thrown, which is time $t = 0$.

Using this scheme, the circuit of Figure 30-5 can be used to generate a solution to the differential equation subject to the desired initial conditions. If all four switches are ganged together so that they all switch at the same time, this solves the differential equation subject to the desired initial conditions. The switches are up to set the initial conditions; then, all switches are moved to the down or run positions, and the output voltage corresponds to the motion of an object subjected to a given $F(t)$.

A SECOND EXAMPLE

Consider another very simple differential equation,

$$\frac{dx}{dt} = -kx \tag{30-4}$$

where k is a constant. The solution is

$$x(t) = A \exp(-kt) \tag{30-5}$$

where $A = x(0)$. But solving this with an electrical analog illustrates one final point. The circuit in Figure 30-6 does the integrating and has provisions to set the initial conditions, but it does not force x to be equal to $-dx/dt$. The solution to this is both trivial and subtle—connect the output to the input. The output x and the input $-dx/dt$ are forced to be equal by the connection (see Figure 30-7).

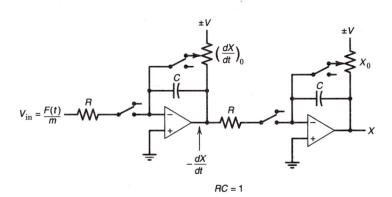

FIGURE 30-5 An analog computer to solve Equation 30-2.

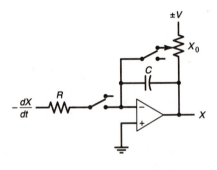

$$RC = 1$$

FIGURE 30-6 A partial solution to Equation 30-3.

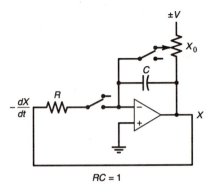

$$RC = 1$$

FIGURE 30-7 An analog computer to solve Equation 30-3.

DISCUSSION

The two examples above illustrate all the basic ideas of an analog computer. Given circuits that can add, subtract, change sign, multiply, divide, integrate, and differentiate (all of which were discussed in Chapter 29) along with ganged switches, potentiometers, and voltage sources for setting initial conditions, and perhaps methods of generating voltages that have certain desired functional forms for driving forces, essentially any set of linear coupled differential equations can be solved.

A great deal of lore about how to do successful and useful simulations using analog computers has developed. For instance, although RC is set to 1 in the preceding examples, setting RC to some other value would simply change the time scale; that is, time is no longer measured in seconds but in units of $\tau = RC$. Thus time can be expanded or contracted. Care has to be taken that the problem is scaled correctly so that the op-amp frequency response, slew rate, maximum output levels are not exceeded. Finally, usable analogies between various physical systems—mechanical systems, fluid dynamics, or heat flow systems—have to be set up to translate the problem of interest into an electrical problem.

PROBLEMS

30-1 Design an analog computer to solve the following system of linear equations:

$$a_1x + b_1y = c_1$$
$$a_2x + b_2y = c_2$$

30-2 Design an analog computer to solve the following differential equation for the damped driven oscillator:

$$m\frac{d^2x}{dt^2} = -b\frac{dx}{dt} - kx + f(t).$$

30-3 Design an analog computer to solve the following differential equations. (These are two coupled oscillators, one of which is driven.) For simplicity, omit the setting of the initial conditions:

$$m_1\frac{d^2x_1}{dt^2} = -b_1\frac{dx_2}{dt} - k_1x_1 + f(t)$$

$$m_2\frac{d^2x_2}{dt^2} = -b_2\frac{dx_1}{dt} - k_2x_2$$

ANSWERS TO ODD-NUMBERED PROBLEMS

30-1 The solution is shown in Figure 30-8.

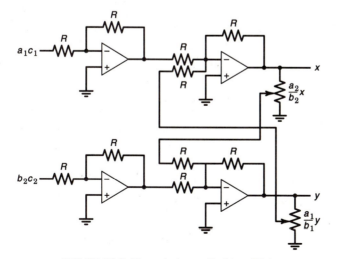

FIGURE 30-8 The solution to Problem 30-1.

30-3 The solution is shown in Figure 30-9.

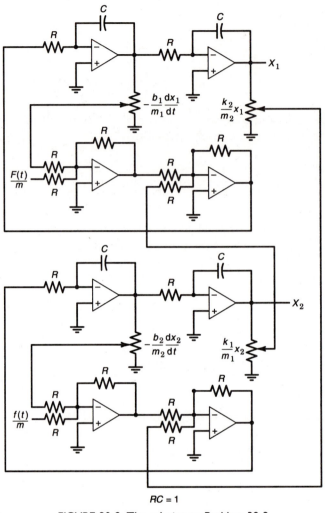

FIGURE 30-9 The solution to Problem 30-3.

Chapter 31

OPERATIONAL AMPLIFIER PARAMETERS

OBJECTIVES

1. Be familiar with the parameters and terminology used to describe operational amplifier performance.

2. Be familiar with some of the trade-offs necessary in op-amp circuit design.

3. Be familiar with an equivalent circuit containing many of the op-amp imperfections.

INTRODUCTION

Despite the rather remarkable performance characteristics of currently available op-amps, none of them is perfect nor, for that matter, can they ever be made perfect. Furthermore, there is a natural tendency to design circuits that demand more and more from the op-amps. At present, there are thousands of different op-amps available, most of them offering a different combination of performance characteristics and price. These op-amps have been designed to fill various niches in the performance spectrum. For instance, they might be characterized as high power or low power, as wide bandwidth or low bias current, as ultrastable, and so on.

The process of building an actual op-amp circuit involves selecting the op-amp to be used. Ultimately, this boils down to a compromise based on various performance specifications, the cost, and the availability. Although the subtleties of making such a decision may be beyond an introductory discussion, the terminology used to describe these parameters is not. This chapter contains a section on how to pick an op-amp, a long list of op-amp characteristics, and a short section relating some of these parameters to errors in op-amp circuit performance. It ends with one large circuit diagram summarizing op-amp imperfections.

There are some problems at the end of this chapter to be solved; but the real test of mastery of this material is to be able to build circuits that perform as well as they are expected to perform.

HOW TO PICK AN OP-AMP

To pick the best op-amp for an application, it is necessary to have a clear idea of the design objectives of the circuit as well as a good understanding of the published specifications describing the op-amps. In a complete design, the following list of circuit parameters need to be considered and, where necessary, specified.

Signal Levels. When picking an op-amp, one must have a clear understanding of the sizes of the input and output voltages and currents. That is, are the inputs to be millivolts or volts? Is the output to be a few volts or hundreds of volts?

Accuracy. Errors in the output of an op-amp are due to many sources. For instance, errors may be the result of offsets, drifts, too little gain, and frequency effects. There must be a clear understanding of how large the errors in the output can be. For instance, must the output be accurate to 1 mV, to 1%, or 0.01%? Furthermore, there must be an understanding of what kind of error is most serious and what kinds, if any, are irrelevant.

Bandwidth Requirement. What frequency range of operation is needed? Is dc operation needed? What is the maximum frequency of interest? Is the transient response of the circuit of interest, or only the steady state?

Environmental Conditions. The environment in which the amplifier will operate includes specifying the temperature range, the electrical noise environment, the power supply stability, ground fluctuations, and so on.

Power Budget. There may well be design limitations on the power available for the op-amp; these considerations may be crucial and are probably the first ones to think of when looking for a suitable op-amp. Obviously, battery-operated equipment places a premium on low-power operation, but there may be other considerations. In particular, how much power can the op-amp draw? Are there any limitations on how many power supply voltages are available and what they are?

Space Budget. There are some occasions when there are limitations on how much space is available for the circuit, or how many components are allowed.

Financial Budget. Better op-amps cost more. Commercial design is always a trade-off between cost and performance. Laboratory design cannot totally ignore costs.

Once the design objectives have been specified, it is possible to look for an op-amp that will meet the objectives. Because of the rapid rate of change in the modern electronic world, the only real current sources of information are the data books published by manufacturers, like Analogic, Analog Devices, National Semiconductors, and Texas Instruments—to name only four domestic producers. Buy, beg, or borrow, some data books and **study them carefully.** Get the full data sheets and application notes from the manufacturer for any op-amp that looks satisfactory. **Read these completely and carefully**—they are a gold mine of information, suggestion, and warnings.

One unfortunate fact is that not all manufacturers define all the parameters exactly the same way. When making a choice between two different manufacturers, be sure to understand the exact definition that each manufacturer uses for any critical parameters.

In many cases, a general-purpose op-amp that is available cheaply from multiple sources will do. In more demanding cases, a special-purpose amplifier may be needed; this will require much more searching of the data books and catalogs. But in a surprising number of instances, a walk to the nearest Radio Shack to buy a 741 will solve the problem.

OP-AMP PARAMETERS

This section contains the definition and a short discussion about the meaning and use of a long list of different op-amp parameters. Since there is no obvious way to organize this list of parameters, they are simply given in alphabetical order. However, because many of these parameters are related, several readings of this section are in order. Finally, although this section is generally applicable to all op-amps, in some cases there are minor differences in the detailed way in which different manufacturers define some of these parameters. The definitions used here were generally taken from the Analog Devices Catalog.

Absolute Maximums

The absolute maximum specifications include the **absolute maximum power supply voltages** that can be used for the op-amp. An **absolute maximum input voltage** is sometimes specified; often, this is the same as the maximum power supply voltage. Exceeding any of these limitations offers the possibility of destroying the op-amp. Sometimes a **maximum current** for one pin of the package is specified.

There is also normally a **maximum ambient temperature** specification. This tells the maximum temperature at which the op-amp can be expected to operate correctly. Even if the op-amp were to operate at higher temperatures, most of its specifications would be strongly degraded by the temperature.

Absolute Maximum Differential Voltage

The **absolute maximum differential voltage** is the maximum voltage difference that can be applied between the two inputs of the amplifier without causing permanent damage to the op-amp. In normal linear differential operation, the feedback maintains the voltage difference between the two inputs to be a small fraction of a volt. However, in certain special cases, such as for comparators, the differential input voltage can become large. The differential input voltage can also become large in normal linear circuits when the operation of the circuit breaks down, for example, when the output is saturated or when the slewing rate is exceeded.

Common Mode Rejection

A voltage that is present at both inputs of an op-amp is called the **common mode voltage**. Thus if the inverting input is at voltage V and the noninverting input is at $V + \varepsilon$, then V is the common mode voltage and ε is the differential input voltage. An ideal op-amp produces an output that is related only to the differential input voltage and that is unrelated to the common mode voltage. Thus where the differential input is zero, the output should be zero no matter what the common mode voltage.

Because of imperfections of the op-amp, common mode input voltages are not totally rejected in the output of the op-amp. The closed loop **common mode gain** is defined as the ratio of the common mode output voltage to the common mode input voltage causing it. (This must be done for the closed loop case since, in general, an op-amp in an open loop situation has its output saturated at one limit or the other.)

Finally, the **common mode rejection ratio, CMRR,** is defined as the ratio of the closed loop differential gain to the closed loop common mode gain. Typical CMRR values are 90 db or better, thus indicating that differential gain is a factor of 50,000 or more greater than the common mode gain. The most serious problem with specifying CMRR is that it is a highly nonlinear parameter. It may vary quite rapidly with common mode voltage, temperature, power supply voltages, the other parameters. Generally, published CMRR specifications refer to low-frequency voltages; usually, CMRR decreases with increasing frequency.

Common Mode Voltage Maximum

Because the CMRR is nonlinearly dependent on the common mode voltage, it is necessary to specify a maximum common mode voltage that can be present and yet still have a CMRR equal to, or better than, the stated value of the parameter. This is the **common mode voltage maximum.** It is the maximum common mode voltage, either above or below ground (power supply common), that is permitted while requiring the CMRR to be equal to, or better than, the stated value of the parameter. It should be noted that this voltage establishes a potential maximum input voltage for a voltage follower circuit.

There may also be an **absolute maximum common mode** voltage specified for the op-amp. As with other absolute maximum specifications, exceeding this limit may destroy the op-amp.

Drift Versus Supply

The input offset voltage, bias current, input offset bias current, and other parameters all vary or **drift** as the supply voltages vary. The published parameters all refer to low-frequency variations in the supply voltages, since the effects of dc variations in the supply are less than those due to the thermal drifts in these parameters. There may well be other much larger effects due to high-frequency variations in the supply voltages. These are generally not specified and are sufficiently unpleasant that the power supply lines should be heavily filtered to eliminate any possible high-frequency variations.

Drift Versus Temperature

The input offset voltage, bias current, and input bias current all change or drift as a function of temperature. These drifts are most often the major source of errors in precision op-amp circuits. As a result, a great deal of the design effort going into op-amps and op-amp circuits is intended to minimize the effects of these drifts. The **temperature coefficients,** or **tempcos,** of these parameters are generally given in the specification sheets. There is no accepted way of defining these temperature coefficients, so that it is necessary to read the data sheets very carefully. Typical methods include specifying them at only the normal operating temperature (25°C), specifying an average over the stated operating temperature range, or specifying the maximum values they have over the operating range. Which choice is made depends on the marketing philosophy of the manufacturer.

Drift Versus Time

Most of the parameters that describe an op-amp change as the components in the circuit age. The most important of these changes have to do with the input offset voltage, bias current, and input offset current. These variations are typically random and rarely accumulate in a linear fashion over time. Thus the expected drift over a 30-day period would not be 30 times the drift over a 1-day period. The usual assumption is to assume a random walk behavior and thus scale expected drift with the square root of time. Thus the maximum drift expected in 30 days would be $\sqrt{30}$ times the maximum drift expected in 1 day.

Full Power Response

Because of slew rate limitations and other related parameters, the large-signal and small-signal characteristics of an op-amp differ substantially. An amplifier can accurately produce small-output signals at a higher frequency than it can for large-output signals. This is directly related to the slew rate. There are at least two common measures of full power response.

The **full power peak response**, f_p, is defined as the maximum frequency at which the amplifier can deliver the rated output voltage and current to a resistive load. Frequently, graphs showing the maximum frequency versus the desired maximum output voltage swing are given in the specifications. Although this is defined in terms of a sine wave, no distortion specifications are included.

A somewhat more precise and restrictive definition is the **full linear response** parameter. This is the maximum frequency for which the op-amp will produce a sinusoidal output at rated power without the op-amp exceeding some specified distortion level when it is used in a circuit with unity closed loop gain.

Input Bias Current

For the op-amp to function, there must be some current flowing in the input lines; the direction of this current depends on the details of the op-amp and the input circuit. This current flows through whatever circuit is connected to the input of the op-amp. Generally, the current flowing in each input is essentially the same, but this is not necessarily true.

A precise definition of the **input bias current**, I_b, is the current that must be supplied to an input by a current source having infinite source impedance in order to drive the output of the op-amp to zero, assuming, of course, that there is no differential or common mode input present. Manufacturers differ over using the average of the two bias currents or the maximum value in their specifications.

Input bias currents drift as a function of temperature, time, and power supply voltages. Although specifications for these drifts are frequently given, the exact conditions under which these specifications and the input bias current, itself are measured allow creative marketing departments to produce specifications that may be misleading. As an example of what might be done, the input bias current of an FET input op-amp roughly doubles for each 10°C increase in junction temperature. It is not unusual for a junction to warm up at least 20°C during the first few minutes of operation. If the input bias current was specified at the instant of turn on, then the normal operating bias current would be roughly four times the specified value.

Input Offset Bias Current

The **input offset bias current, I_{os},** is the difference between the two input offset currents. Since the two inputs of an op-amp are usually symmetrical, the input offset bias current tends to be a small, fairly constant fraction of the input bias current; 10% of the input bias current is a reasonable "ball park" valve for an initial design. To the extent that the two input bias currents track each other, the variations in the input offset bias current are simply a constant fraction of the variations in the input bias current. However, for very precise work, estimates of the temperature, time, and power supply drifts in the input offset bias current will have to be made. As stated above, the definitions of these parameters by different manufacturers may *not* be the same.

Input Impedance

There are two different input impedances: the impedance between the two inputs, and the impedance between an input and the circuit ground (power supply common).

The **differential input impedance, Z_i,** is the small-signal impedance measured between the two inputs. To a good approximation, this input impedance can be modeled as a resistor R_i in parallel with a capacitor C_i. Because of the frequent nonlinear nature of the inputs to an op-amp, good specifications will include complete details, often with a schematic of the test circuit under which these parameters were measured.

The **common mode input impedance, Z_{cm},** is impedance measured between an input and the power supply common. To a good approximation, this can also be modeled as a resistor R_{cm} in parallel with a capacitor C_{cm}. As for the differential case, a full specification of the conditions under which these parameters were measured is often given. In addition, since the inverting and noninverting inputs may differ, it is necessary to know which one is being measured.

The common mode input impedance for the inverting input is rarely important because of the feedback in the usual circuits. However, the common mode input impedance for the noninverting input sets the upper limit on the closed loop input impedance for noninverting circuits—see the discussion that follows and Chapter 29.

Input Offset Voltage

Because of the imperfections in op-amp construction, an op-amp will usually have a nonzero output even when there is no differential input voltage. The input offset voltage is a measure of this effect. In particular, the **input offset voltage, V_{os},** is defined as the voltage required at the input of the op-amp from a voltage source having zero source impedance to drive the output of the op-amp to zero. A complete specification of the input offset voltage will include specifying the temperature and power supply voltages at which it was measured. The input offset voltage drifts with time, temperature, and power supply voltages. A complete set of specifications will include coefficients for all of these drifts. Most amplifiers have provisions for adjusting the initial offset to zero using an external potentiometer.

Input Noise

The **input noise voltage, V_n,** is defined as the equivalent input differential noise voltage required to produce the observed output noise signal when the input is terminated in a zero-impedance source. This is analogous to the input offset voltage definition, only it is for a noise voltage.

The **input noise current, I_n,** is defined as the equivalent input noise current at either input that is required to produce the observed output noise signal when the input is terminated in a large-input impedance. This is analogous to the input bias current definition.

Although the definitions of the input noise voltage and currents are similar to those for the offset voltage and bias current, noise is more complex because bandwidth must also be considered in dealing with noise. Depending on the type of amplifier, noise may have different frequency characteristics; for instance, $1/f$ noise, resistor noise, and junction noise all have different frequency distributions. As a result, a complete specification of noise for an op-amp will involve several different noise specification.

Open Loop Gain

The **open loop gain, A,** is the ratio of the change in the output voltage of the op-amp to the change in the differential input voltage that caused it. The nominal open loop gain of most op-amps is specified for dc; however, frequently, a plot of open loop gain as a function of frequency is provided for an amplifier. At least two values of the open loop gain are usually given: the first is the minimum, the value the manufacturer guarantees the op-amp will reach; the second is a typical value that is expected from any average sample taken off the shelf.

Output Resistance

The **Thevenin equivalent output resistance, R_o,** for the op-amp is usually specified. While it might be more accurate to specify an impedance, usually the resistance specification is given, and any phase shifting effects are covered by giving a graph of phase shift versus frequency as part of the Bode plot for the amplifier.

Overload Recovery

For an op-amp that is operating in its linear region, the output changes essentially instantaneously when the input changes, aside from a delay of a few nanoseconds. However, when an op-amp is driven into saturation by an input signal that is too large, and when the input signal is then removed, the output does not start to change immediately, but only after a characteristic delay time. The **overload recovery time** is a measure of this characteristic delay time. The normal definition is the time required for the output voltage to recover to the correct voltage for linear operation from a saturated condition caused by a 50% overdrive.

Power Supply

Besides the absolute maximum voltage specification, there may also be a specification for the typical power supply voltages or a normal range of voltages given. There may also be a specification for power consumption. This is a somewhat difficult specification because it depends, in part, on how much power is being delivered to the output. Sometimes, a specification for how much power the amplifier draws when working into an open circuit is given. The usual assumption is that this power consumption is more or less constant, and the total power consumption can be found by adding the power delivered to the output load to this quiescent power.

Rated Output

The rated output voltage and rated output current are related to the peak values of the voltage and current that can be simultaneously supplied by the amplifier. It is not useful to specify the maximum value that is found when many op-amps of a particular type are measured because the chances of any particular op-amp reaching this maximum would be very small. What is useful is to specify the minimum value that any given op-amp is guaranteed to reach. Thus the definitions of these two quantities are somewhat convoluted.

The **rated output voltage, V_{max},** is defined as the minimum guaranteed value of the peak output voltage that can be obtained at the rated current output before clipping or excessive nonlinearity occurs. A complete definition includes defining "excessive" and specifying the load impedance of the output.

The **rated output current, I_{max},** is defined as the minimum guaranteed value of current supplied at the rated output voltage before clipping or excessive nonlinearity occurs.

Any particular op-amp may be able to generate larger output voltages and currents simultaneously, but it is guaranteed that it will be able to generate at least V_{max} and I_{max} simultaneously. An op-amp may be able to generate more than I_{max} output current when the output voltage is less than V_{max}, and so forth. Obviously, an op-amp will produce the maximum output current when it is working into a short circuit; likewise, it will produce the maximum output voltage when it is working into an open circuit. Finally, it should be noted that most op-amps are protected against damage when the output is short circuited to ground, and many are protected against shorts to the power supplies.

Settling Time

The **settling time,** t_s, is defined as the time required after the application of a step function input voltage for the output to rise and remain within a specified error band symmetrical about the final value. Although the various frequency-related parameters, settling time, full power response, and slew rate are all related, one of them cannot be predicted given the others; hence, to some extent they are independent parameters. These various terms are illustrated in Figure 31-3.

Slew Rate

The **slew rate,** dV_o/dt_{max}, is the maximum rate of change of the output voltage for a large change in the input voltage. This is usually measured in volts per microseconds.

Unity Gain Small-Signal Response Frequency

The **unity gain small-signal response frequency,** f_u, is the frequency at which the small-signal open loop gain falls to 1 (or 0 db). It is necessary to measure this with small signals, since the slewing rate limitation of the op-amp will typically limit the frequency response for large-output signals.

ERROR SOURCES AND CONTROL

One of the major sources of error in op-amp circuits is insufficient loop gain. **Loop gain** is defined as the difference between the open loop gain and the closed loop gain at the frequency of interest (see Figure 31-3). If the gains are measured in db, then the loop gain is truly the difference; if the gains are measured as multiplicative numbers, then the loop gain is the ratio of the open loop gain to the closed loop gain.

In general, the overall accuracy of an op-amp circuit depends on the loop gain. In particular, the independence of the gain of the circuit from the gain of the op-amp itself, the suppression of noise, and distortion and other measures of accuracy all depend on the loop gain. High accuracy requires high loop gain; to a good approximation, the accuracy is 1/(loop gain). This was illustrated for the case of the inverting amplifier in the problems for Chapter 28, where it was shown that the departure of the gain from the nominal value of R_f/R depends on the loop gain. In particular, a loop gain of 100 means that the gain will differ from R_f/R by 1%; a loop gain of 10000 means that the gain will differ by 0.01%.

The desire for accuracy interacts with the desired frequency response of an amplifier since the loop gain rapidly decreases with frequency (refer to Figure 31-3). Thus to obtain an amplifier with a frequency response that is uniform to 0.01% at even a moderately high frequency will require either a very high gain, high-frequency op-amp or several high-frequency op-amps running in series.

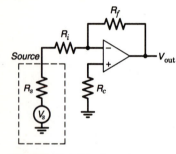

FIGURE 31-1 The inverting amplifier.

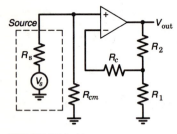

FIGURE 31-2 Noninverting amplifier.

A second major source of error in op-amp circuits is due to input offsets and the drifts of these offsets. An inverting amplifier is shown in Figure 31-1, and a number of expressions related to these errors follow. Figure 31-2 shows the noninverting amplifier configuration, and similar expressions are also given. The meaning of the expressions and some of the terms used in them are discussed following both sets of results. For the inverting amplifier, the following results hold true. When

$$R_c = 0 \quad \text{and} \quad R_s \ll R_i$$

it can be shown that

$$V_{out} = -\frac{R_f}{R_i}\left[V_s + V_{os}\left(\frac{R_f + R_i}{R_f}\right) + I_b R_i\right] \tag{31-1}$$

Signal / Input error

However, if $R_c = R_i R_f/(R_f + R_i)$, then this becomes

$$V_{out} = -\frac{R_f}{R_i}\left[V_s + V_{os}\left(\frac{R_f + R_i}{R_f}\right) + I_d R_i\right] \tag{31-2}$$

Signal / Input drift error

Also, the closed loop input impedance is approximately given by

$$R_{in} = R_i$$

Again, for the case where $R_c = 0$,

$$V_{out} = \frac{R_2 + R_1}{R_i}[V_s + V_{os} + I_b R_s] \tag{31-3}$$

Signal / Input error

For the case where $R_c = R_s - [R_1 R_2/(R_1 + R_2)]$,

$$V_{out} = \frac{R_2 + R_1}{R_i}[V_s + V_{os} + I_d R_s] \tag{31-4}$$

Signal / Input drift error

The input impedance is approximately given by

$$R_{in} = R_{cm}$$

The preceding expressions allow an easy comparison between offset errors, drift errors, and signal voltages. Thus if the typical inputs are 1 V and 0.1% accuracy is desired, then the total input drift error must be less than 1 mV.

Since most op-amps have provisions for adjusting the initial input offset voltage to zero, only the drift in the input offset voltage is of interest once the adjustment has been made. As a result, V_{os} need only be an estimate of the drift in the input offset voltage.

In Equations 31-1 and 31-3, one of the error terms is due to the bias current I_b flowing through an input resistor. In either configuration, R_c does not affect the gain. By picking the value of R_c correctly, however, the effects of the input bias current are canceled, and only the input offset bias currents, indicated as I_d in Equations 31-2 and 31-4, need to be considered.

In the inverting case, the effects of the input bias current can be further reduced by decreasing R_s. But this increases the loading on the signal source and introduces a new error. Thus a conflict exists between high input impedance and low offset errors due to input currents. As a result, the noninverting configuration is often preferred for situations that require high input impedances. Even low-cost general-purpose op-amps have $R_{cm} = 10^7$ Ω, and values up to 10^{11} Ω are available for FET types.

Of course, the easiest way to eliminate the effects of the input offset voltage and input bias currents and their associated drifts is to use a blocking capacitor because these effects are either dc or of very low frequency. Thus if a circuit does not need response extending to dc, a judiciously selected capacitor will eliminate many of the problems.

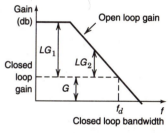

FIGURE 31-3 Gain definitions LG_1 and LG_2 are the loop gains at two different frequencies, respectively.

FREQUENCY CONSIDERATIONS

The Bode plot in Figure 31-3 illustrates the various gain definitions and their frequency dependance. In particular, this graph illustrates the decrease in loop gain as the frequency increases. This decrease is the reason why the performance of an op-amp circuit often deteriorates as the frequency increases.

The graph in Figure 31-4 illustrates the definitions of the settling time and a number of related terms.

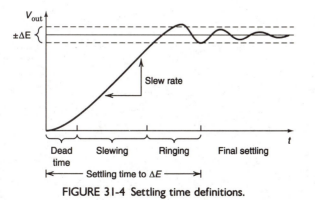

FIGURE 31-4 Settling time definitions.

A FINAL EQUIVALENT CIRCUIT

Finally, for those who like circuit analysis, the following equivalent circuit for an op-amp is included (see Figure 31-5). This equivalent circuit includes most of the imperfections that have been discussed in this chapter. The usual approach to estimating the effect of imperfections of the op-amp on real circuits is to deal with them one at a time—that is, set all but the one of interest to zero. This makes the calculations feasible.

PROBLEMS

31-1 For an op-amp circuit that has a nominal closed loop gain of 100, a nominal output of 10 V, a drift in the offset bias voltage of 1.0 mV/°C, and which operates over a temperature range of 20°C, what is the expected accuracy? What would the answer be if the drift were 10 μV/°C?

31-2 For the situation described in Problem 31-1, if the input resistance is 10k ohms (the sum of R and R_s) and if the drift in the input bias current is 1 μA/°C, what is the error due to the drift in the bias current? If the drift is 0.1 μA/°C?

31-3 If the output of an op-amp is a sawtooth voltage (triangles) with a peak-to-peak voltage of 5 V, what is the maximum frequency the op-amp can operate at if its slewing rate is 1 V/μS?

31-4 Why would two identical op-amps, each being used as an inverting amplifier having a closed loop gain of 10 and being placed in series, have a better frequency response than just one of these op-amps operating in a circuit with a gain of 100?

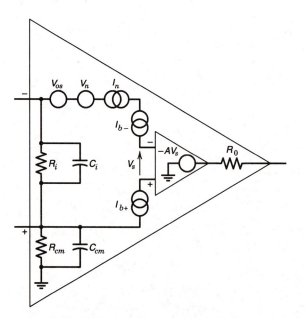

FIGURE 31-5 A grand equivalent circuit.

31-5 An amplifier circuit has an input of about 10 mV, an effective input resistance of 10k ohms, and is used over a 40°C operating temperature range. If this amplifier is to have a 1% accuracy, what are the limits on the drift in the input offset voltage and input offset bias current?

ANSWERS TO ODD-NUMBERED PROBLEMS

31-1 20% and 0.2% accuracy

31-3 100 hz

31-5 A total offset drift of less than 0.1 mV, which corresponds to a drift of less than 2.5 μV/°C; also, a total drift in input offset bias current of 10 μA, which corresponds to a drift of less than 0.25 μA/°C

Chapter 32

THREE SPECIAL-PURPOSE AMPLIFIERS

OBJECTIVES

1. Become familiar with the nature of and uses of instrumentation amplifiers, stabilized amplifiers, and isolation amplifiers.

INTRODUCTION

Three very useful amplifier configurations based on one or more op-amps and auxiliary circuitry are introduced in this chapter. These amplifiers are useful in situations that demand a performance this is unobtainable from a single op-amp. These three configurations include the instrumentation amplifier, the stabilized amplifier, and the isolation amplifier.

The purpose of this chapter is to provide an awareness of these circuits that will prove useful in special situations. For each of these three categories of amplifiers, only the most general generic version is discussed. The successful selection and use of one of these amplifiers requires an extensive study of the data sheets provided by the manufacturer.

INSTRUMENTATION AMPLIFIER

As was pointed out in the previous chapter, there is a trade-off between high input impedance and low drift due to the input bias current in the inverting configuration. The noninverting configuration has better input characteristics in this regard and is frequently the best choice to use. However, if true differential operation is needed, both the inverting and noninverting inputs must be used. The instrumentation amplifier is designed to permit true differential operation with a very high input impedance in each input lead.

The typical instrumentation amplifier is a committed gain differential amplifier that has high input impedance, excellent common mode rejection, stable gain, high accuracy, low offset and drift, and low effective output impedance. In short, it is just a little bit closer to the ideal op-amp than is possible

284

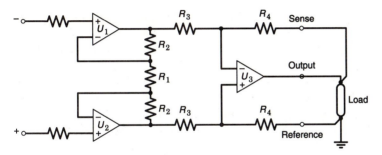

FIGURE 32-1 The three op-amp instrumentation amplifier.

to obtain with a simple op-amp. The **committed gain** means that the gain is either preset (usually in the range of 1 to 1000) or can be set to one of several values by connecting two pins together, or can be set by providing only one external resistor.

The classical instrumentation amplifier is built from three op-amps. Two op-amps are used as buffers on the inputs to provide the high input impedance required. This configuration is shown in Figure 32-1. The gain G for this circuit is given by

$$G = \left(1 + \frac{2R_2}{R_1}\right)\left(\frac{R_4}{R_3}\right) \tag{32-1}$$

By using high-quality FET op-amps and using all the tricks of modern technology (laser trimming, etc.), the resistors and the entire circuit can be very closely matched to give excellent results. Of course, there are many themes and variations on this configuration. Delving into them goes far beyond an introduction. Several layers of additional complexity are presented in Horowitz and Hill (see Appendix A for the full reference). Analog Devices (also see Appendix A) makes a number of these amplifiers—the AD 524 is a good example.

THE STABILIZED AMPLIFIER

Sometimes it is necessary to have ultrastable circuits, circuits that have drifts of the order of 0.1 μV/°C and 1 μV/year. For instance, electrometers or integrators may need such stability. To meet these goals, several different kinds of stabilized amplifiers have been designed.

The traditional approach is the **chopper stabilized amplifier.** In this amplifier, the input dc signal is converted to ac, the ac signal is amplified, and the dc drifts are blocked with a capacitor. Then, the ac signal is converted back into dc—(see Figure 32-2). In the original versions, the conversion to ac

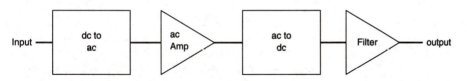

FIGURE 32-2 A block diagram of a chopper stabilized amplifier.

was done by a mechanical switch that chopped or switched the input on and off at regular intervals—hence, the name. Several sets of switches could be driven in synchronism by one motor: one set chopped the input and another set was used to reconvert the ac output to dc again. Mechanical switches take considerable power to operate, and they tend to be noisy. As a result, designing and using these amplifiers was moderately difficult.

In more recent times, several additional methods of chopping the input signals have been tried. Optical choppers are used (the input is converted into light, the light is chopped, and then the light is converted back into an electrical signal) as well as choppers based on solid state switches—FETs. Again, the variations and complexities of these circuits quickly take us beyond the level of an introduction.

Note that an FET is a field effect transistor. FETs can be made to operate as electrically controlled off–on switches. High-quality FET switches that can be controlled by TTL or CMOS logic signals are available in integrated circuits or can be included with other circuitry in one integrated circuit (IC). These switches are becoming ubiquitous in IC circuits. See Chapter 45 for more details on FETs.

With the development of high-quality FET switches together with the ability to combine large amounts of analog and digital circuitry into a small space, many other possibilities for stabilizing an amplifier have become available. One possibility uses the following scheme:

1. Ground the input of the op-amp and measure the output due to the offset.
2. Generate an appropriate input to negate this output.
3. Sum the signal and this correcting input to generate a zeroed output.
4. Repeat the steps above fairly rapidly so that the zeroing tracks any drifts in the input.
5. Have two complete amplifier and zeroing chains, and set up the switching on the input and output so that one of the chains is measuring signal while the other is measuring its offsets at any instant of time.

If all this is done correctly, the input is unaffected by the switching; that is, it always appears to be connected to the input of an op-amp circuit. Furthermore, the output always appears to come from an amplifier that has zero offset errors. As usual, more details would take us beyond an introductory discussion, but it should be noted that this type of circuit is fairly new. As a result, the details and capabilities of these circuits are changing rapidly. This type of circuit is sometimes called a **commutating auto-zero amplifier, CAZ**. A typical unit of this type is the AM 7600 made by Datel-Intersil.

ISOLATION AMPLIFIER

The isolation amplifier was designed to provide a solution to cases where there can be **no electrical connection** between the input and the output sides of the circuit, and the circuit must be able to withstand very high common mode voltages. This is often needed when a differential signal must be measured in the face of a very high common mode voltage, for instance, when

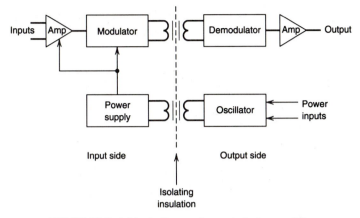

FIGURE 32-3 A block diagram for an isolation amplifier.

looking for small dc offsets between two different 13.6-kV power lines. Conventional op-amps cannot withstand such high common mode voltages.

Although making such a measurement may seem impossible on first glance, it really is not too hard to do. There are two problems that must be solved: (1) getting power for the input side of the circuit, and (2) getting the signal from the input side to the output side. There is an obvious solution to getting power to the input side of the circuit, which is absolutely independent of the output side of the circuit (ground and all). This is to use batteries to power the input side. Although this is done for some isolation amplifiers, by modern standards it is relatively crude. The block diagram in Figure 32-3 illustrates one modern design for an isolation amplifier. The features of this circuit are the following:

1. Power is supplied to the input side of the circuit by means of a very well-insulated transformer which bridges the input and output sides of the circuit. By operating at a high frequency, typically more than 100 kHz, much smaller capacitors and transformers are needed.
2. The output is converted to an ac signal by the modulator, transmitted to the output side of the circuit by means of another transformer, and then the ac signal is converted back to dc by the demodulator. Any final amplification needed is done on the output side. There is an alternate version where the output of the input side is converted to a light signal by means of a light-emitting diode and then converted back into an electrical signal on the output side by a photodiode. The output of the diode can be further amplified if necessary.

By this means, total isolation between the input and output sides of the circuit can be obtained. The quality of the isolation is determined by the quality of the insulation used between the two sides of the circuit. Typically, modern isolation amplifiers are rated for several thousands of volts. Again, a typical device of this nature is the AD277 manufactured by Analog Devices.

Chapter 33

FORMAL AMPLIFIER THEORY

OBJECTIVES

1. Have a knowledge of the basic concepts of feedback.

2. Know the advantages of negative feedback.

3. Know the conditions necessary for amplifier circuits to be stable.

4. Know what compensation means.

INTRODUCTION

A great deal of circuit theory has been developed for treating amplifier circuits. Although most of this theory is too formal and too advanced to be of use in an introduction, much of the terminology used to describe amplifier circuits comes from this circuit analysis. In addition, insight into the features of amplifier circuits can be gained from a quick study of formal amplifier theory. Therefore, a very brief introduction to several of the topics of formal amplifier theory is given here.

This chapter presents four distinct but related topics. First, there is a formal description of amplifier feedback circuits, then there is a short discussion of some of the advantages of negative feedback, which is followed by a discussion of amplifier stability. Finally, we examine the related topic of amplifier compensation.

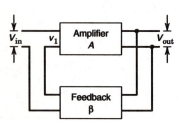

FEEDBACK

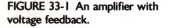

FIGURE 33-1 An amplifier with voltage feedback.

The amplifier in Figure 33-1 has an open loop gain A, and a fraction β of the output voltage is fed back to the input of the amplifier; β is sometimes called the **feedback fraction.** Analysis of this circuit gives

$$v_1 = v_{in} + \beta v_{out} \tag{33-1}$$

But

$$v_{out} = Av_1$$

$$v_{out} = Av_{in} + \beta Av_{out}$$

and finally

$$v_{out} = \frac{A}{1 - \beta A} v_{in} \tag{33-2}$$

The term βA is called the loop gain; this is the same loop gain discussed in the preceding chapter. Both A and β may be complex numbers; that is, there are both amplitude and phase relationships hiding in these expressions. For now, possible changes in these phases will be ignored.

If the loop gain is very large compared to 1, that is, if $\beta A >> 1$, this expression becomes

$$v_{out} \approx -\frac{1}{\beta v_{in}} \tag{33-3}$$

As an example, consider the noninverting op-amp circuit shown in Figure 33-2. For this circuit, the feedback circuit is just a voltage divider and

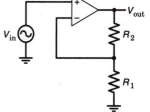

FIGURE 33-2 A noninverting amplifier circuit.

$$\beta = -\frac{R_1}{R_1 + R_2} \tag{33-4}$$

(The minus sign occurs because this term decreases the voltage across the inputs of the op-amp.) In the approximation that the gain A of the op-amp is very large, this gives

$$v_{out} \approx \frac{1}{\beta v_{in}}$$

$$\approx \left(\frac{R_1 + R_2}{R_1} \right) v_{in}$$

$$\approx \left(1 + \frac{R_2}{R_1} \right) v_{in} \tag{33-5}$$

just as was obtained before.

The feedback is said to be **negative** if its effect is to decrease the output from what it would have been without the feedback. The feedback is said to be **positive** if its effect is to increase the output from what it would have been without the feedback. For the noninverting amplifier, the feedback is negative, as it is for all the op-amp configurations discussed in Chapter 29.

For the case discussed above, the signal fed back to the input of the amplifier is proportional to the output voltage; this is called **voltage feedback.** It is possible to arrange the circuit so that a voltage proportional to the current flowing in the output circuit is fed back into the input of the circuit; this is called **current feedback.** Furthermore, it is possible to combine both current and voltage feedback in the same circuit. Naturally, the analysis of these situations gets very messy; however, no new concepts emerge from this analysis, so nothing further will be stated here.

A word of warning to complete this section: There is a sign convention built into Equations 33-1 and 33-2. The convention used here is that β is negative for negative feedback; thus, $-\beta$ is positive for negative feedback. Other authors use other conventions. When reading other texts, be sure to understand the sign conventions to prevent confusion.

ADVANTAGES OF NEGATIVE FEEDBACK

There are many advantages to using negative feedback in an amplifier. A few of the most important of these advantages are discussed in this section.

Using negative feedback gives **improved gain stability.** If the loop gain is large enough, the gain of the overall circuit depends only on β, which can be set by high-quality, stable resistors. This means that the gain of the overall circuit does not depend on the exact value of A for the particular op-amp, nor does it vary as the op-amp ages and changes in its gain occur. One facet of this situation has been examined numerically in the problems for Chapter 28.

Negative feedback gives **improved linearity.** The open loop gain of most op-amps varies as a function of the input voltage. This means that the transfer function, a plot of the output versus the input, is not a straight line—see Figure 33-3. This is typical of all solid state amplifiers without feedback. However, as long as the gain is large enough, the gain of the amplifier circuit with feedback will be constant. This means that the transfer function will be linear but will have a much smaller slope. This point and the previous point are different ways of saying roughly the same thing.

Negative voltage feedback **increases** the apparent **input impedance** of the amplifier. This is because the voltage fed back to the input tends to reduce the current flowing in the input leads. This increase was demonstrated for the noninverting amplifier and the voltage follower (see Chapter 29).

Negative voltage feedback **decreases** the apparent **output impedance** of the amplifier. This has also been demonstrated for the cases of the noninverting amplifier in Chapter 29. In a certain sense, this may be thought of as a generalization of the reduction of noise and distortion (see below).

An amplifier used in a circuit with negative feedback can be used over a **wider frequency range** than if there were no feedback. This effect has also been discussed in Chapters 29 and 31.

The use of negative feedback **reduces the effects of noise and distortion** generated internally in the amplifier circuit. In Figure 33-4, an amplifier is

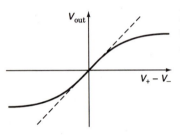

FIGURE 33-3 The transfer function for a typical op-amp showing a nonlinear open loop response.

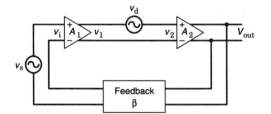

FIGURE 33-4 A model for investigating the effects of negative feedback on a noise source inside an amplifier.

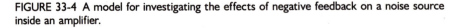

shown schematically as two stages, the first with a gain of A_1 and the second with a gain of A_2, with a voltage generator V_d to simulate either a noise or a distortion source between the two stages. The reduction in the effects of the noise and distortion can be seen by analyzing this circuit.

The first step of the analysis is to write down the expressions for the inputs and outputs of each of the amplifiers. These are

$$v_2 = v_1 + v_d$$

$$v_{out} = A_2 v_2 = A_2(v_1 + v_d)$$

$$v_i = v_s + \beta v_{out}$$

$$v_1 = A_1 v_i$$

Next, eliminate the unknown terms, first v_i and then v_1:

$$v_i = \frac{v_1}{A_1}$$

$$\frac{v_1}{A_1} = v_s + \beta v_{out}$$

$$v_1 = A_1 v_s + \beta A_1 v_{out}$$

$$v_{out} = A_2(A_1 v_s + \beta A_1 v_{out}) + A_2 v_d$$

$$v_{out} = \frac{A_1 A_2 v_s}{1 - \beta A_1 A_2} + \frac{A_2 v_d}{1 - \beta A_1 A_2} \tag{33-6}$$

To get some feel for what this means, let $A_1 = A_2 = 100$ and $\beta = -0.1$; then

$$v_{out} = 9.99 v_s + 0.1 v_d$$

Thus the noise and distortion are amplified less than the signal and much less than A_2, the gain of the portion of the op-amp that follows the noise source. Thus the negative feedback reduces the effects of the internally generated noise. This conclusion is not true if the distortion and noise come from the input side of the first stage of the amplifier. This means that great care spent on a low-noise, low-distortion, high-gain front end of an amplifier results in a very good overall amplifier when negative feedback is used. It should be noted that the front end of an amplifier generally has to deal with the smallest signals, so that the effects of nonlinearities are minimized.

STABILITY

An amplifier that has positive feedback has $\beta A > 0$. In particular, it is possible that βA may be equal to 1; and, as Equation 33-2 indicates, something unusual happens then. In this situation, the amplifier acts as an oscillator and is not useful as an amplifier at all. Even if $\beta A \neq 1$ but is only near to 1, the amplifier may show unwanted ringing and frequency-related effects. In either situation, the amplifier is said to be unstable. In general, amplifier instability is a serious bug in a circuit and should be avoided.

An amplifier with negative feedback always has βA far from 1, so it should always be stable. Unfortunately, the distinction between positive and negative feedback is not as clear as it has been implied thus far. The amplifier has a phase shift that varies as a function of frequency—see the Bode plots in Figure 33-5. Thus when the full complex nature of Equation 33-2 is considered, it may be possible that an amplifier that has massive negative feedback at dc may have positive feedback at some high frequency and may even spontaneously oscillate.

As a result of all this, a certain amount of attention has to be paid to gain and phase shifts as a function of frequency to make sure that an amplifier circuit is completely stable and, hence, behaves properly. Considerable effort has been invested in both the theoretical and experimental investigations of amplifier instability to develop circuits and criteria that can be used with little additional analysis and yet produce satisfactorily stable amplifier circuits. The three most common criteria for stability will be discussed here in order of their complexity.

The simplest criterion is based on the Bode plot. Recall that the Bode plot is a plot of both the gain (in db) and the phase shift (in degrees) of an amplifier versus the logarithm of the frequency. The Bode plots for two op-amps are shown in Figure 33-5. The Bode plot shown in the upper portion as a solid line never has a slope greater than 20 db/decade; the plot shown as a dashed line has a considerably greater slope. This second amplifier may result in unstable amplifier configurations. The simplest stability criteria is that an op-amp will be stable if the slope of its Bode plot is never greater than 30 to 40 db/decade before it reaches the unity gain level (that is, the point where the gain = 1, or 0 db gain).

However, this criterion is unnecessarily severe. The op-amp whose Bode plot is shown as dashed lines in Figure 33-5 has a much better high-frequency response than the one shown as solid lines; if stability can be maintained, it will make a better high-frequency amplifier. The following criteria are less restrictive, but somewhat harder to apply.

The second criterion is based on the Bode plot of βA, shown in Figure 33-6. For an amplifier with nominal negative feedback, βA is a negative real number, that is, it starts out at low frequencies with a phase shift of $-180°$. As the

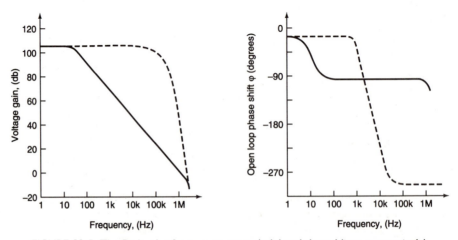

FIGURE 33-5 The Bode plot for two op-amps (solid and dotted lines, respectively).

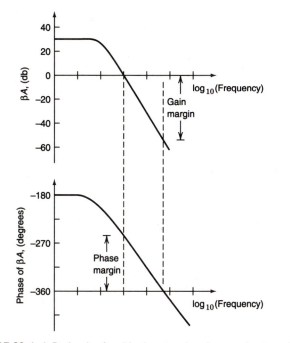

FIGURE 33-6 A Bode plot for βA, showing the phase and gain stability.

frequency increases, the phase shift increases; if it reaches $-360°$, then βA is a positive real number and positive feedback occurs. If βA comes near to the value of 1, the amplifier is unstable. Therefore, the amplifier may be unstable if the unity loop gain point occurs at a higher frequency than that at which the phase of βA reaches $-360°$. Thus the possibility of instability can be investigated from the Bode plot.

Figure 33-6 is a Bode plot for a stable amplifier, showing the phase stability margin and the gain stability margin. These margins are essentially the distance on these plots from the danger point of $\beta A = 1$. An amplifier is considered to be stable if the gain stability margin is greater than 10 db and the phase margin is more than $50°$.

The final criterion is the Nyquist criterion. A Nyquist plot is a plot of the real and imaginary parts of βA for all frequencies. This is displayed as a locus of points on the complex plane. Figure 33-7 is a typical Nyquist plot—note the asymmetrical scales.

The Nyquist criterion is simple. The amplifier is unstable if the point $\beta A = 1 + j0$ is encircled by the plot. An amplifier is stable if this point is outside the Nyquist plot. This criterion requires considerable information about an amplifier, but it gives the most accurate information for borderline cases.

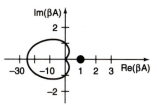

FIGURE 33-7 A Nyquist plot for a stable amplifier. This plot requires many measurements; m = imaginary, Re = real.

AMPLIFIER COMPENSATION

The high-frequency falloff in gain for an op-amp is due to various internal capacitors that form RC low-pass filters. Starting at some low frequency, as the frequency increases, first one RC filter starts to have an effect, and the gain starts to fall off at 6 db/octave—see Chapter 9 for a quick review of the low-pass filter. (A low-pass filter has a response that eventually falls off as $1/\omega$ as the frequency increases. This falloff corresponds to a factor of 2 for every octave, or 6 db/octave or 20 db/decade.) As the frequency increases, another *RC* may set in; now, the gain falls at 12 db/octave, and so on. Both the phase shifts and the attenuation (in db) add for the various *RC* low-pass filters. A problem occurs if the phase shift of the amplifier reaches 180° before the loop gain reaches 0 db.

The worst case occurs when an op-amp is used in a voltage follower, since the loop gain is the open loop gain. Possible Bode plots for two op-amps are shown in Figure 33-5 above. Even though these amplifiers both have the same unity gain frequency, they have different Bode plots. The amplifier represented by the dashed curve would be unstable when it is used in a voltage follower.

The solution to making this amplifier absolutely stable is to introduce one capacitor that starts cutting into the gain at such a low frequency that the open loop gain falls to 0 db before the phase shift of the amplifier reaches 180°. This amplifier will be stable when it is used as a voltage follower, which is the worst case for an amplifier used with negative feedback. This process of introducing a capacitor to produce an absolutely stable amplifier is called **compensation.**

One unfortunate result of compensation is that the high-frequency response of the compensated amplifier is poorer than that of the uncompensated amplifier. For high-frequency operation, uncompensated amplifiers are available. However, when one uses them, the possibility of an unstable amplifier always exists. Most common op-amps are compensated so that they will be unconditionally stable when they are used in all normal circuits. However, uncompensated versions of most of the common op-amps are available—the 748 is the uncompensated version of the 741.

Chapter 34

COMPARATORS AND SCHMITT TRIGGERS

OBJECTIVES

1. Become familiar with comparators and Schmitt triggers.

2. Become familiar with the parameters describing comparators.

3. Be able to design Schmitt trigger circuits having the desired threshold and hysteresis.

INTRODUCTION

The two most important circuits for interfacing between the digital and the analog worlds of electronics are introduced in the next two chapters. This chapter deals with comparators, that is, decision-making circuits. The next chapter deals with a circuit that generates an analog voltage proportional to a digital number.

There is some ambiguity about where this chapter, by its very nature, should go—in the analog section, in the digital section, or in a special section for interface circuits. In any case, some knowledge of both op-amp circuits and of digital logic is needed as preparation for mastering chapter.

COMPARATORS

In the simplest terms, a comparator is a circuit that tells which of two voltages is larger. This implies a circuit with two analog signal inputs and a single digital output whose state reflects the relative values of the input. Thus for the circuit shown in Figure 34-1, V_{out} will be a logical True when $V_1 > V_2$ and will be a logical False when $V_1 < V_2$. As usual, for simplicity no mention has been made of the power supply lines necessary for the operation of the circuit.

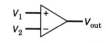

FIGURE 34-1 A comparator.

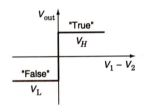

Ideal case

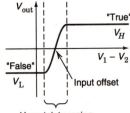

Typical case

FIGURE 34-2 Transfer functions for an ideal and a typical comparator.

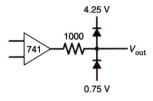

FIGURE 34-3 An op-amp with diode clamps on the output.

Several of the most important specifications describing a comparator are the following:

The **output voltage levels**, that is, what output voltage corresponds to a True and what voltage corresponds to a False. This may be stated in terms of compatibility with certain logic families; for instance, TTL compatible implies a Low output near 0 and a High output near 5 V.

The **speed of response**, that is, how long it takes the output to change from True to False, or vise versa, when the inputs are changed. Typical values range from 40 to 100 nsec, with the fastest comparators needing only a few nsec.

The **input offset voltage** and **uncertainty region** are measures of the possible error in the decision making. These two parameters are best illustrated in terms of the transfer function for a comparator. The transfer function is simply the plot of the output versus the difference between the two inputs. The plot in Figure 34-2 shows the transfer function for an ideal comparator and for a typical comparator.

The **input offset voltage** is how far the decision point is offset from the true zero $V_1 - V_2$. Ideally, this would be zero. The **uncertainty region** is a measure of the range of input voltages for which the output is neither a True nor a False. Ideally, this range would be zero. The uncertainty region is inversely related to the gain of the amplifier in the comparator.

Just as the important specifications above are similar to those needed for op-amps, a complete set of specifications for a comparator includes a number of additional specifications all of which are similar to those for op-amps. These other specifications include values for the input and output resistances (impedances), input bias current, maximum differential input voltage, maximum common mode voltage, power supply voltages and power requirements, packaging information, and cost.

However, in general, the selection of a comparator is much easier than the selection of an op-amp. There really are only three major considerations: (1) compatibility of the output with the logic family being used; (2) speed; and (3) an evaluation of the possible errors in the input circuit; this includes the effects of the input offset, the uncertainty region, and perhaps the input bias current and any possible loading effects of the input.

On some level, an op-amp operating without feedback functions as a comparator. Because of its very high gain, it has a very small uncertainty region— for the 741 this is 50 μV or less. There are two problems with using an op-amp as a comparator: First, the output levels are not necessarily compatible with the logic family being used; and second, the speed of response may not be good enough.

The first of these problems can be easily solved. For instance, the circuit in Figure 34-3 shows a 741 with its output clamped between 0 and +5 V by the two diodes. Thus the output of this circuit is compatible with most common logic families, TTL, CMOS, and the like.

However, it generally seems as though speed is of primary interest in circuits involving comparators. Normal run-of-the-mill comparators switch between True and False in times of the order of 50 nsec, whereas a typical op-amp would take a few leisurely microseconds. Ordinary comparators are built with lower gain but higher speed amplifiers than the usual op-amps. Since they are not intended to be used as amplifiers with negative feedback, no compensation is needed to ensure stability.

At typical comparator is the LM311 made by National Semiconductors. It has the following basic specifications: 200 nsec response time, 3.0 mV maximum offset voltage, 0.1 μA input bias current, and TTL-compatible outputs.

Comparators are used whenever decisions have to be made. They are a central component in the construction of any analog-to-digital converter, of which the digital voltmeter is the most common example. Whenever something must be turned on or off depending on the level of some other signal, comparators are used. For instance, the small circuit that decides when to turn on or turn off the street lights involves a solar cell to convert the light intensity into a voltage and a comparator which makes the decision. Even the transmission of digital signals over long distances, 10 feet or more for ordinary TTL signals, often requires the use of comparators. For this use, special comparators called line receivers are often used.

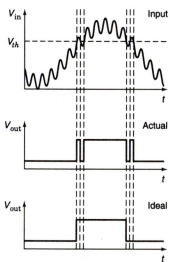

SCHMITT TRIGGER

The very speed of a comparator often causes a problem. The nature of this problem can best be explained by the simple illustration shown in Figure 34-4. FIGURE 34-4 A false response In this example, the input to the comparator is a relatively slow signal plus from a comparator. some high-frequency noise. Of course, the comparator does not know about the distinction between the signal and the noise; it sees only the sum of the two, which is shown in the upper part of the figure. The ideal output would reflect only the signal and somehow reject the effects of the noise; this ideal response is shown in the lower part of the figure. However, the comparator simply responds to the actual input; the output the comparator generates owing to the combination of the signal and noise is shown in the middle of the figure. Whereas the ideal output would have only one positive going pulse, the actual output has three pulses. If the output of the comparator were being counted, the result would be incorrect, and incorrect by a factor that depends on the details of the noise. In other words, the result of the counting would be meaningless.

This is a fairly general problem. Any time a slowly varying signal is the input to a comparator, the output may have a number of false transitions because of the effects of small noise signals. In many instances, these extra transitions are not a problem, but in others they are.

Simply slowing down the comparator so that it cannot respond to the high-frequency noise signals is not always a very good solution because there may be situations where the real signals sometimes vary slowly and sometimes vary rapidly. A better solution is to somehow take advantage of the fact that generally the signals have large amplitude changes and the noise has a relatively small amplitude. The Schmitt trigger circuit takes advantage of this distinction.

The basic idea behind a Schmitt trigger is the use of positive feedback around a comparator so that when the output switches from one state to another, the threshold at which the next transition will occur also changes. Thus once the input has crossed the threshold, it must change back by some minimum voltage called the **hysteresis voltage** before it crosses the threshold again. This will be explained two more times below, first in words and then via a fairly complete analysis of a Schmitt trigger circuit.

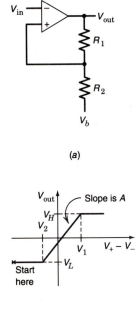

(a)

(b)

FIGURE 34-5 A Schmitt trigger circuit and its transfer function.

Consider the case of a normal comparator with a slowly varying input that is approaching the threshold from below. When the input crosses the threshold, the output switches from one state to the other; then, if the input decreases by a tiny amount due to noise, the effective input crosses the threshold again and the output switches back. Thus, as illustrated in Figure 34-4, small-noise signals superimposed on a slowly varying signal can cause the output of the comparator to switch back and forth many times. For the Schmitt trigger, however, there is some positive feedback so that as the input signal crosses the threshold in the upward direction for the first time, not only does the output change but also the threshold is lowered by an amount V_{hyst} so that, even if the input then decreases by a small amount because of noise, the output does not change. Only when the input decreases by an amount greater than V_{hyst} will the output switch back. By picking the value of V_{hyst} to be some compromise between the size of the typical noise signal and the size of the expected input signal variations, the Schmitt trigger circuit can be designed to minimize the response to the noise and yet respond correctly to the expected signals. Of course, V_{hyst} is the hysteresis voltage.

ANALYSIS OF A SCHMITT TRIGGER CIRCUIT

A Schmitt trigger built from an op-amp is shown in Figure 34-5. Positive feedback from the output to the noninverting input is provided by the voltage divider. The input voltage V_{in} goes to the inverting input. For a large positive input, the output of the comparator will be low, V_L. As the input become less positive, at some point it crosses a threshold where the output switches from V_L to V_H. This causes the noninverting input voltage to increase suddenly and, as a result, V_{in} is no longer near the threshold. This positive feedback is what makes the hysterisis. The values of the resistors and V_b can be varied to adjust the threshold and the hysteresis.

The analysis of the Schmitt trigger circuit is somewhat complicated and confusing. This analysis is first carried out in great detail; then, some approximations and simplifications are made to show that the solution really does what is advertised.

The circuit to be analyzed and its transfer function are shown in Figure 34-5. This circuit is built from an op-amp having an open loop gain A and two saturation output voltages V_H (High output) and V_L (Low output). No assumptions will be made about A at this point, since A for typical comparator circuits may only be 1000 or less. Likewise, no assumptions about the signs and magnitudes of V_H, V_L, and V_1, and V_2 are made other than the fact that

$$V_L < V_H \quad \text{and} \quad V_2 < V_1$$

However, the relationship between the input and the output of the op-amp means that

$$A(V_1 - V_2) = V_H - V_L \tag{34-1}$$

One point of confusion is the sign of the input voltage V_{in} versus the sign of the voltage plotted in the transfer function. V_+ is the voltage of the noninverting input with respect to ground; V_- is the voltage of the inverting input. The transfer function shows the output versus the difference in the two input voltages V_+ and V_-. The actual input voltage to the circuit V_{in} is connected to the inverting input, that is, $V_{in} = V_-$. Thus a very large positive value of V_{in} corresponds to a large negative value of $V_+ - V_-$ and is far to the left on the transfer function graph. Since the entire analysis depends on the signs of various terms, try to keep this straight from the very beginning.

First, a qualitative analysis of the operation of the circuit shown in Figure 34-5 involves five steps:

1. Start with V_{in} very positive, that is, $V_{in} >> V_+$, so that the output is low. That is, $V_{out} = V_L$. This is the point marked "Start" on the graph of the transfer function.

2. Let V_{in} start to decrease slowly. This corresponds to sliding to the right along the transfer function. At first, nothing happens. But when

$$V_+ - V_{in} = V_+ - V_- = V_2 \qquad (34\text{-}2)$$

the output voltage starts to increase. This corresponds to having reached the corner of the transfer function and starting to slide up the slope.

3. Because V_{out} increases and this is coupled back to V_+ via the voltage divider (the positive feedback), V_+ increases and the output increases even more. Thus a small change in the input causes the operating point to slide slightly up the slope, and the corresponding change in the output voltage causes the operating point to slide even farther up the point. This means that there is a further increase in V_{out} and the analysis cycle must continue. Either this process will converge to some final value on the slope or else the output will switch suddenly to the other extreme. Mathematically there are two possible cases. If

$$\frac{AR_2}{R_1 + R_2} < 1 \qquad (34\text{-}3)$$

then nothing special happens. But if

$$\frac{AR_2}{R_1 + R_2} > 1 \qquad (34\text{-}4)$$

then the output switches to V_H. This is the case of interest.

4. Because V_{out} has gone from V_L to V_H, V_+ has increased substantially. Therefore, $V_+ - V_{in}$ is now far more positive than it was just before these changes started. Thus the operating point is now far to the right of the right-hand corner of the transfer function. Therefore, a slight increase in V_{in} will make no difference in V_{out} once the threshold has been crossed. Remember, the transition was reached because V_{in} was decreasing, so a slight noise ripple on V_{in} will not cause multiple transitions.

5. Thus V_{in} must increase significantly before the output will switch back to the original state. As a result, hysteresis has been introduced to the circuit.

This analysis can be made quantitative. It can be shown that (do this as an exercise)

$$V_+ = V_b + (V_{out} - V_b)\frac{R_2}{R_1 + R_2} \tag{34-5}$$

It is assumed that

$$\frac{AR_2}{R_1 + R_2} > 1$$

(This is not necessarily trivial for very-high-speed comparators.)

There are two thresholds or two trigger points. These are $V_{th}(U)$, the upper threshold, which occurs where V_{out} switches from V_H to V_L. This occurs when V_{in} increases to

$$V_+ - V_{in} = V_1 \tag{34-6}$$

and is approached from the right on the transfer graph.

The second threshold is $V_{th}(L)$, the lower threshold, which occurs when V_{out} switches from V_L to V_H. This occurs when V_{in} decreases to the point where

$$V_+ - V_{in} = V_2 \tag{34-7}$$

and is approached from the left on the transfer graph.

Combining these various terms allows the evaluation of the two thresholds. For the approach to the upper threshold, $V_{out} = V_H$ and

$$V_{th}(U) = -V_1 + V_b + (V_H - V_b)\frac{R_2}{R_1 + R_2} \tag{34-8}$$

whereas for the approach to the lower threshold, $V_{out} = V_L$ and

$$V_{th}(L) = -V_2 + V_b + (V_L - V_b)\frac{R_2}{R_1 + R_2} \tag{34-9}$$

Note that V_1 and V_2 are determined by the gain and the input offset of the op-amp. Clearly, $V_{th}(U) > V_{th}(L)$.

Finally, the hysteresis can be calculated as follows:

$$V_{hyst} = V_{th}(U) - V_{th}(L)$$

$$= -V_1 + V_2 + (V_H - V_L)\frac{R_2}{R_1 + R_2}$$

$$= (V_H - V_L)\left(\frac{R_2}{R_1 + R_2} - \frac{1}{A}\right) \tag{34-10}$$

Although the full expressions given above may be needed to design a real Schmitt trigger circuit, some approximations make the general structure of these results clearer. If you assume that the gain of the comparator (op-amp) is very large, then the term $1/A$ can be ignored. This also means that $V_1 = V_2$. Also assume that there is no input offset voltage; this means that $V_1 = V_2 = 0$. Then, the expressions become

$$V_{hyst} = (V_H - V_L)\frac{R_2}{R_1 + R_2} \qquad (34\text{-}11)$$

$$V_{th}(U) = V_b + (V_H - V_b)\frac{R_2}{R_1 + R_2} \qquad (34\text{-}12)$$

and

$$V_{th}(L) = V_b + (V_L - V_b)\frac{R_2}{R_1 + R_2} \qquad (34\text{-}13)$$

These results show that to design a Schmitt trigger, the output voltage swing and the desired hysteresis determine the resistor ratio (Note that only the ratio is determined so that one of the values is still free and may be chosen to satisfy any other constraint.) Then, V_b is adjusted to give one or the other of the two threshold voltages—the second is determined by the first and by the hysteresis.

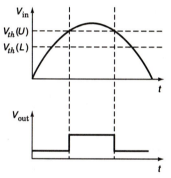

DISCUSSION OF THE SCHMITT TRIGGER

If a sine wave were input to a Schmitt trigger and the input and output observed on the same time scale with an oscilloscope, a display similar to the one shown in Figure 34-6 might be obtained.

FIGURE 34-6 The response of a Schmitt trigger circuit to a sine wave input.

Some comparators and/or line receiver circuits have hysteresis built into them so they do not require external feedback networks to cause the hysteresis. The schematic symbol shown in Figure 34-7 is sometimes used for these circuits; the little box with tails is included to designate the existence of hysteresis.

FIGURE 34-7 A comparator with hysteresis.

PROBLEMS

34-1 Assuming an op-amp with infinite gain, no offset, and output voltage levels of +15.0 and −15.0 V, design a Schmitt trigger that has a 2-V hysteresis and a lower threshold of 2.5 V.

34-2 Given a comparator with output levels of +5.0 V and −0.75 V and a gain of 1000 and no offset, design a Schmitt trigger circuit that has a hysteresis of 100 mV and a lower threshold of 2.5 V.

34-3 Given an unknown positive input voltage, a 10 V reference voltage, and assorted resistors, comparators, and logic units, design a circuit whose output is True when $3.0 < V_{in} < 4.0$ V.

34-4 Given an op-amp with output levels of −0.75 V and −1.5 V and a gain of 12, design a Schmitt trigger circuit that has a hysteresis of 50 mV and a lower threshold of −75 mV. Note that in this case, the finite gain of the op-amp cannot be ignored and some assumptions about V_1 and V_2 will have to be made.

ANSWERS TO ODD-NUMBERED PROBLEMS

34-1 picking $R_2 = 1$ kΩ, then $R_1 = 14$ kΩ and $V_b = 3.75$ V

34-3 one possible solution given in Figure 34-8

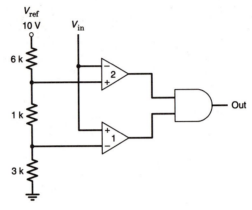

FIGURE 34-8 A solution to Problem 34-3.

PART FIVE

SYSTEMS

This part presents an introduction to several of the important electronic systems that involve digital and analog circuits as well as discrete components, in some cases. The purpose of this part is to provide some familiarity with these circuits, which are common and useful, as well as to show how many of the simple circuits discussed earlier in this book can be combined into useful systems.

The nature of this material means that this discussion must be very general. Too much detail would require restricting the discussion to one particular circuit without any benefit being gained from such specificity. Likewise, the nature of this material also precludes exercises and problems, since the only valid problem would be "design, build, and make work a circuit that will...", and this belongs in a laboratory.

The first two chapters, Chapters 35 and 36, of this section discuss the two circuits that form the basis for most digital control and measurements, including the digital voltmeters as the simplest example. Chapter 37 introduces some of the circuits that are used in signal generators, music synthesizers, and many other places. Chapter 38 presents a short introduction to power supplies, their nature, purpose, and design, as well as much of the terminology that is used with respect to them. Finally, Chapter 39 describes an analog-to-digital converter that is interfaced to a hypothetical computer system.

PREREQUISITES

CHAPTER 35 DIGITAL-TO-ANALOG CONVERTERS
1. A knowledge of simple op-amp circuits—Chapter 29.
2. A knowledge of binary number systems—Chapter 54.
3. A knowledge of simple digital electronic concepts—Chapter 19.
4. A knowledge of basic circuit analysis—Chapter 3.
5. FET switches are mentioned but an understanding of them is not necessary to master this chapter.'

CHAPTER 36 ANALOG-TO-DIGITAL CONVERTERS
1. A familiarity with digital-to-analog converters—Chapter 35.
2. A familiarity with comparators—Chapter 34.
3. A familiarity with flip-flop circuits—Chapter 21—and simple digital logic—Chapter 19.

CHAPTER 37 VOLTAGE CONTROLLED OSCILLATORS AND
FUNCTION GENERATORS

1. A familiarity with a number of op-amps circuits—Chapter 29.
2. A familiarity with comparators—Chapter 34.
3. A familiarity with the astable multivibrator—Chapter 23.
4. A knowledge of RC circuits—Chapter 12.

CHAPTER 38 POWER SUPPLIES

1. A familiarity with basic circuit analysis—Chapter 3.
2. A familiarity with simple ac circuit analysis—Chapter 8.
3. A familiarity with transformers—Chapter 11.
4. A familiarity with op-amp circuits—Chapter 29.
5. A knowledge of diodes—Chapter 41—and simple transistor
circuits—Chapter 43—is assumed in the discussions.

CHAPTER 39 A COMPUTER INTERFACE FOR AN A TO D

A familiarity with a whole list of topics:

1. Analog to Digital Converters—Chapter 36.
2. Computers—Chapter 27.
3. Bus Drivers—Chapter 25.
4. Logic Circuits—Chapters 22 and 25.
5. One-Shots—Chapter 23.

Chapter 35

DIGITAL-TO-ANALOG CONVERTERS

OBJECTIVES

1. Understand the two main designs for digital-to-analog converters.

2. Become familiar with some of the variations on the basic D/A converters.

3. Become familiar with the parameters used to describe D/A converters.

INTRODUCTION

The **digital-to-analog converter**, known as the **D/A converter** (read as D-to-A converter) or the **DAC**, is the second of the major interface circuits, that form the bridge between the analog and digital worlds. DACs are the core of many circuits and instruments, including digital voltmeters, plotters, oscilloscope displays, and many computer-controlled devices. In this chapter, the operation of a DAC is described, and two basic DAC circuits are analyzed in some detail. Then a number of variations on the basic circuit are discussed. The chapter ends with a listing of the various parameters that are used to describe the operation of DACs.

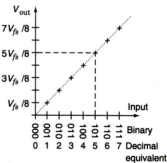

THE PURPOSE OF A DIGITAL-TO-ANALOG CONVERTER

The purpose of a DAC is to produce an output voltage that is proportional to a digital binary number that is input to the circuit. For instance, a perfect DAC with a three-bit binary number as its input would have eight distinct output voltages given in Table 35-1. This output is plotted in Figure 35-1. For this example, V_{fs} stands for the **full-scale voltage**, even though the output never quite reaches this value. For a real DAC, this may be a fixed voltage, or it may be a voltage that can be externally adjusted (see below). For this three-bit example, a change in the **least significant bit**, **LSB**, corresponds to a volt-

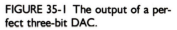

FIGURE 35-1 The output of a perfect three-bit DAC.

TABLE 35-1 Response of a Perfect Three-Bit DAC

Binary Input	Decimal Equivalent	Output Voltage
000	0	0
001	1	$1/8\ V_{fs}$
010	2	$1/4\ V_{fs}$
011	3	$3/8\ V_{fs}$
100	4	$1/2\ V_{fs}$
101	5	$5/8\ V_{fs}$
110	6	$3/4\ V_{fs}$
111	7	$7/8\ V_{fs}$

age step of $V_{fs}/8$. This is the smallest voltage step the output can make. As the obvious generalization of this, the minimum step of a n-bit A/D is given by

$$\text{Minimum step} = \frac{V_{fs}}{2^n} \tag{35-1}$$

The **resolution** of a DAC is the value of the smallest possible voltage step expressed as a percentage of the full-scale value. Thus the resolution of a D/A is simply the value $1/2^n$ expressed as a percentage. Table 35-2 shows the resolutions obtained for various numbers of bits.

   With current technology, it is possible to produce cheap integrated eight-bit DACs. Units with 16-bit resolutions are available for a reasonable price; and units claiming to have even better resolution are on the market. Using a unit with a 16-bit resolution is not a trivial exercise; for instance, a 16-bit device with a full-scale voltage of 1 V has a minimum step corresponding to 15 μV. Such a small voltage can easily be masked by the larger noise voltages occurring in a circuit. The voltages generated by dissimilar metals in contact with each other in conjunction with either a thermal gradient or a little moisture can easily dwarf this minimum step. Using very-high-resolution devices requires lots of knowledge, skill, and almost-clean-room techniques.

ACTUAL D/A CIRCUITS

Modern DACs are based on two circuit elements. One is the voltage-to-current converter, or the summing amplifier; see Chapter 29 for a review of this cir-

TABLE 35-2 Resolution of DACs as a Function of the Number of Bits

Number of Bits n	$1/2^n$	% Resolution
2	0.25	25
3	0.125	12.5
4	0.0625	6
8	0.0039	0.4
10	0.00098	0.1
12	2.44×10^{-4}	0.02
14	6.1×10^{-5}	0.006
16	1.5×10^{-5}	0.0015
20	9.5×10^{-7}	0.0001

cuit. The second circuit element is the FET switch; see Chapter 45. These electronic off-on switches controlled by a digital signal have a very low on resistance (10 to 100 ohms) and very high off resistances (multimegohms or more). In the following schematics, the digital lines are not shown and the switches are indicated as ordinary mechanical switches. But in actual situations, these switches would be FETs. (Of course, it was not too many years ago that these switches would have been relays operated by the digital signals.)

A very simple DAC circuit is shown in Figure 35-2. In the Figure, **MSB** stands for most significant bit, and **LSB** stands for least significant bit. In addition, the bits in the figure are numbered, bit 1 being the MSB and bit n the LSB.

The circuit shown in Figure 35-2 is the simplest possible DAC. In essence, each switch/resistor combination forms a current source that generates a current $V_{ref}/2^n$ which is injected into the summing point of the current-to-voltage converter; the summing point is kept at ground by the feedback loop around the op-amp. The output of the op-amp is given by

$$V_{out} = I_t R \qquad (35\text{-}2)$$

where I_t is the total current flowing into the summing point. For instance, if bits 1, 3, and 6 were on, the output voltage would be given by

$$V_{out} = R \times V_{ref}\left(\frac{1}{2R} + \frac{1}{2^3 R} + \frac{1}{2^6 R}\right)$$

$$= R \times V_{ref}\left(\frac{1}{2R} + \frac{1}{8R} + \frac{1}{64R}\right)$$

$$= V_{ref}\frac{41}{64}$$

The scheme illustrated in Figure 35-2 yields perfectly serviceable DACs; but it has several disadvantages and, as a result, it is rarely used in modern DAC design. First, many different resistor values are needed. Each of these resistors must be carefully controlled to ensure the ultimate precision of the DAC; this creates problems in trimming the circuit to make it operate within design specifications. Second, the resistors span a very large range, running from R to $2^n R$. Very high and/or very low resistance values present technical problems in making integrated circuits. Finally, the most significant bit switch turns off and on a relatively large current; capacitive effects in this switch

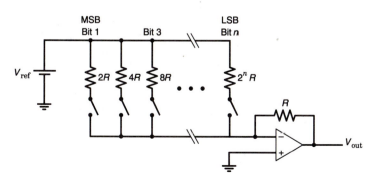

FIGURE 35-2 A simple DAC circuit. Each switch is controlled by a digital input, which is not shown. Power leads are also not shown.

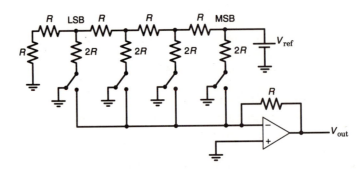

FIGURE 35-3 An R-2R DAC. Again the power lines and the digital input lines are not shown.

and the other switches (which are all in parallel) can limit the slewing rate of the DAC output and, hence, the time response of the DAC. A scheme in which only a few resistor values are needed and in which the capacitive effects of the switches corresponding to the less significant bits are minimized would result in a better DAC.

The circuit shown in Figure 35-3, called the **R–2R network**, has the desired advantages. It requires only one resistor value (since $2R = R + R$), so the automatic resistance trimming equipment need only trim one value. The effects of the capacitance of the less significant switches is decreased by the structure of the network. Although the specific case shown has only four bits, it could easily be extended to any number of bits with no additional complexity except more sections. The only thing not immediately obvious about this circuit is how it works.

The analysis of this circuit is relatively easy, but it does require some insight and a strategy that may almost seem like a trick. There are three main steps in the analysis. The circuit has been redrawn in Figure 35-4 with names given to each node to aid in the analysis. The first step in the analysis requires realizing that the resistance from any of the nodes, A, B, C, or D, to ground looking backward, that is, away from the reference source, is R; this value does not depend on the settings of the switches. The demonstration of this fact requires several substeps:

1. Remember, the summing junction of the op-amp is a virtual ground.
2. Starting at junction D, break the line (conceptually) between D and C. Now, there are two parallel paths to ground. The resistance looking back-

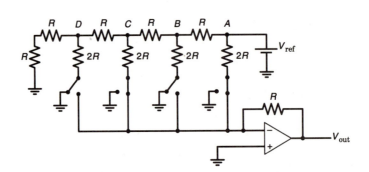

FIGURE 35-4 An R-2R network with several switches closed and each node named.

ward (horizontally) is $2R$. The resistance looking downward through the switch is $2R$, no matter what the position of the switch, since the summing point is also ground. Thus from D there are two parallel paths to ground, each having a resistance of $2R$. Hence, the resistance from D to ground is R.

3. Next, reconnect junction C and D, and break the line between C and B. Now, do a similar analysis from the point of view of junction C. Looking from C toward D, there is a resistance R between C and D, which is in series with the resistance R from D to ground. Thus this path from C has a total resistance of $2R$. Just as described above, the resistance looking downward through the switch connected to C is $2R$, no matter what the switch setting. Hence, just as for D, there are two parallel paths from C to ground, and each has a resistance of $2R$. Thus the total resistance from C to ground is R, no matter what the setting of switches C or D.

No more. Obviously, this process can be repeated as many times as necessary. Looking backward from each of the junctions, the resistance to ground is R.

The second step in the analysis is to understand how the resistor chain operates as a voltage divider. Each node has a voltage that is one-half the voltage of the node to its right. To understand this, consider point D as the midpoint of a voltage divider. The voltage is $V_C/2$, as Figure 35-5 should make clear. Thus the voltage at point D is one-half the voltage at point C. However, since the resistance to ground at any node looking backward is R, this analysis could have been applied to any node. Thus the voltage at point C is one-half that at point B, and so on. As a result, reading from right to left, the voltage at each junction decreases by another factor of 2.

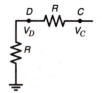

FIGURE 35-5 Point D of Figure 35-4 viewed as a voltage divider.

The final step in the analysis is to consider the effects of the switches. Each switch controls a current of $V/2R$, where V is the voltage at the node to which the switch is attached. Thus, since $V_A = V_{ref}$, if the switch corresponding to the MSB is closed, a current equal to $V_{ref}/2R$ is injected into the summing point. Since $V_B = V_A/2$, the switch connected to node B controls a current of $V_{ref}/4R$. As was shown above, each node has a voltage equal to one-half the voltage of the node to the right. Thus, each switch controls a current one-half as large as the one to the right, and the least significant switch controls a current of $V_{ref}/2^n R$. The total current injected into the summing point is the appropriate sum of currents to represent the binary number being input to the converter. For the particular example shown in Figure 35-4, the total current injected into the summing junction is

$$V_{ref}\left(\frac{1}{2R} + \frac{1}{8R}\right) = V_{ref}\frac{5}{8}$$

VARIATIONS ON DAC CIRCUITS

Thus far, only the basic DAC circuit has been discussed. In this section, several variations on this circuit will be described without the benefit of any circuit details. The purpose of this section is to provide an awareness of the terminology used to describe these converters.

The basic circuits discussed above are based on the assumption that the inputs are binary numbers. However, human beings do not like binary numbers; they prefer decimal numbers. As was discussed in the chapter on binary numbers, there are several schemes for coding decimal numbers within the guise of digital electronics and binary numbers. The most common scheme is called BCD, for binary coded decimal, and uses 10 of the possible 16 values that can be coded by four binary bits. It only takes a little ingenuity to work out a weighting scheme in a variation on the R–$2R$ network that results in a converter producing an output proportional to a multidigit BCD number.

For some DACs, the value of V_{ref} is fixed inside the DAC, whereas for others V_{ref} can be supplied externally. If V_{ref} can be supplied externally, the DAC is said to be a **multiplying** DAC, since the output is a product of V_{ref} and the binary number. If V_{ref} may be of either sign, then the DAC is said to be a **two-quadrant** multiplying DAC. Multiplying DACs can be used as digital attenuators or digital volume controls. They are frequently used in this fashion in computer-controlled equipment.

Again, as was discussed in the chapter on binary numbers, there are several ways in which signed numbers can be coded in binary. Again, it only takes a little ingenuity to devise a circuit that will invert the sign of the output of a DAC depending on the status of one binary bit—the circuit in Figure 35-6 is one possible example. DACs that are designed to accept signed binary inputs are also said to be **two-quadrant** DACs. Finally, if both V_{ref} and the input binary number are of either sign, the result is a **four-quadrant** multiplying DAC.

Some DACs are constructed to be used as current sources rather than as voltage sources. Conceptually, they do not different from the ones discussed above, except that the op-amp circuit is arranged to function as a current source rather than as a voltage source.

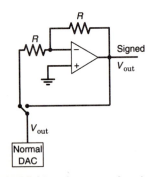

FIGURE 35-6 One way of implementing a signed binary input. The switch is controlled by the sign bit. If the switch is up, the signed output has the opposite sign than V_{out}. If the switch is down, both have the same sign.

DAC SPECIFICATIONS

Further consideration of the details of DAC design or DAC variations go beyond the scope of an introduction. However, a short summary of the terminology used to describe these circuits is completely appropriate here. The remainder of this chapter is devoted to an alphabetical discussion of some of the main terms used in the literature to describe DACs.

Accuracy

There are many ways of expressing the **accuracy** of DACs. Typical specifications state the maximum departure of the output from its theoretical value either as parts per million (ppm), as a percentage of the full-scale value, or as a fraction (or multiple) of the LSB step. Other measures of the accuracy are discussed below.

Compliance Voltage Range

A DAC that outputs a current can only provide some maximum voltage, that is, work into a maximum resistance, before its operation breaks down. This maximum voltage is called the **compliance voltage range**.

Glitches

When the input of a DAC changes from 01111 to 10000, four switches open and one closes. If the switches have different opening and closing speeds, the output may start to go to some incorrect value (perhaps corresponding to 11111 or 00000), then reverse and settle to the correct value. The short incorrect voltage spike, or dip, is called a **glitch**.

Glitches can often cause incorrect circuit behavior; for instance, a fast comparator can respond to the glitch and have its response latched in a flip-flop. In general, glitches can be very serious, and they are very hard to remove by filtering the output. DAC specifications may discuss glitch-free operation or deglitched outputs.

Linearity

The **linearity** error is a measure of the maximum departure of the output of the DAC from a straight line. The line can be either a best-fit line, or a straight line passing through the end points after calibration. Linearity errors are measured in ppm, percent of the full-scale value, or (sub)multiples of an LSB step.

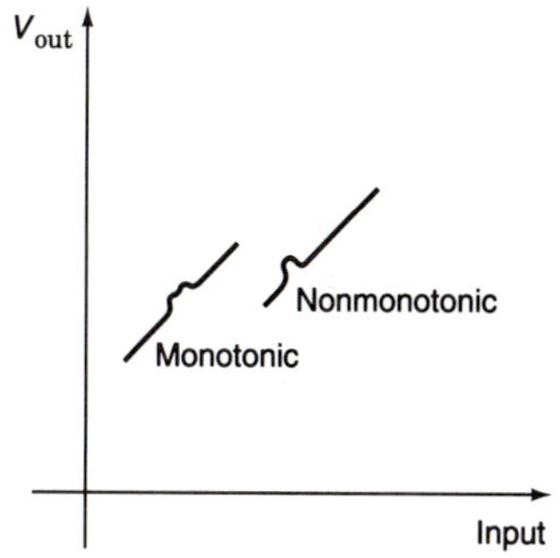

Monotonic

A DAC is said to be **monotonic** if its output increases or remains the same for an increase in the digital input over the entire range of operation. (A DAC is not monotonic if at any point in its operating range, an increase in the digital input causes the output to decrease.) Figure 35-7 illustrates the meaning of nonmonotonic behavior. Short segments of two possible DAC outputs are plotted; both show substantial nonlinearities, but one is monotonic whereas the other is not. Nonmonotonic behavior is often far more serious than simple nonlinearities because many searching algorithms assume monotonic behavior.

FIGURE 35-7 Illustration of monotonic and nonmonotonic outputs.

Other Parameters

Many other terms are used to describe DACs, such as offset, slewing rate, settling time, stability, and temperature coefficients—all of which are essentially the same as for those used to describe op-amps. No further comments need to be made here about these terms.

WHERE TO GET MORE INFORMATION

The best sources of information about DACs is the manufacturer's literature. The major manufacturers all have extensive catalogs devoted to DACs and to related circuits; these often include substantial tutorial sections on the choice and operation of DACs. Analog Devices publishes the **Analog–Digital Conversion Handbook**, which is an excellent introduction to DACs, various ADCs, as well as op-amps and related circuits—see Appendix A for the full reference.

Chapter 36

ANALOG-TO-DIGITAL CONVERTERS

OBJECTIVES

1. Become familiar with the purpose and specifications of analog-to-digital converters.

2. Become familiar with five different A/D converter designs.

3. Become acquainted with the need for signal condition.

INTRODUCTION

This chapter is an introduction to the various techniques for making **analog-to-digital converters**. The analog to digital converter, **A-to-D converter**, **ADC**, or **A/D converter**, is the basis of the digital voltmeter as well as many other digital instruments. There are a number of different designs for making A/D converters, of which the most important are discussed here. In this introduction to A/Ds, only the most basic versions of each of these designs are covered. Many variations of the basic circuits exist; details on these variations can be found in the literature or in the data sheets from the manufacturers.

The terminology and parameters used to describe the performance of an A/D converter are essentially the same as those for a digital-to-analog converter. These are discussed in Chapter 35, and that discussion will not be repeated here.

The various types of A/D converter differ in their speed, potential resolution and accuracy, complexity, and cost. Obviously, with the present ability to combine more and more circuitry, including both analog and digital functions, into a single IC, complexity is no longer a major consideration—cost, speed, accuracy and accuracy-related considerations are the most important. After brief discussions about the purpose of A/D converters and signal conditioning, five different designs will be introduced and evaluated on the basis of these criteria.

THE PURPOSE OF THE A-TO-D CONVERTER

The purpose of an A/D converter is to produce a digital binary number that is proportional to an analog voltage input. The entire discussion concerning DACs of Chapter 35 could easily be repeated here with the inputs and outputs interchanged. This will not be done, but it might be a good time to review that discussion. There are many variations on possible output codes (binary, BCD, gray code, etc.), as well as how to deal with both polarities of inputs. These variations will all be ignored in this chapter. It is simply assumed that the output is in a useful code and that some method is present to invert the input if it is negative.

SIGNAL CONDITIONING

All efforts to make accurate measurements are dogged by noise. This noise may be high frequency or low frequency; but, in general, it is so serious that simply connecting a high-speed A/D to an unconditioned input gives results that vary continuously by large amounts because of the noise signals—in short, the measurement gives nonsense results. The process of making highly accurate measurements involves trying to prevent noise from getting into the system and then somehow making the measurement system less sensitive to the noise than it is to the signal. The process of attempting to eliminate noise from the signal is called **conditioning** the signal.

There is no universal way to condition a signal. For instance, if the goal is to measure very-high-frequency signals with an oscilloscope, then a high-pass filter which eliminates any low-frequency noise will be a reasonable signal conditioning step. If low-frequency signals are being measured, such as dc or very slowly varying signals, an appropriate form of signal conditioning is to average the signal over some time period. Averaging will tend to reduce the effects of very-high-frequency noise, with little effect on the slowly varying parts of the signal. Remember, averaging is another name for introducing a low-pass filter with a very low cutoff frequency into the circuit.

One particularly troublesome type of electrical noise is 60-Hz noise, which prevades the entire globe, even in those regions far from civilization. Any attempt to make an accurate measurement must include some efforts to control the effects of 60-Hz noise, or **hum** to use the old high-fidelity term. If the averaging time is an exact multiple of 1/60 of a cycle, the effects of 60-Hz noise will be greatly reduced. In the following examples, no further comments will be made about signal conditioning, but some form must be present in any real measuring system. It should be noted that the VOM effectively averages 60-Hz noise by its very nature.

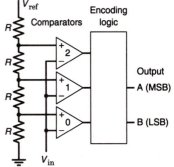

FLASH ENCODERS

The fastest type of A/D is a **flash encoder** or **parallel encoder**. A very simple version of a flash encoder is shown in Figure 36-1. For simplicity, a two-bit flash encoder has been drawn. This involves only three comparators and has very poor resolution. More interesting versions have 8, 16, or more compara-

FIGURE 36-1 A two-bit flash encoder.

tors and correspondingly better resolution. The extension of this circuit to a higher resolution is obvious, but is too tedious to be drawn here.

The operation of this circuit is quite simple. The input signal V_{in} is connected to the inverting input of each of the three comparators. The noninverting input of each of the comparators is connected to a voltage divider chain. Since all of the resistors in the chain are equal, the noninverting input of comparator 0 gets $\frac{1}{4}V_{ref}$, comparator 1 gets $\frac{2}{4}V_{ref}$, and comparator 2 gets $\frac{3}{4}V_{ref}$. For any given input voltage, all of the comparators below a certain level will be False and all the comparators above that level will be True. For instance, if V_{in} were $0.6V_{ref}$, then comparators 0 and 1 would be False and comparator 2 would be True. The encoding logic converts the comparator pattern into the appropriate two-bit binary output. For the example, here, the output would be a binary 2, or bit A would be True and bit B would be False.

Obviously, this design can easily be extended to three bits (eight comparators), four bits (16 comparators), and so on. The repetitive nature of the circuit lends itself to the manufacturer of an IC. Commercial IC units having up to 256 levels are available, but greater resolutions are not available at present.

For any given resolution, this particular type of A/D is the fastest possible. Conversion speeds of about 30 nsec are obtained for eight-bit models (256 level resolution). However, because of the number of circuits needed, this type of converter can not be made with very high resolution. This particular type of A/D is very useful where moderate resolution and very high speeds are needed. A typical use is to digitize television pictures in real time.

COUNTDOWN A-TO-D CONVERTERS

A number of different A/D designs are based on digital-to-analog converters, some logic, and a display. Several of these are discussed in this and the next section.

The **countdown A-to-D converter** is a very simple A/D that is built from a DAC, a counter, a clock, a comparator, and a little logic. A block diagram of the circuit is shown in Figure 36-2. The circuit operates as follows:

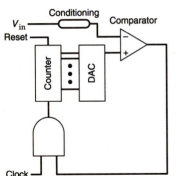

FIGURE 36-2 Part of a countdown A/D converter. The display is not shown.

1. The counter is set to zero at the start; this means that the DAC output is also zero.
2. The counter starts to count upward; the DAC output increases as the counter counts.
3. At some point, the DAC output becomes greater than the conditioned input voltage. At this point, the comparator changes state, and this triggers the logic portions of the circuit.
4. The triggering of the logic causes the following actions to take place: The counter is stopped, and the counter contents are latched and sent to the display. Next, the counter is reset to zero, and the whole process starts over again at step 1 above.

This kind of A/D is very simple, and its resolution is limited only by the quality of the DAC. Because of its simplicity, this was a very popular form of A/D in the early days of modern electronics. However, this type of A/D may be very slow if there are many bits in the DAC. For instance, if a 16-bit DAC were used, it might be necessary to wait for the counter to count up 30,000 or more steps before an answer is obtained.

A slight modification of this circuit, called the **tracking A-to-D**, permits a somewhat better performance. An up/down counter is used and the comparator and the count direction logic are arranged so that when $V_{DAC} > V_{in}$, the counter counts down; and when $V_{DAC} < V_{in}$, the counter counts up. With this arrangement, the DAC output will continuously track the input voltage, and the display can continuously display the output of the counter. As long as the circuit is successfully tracking the input voltage, the display will continuously change by ±1 in the last digit. This causes no problems unless it happens to be a change from 2.99 to 3.00 or the like. As long as the input changes slowly, this circuit will continually track the input. If the input changes by a large amount, it will be some time before the output starts to track the voltage again.

SUCCESSIVE APPROXIMATION A/Ds

The **successive approximation A/D** is substantially faster than the countdown type of A/D, but it is also considerably more complex. A partial schematic is shown in Figure 36-3. The operation of this A/D goes as follows:

1. The cycle starts with all the flip-flop bits set to zero.
2. On the first cycle, the MSB of the DAC is turned on, and the following decision is made: If $V_{DAC} > V_{in}$, then the bit is turned off; If $V_{DAC} < V_{in}$, the bit is left on.
3. When the decision has been made for the first bit, go to the next most significant bit, turn it on, and make a similar set of decisions.

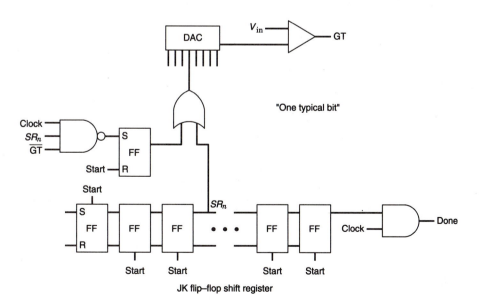

FIGURE 36-3 Part of a successive approximation A/D. The logic for only one bit is shown, and no display logic is shown.

4. Repeat this process until all the bits are done.
5. Transfer the status of the flip-flop register into the output register and start over.

This particular type of A/D takes n cycles, where n is the number of bits in the final result. The speed at which it can proceed is determined by the settling time of the DAC for the MSB. By making the circuit somewhat fancier, it is possible to have one clock speed used for making decisions for the first one or two bits, and then a faster clock speed for the rest of the decision tree, since all the other bits will settle faster than the MSB.

Although the logic of this particular type of A/D may seem quite complex, it is no real challenge to put it all on one IC. Thus single ICs with all the logic for successive approximation A/Ds exist. These can be combined with fast comparators, clocks and DACs to form complete analog-to-digital converters.

SINGLE-SLOPE INTEGRATING A/Ds

A schematic diagram of a **single-slope integrating A/D** is shown in Figure 36-4. A number of switches are shown in the diagram. Although these are shown as simple switches, they really are FET switches driven by the logic portion of the circuit. To keep the diagram simple, not all the connections are shown.

The operation of this circuit is as follows:

1. The capacitor is discharged and the counter is set to zero.
2. The current source starts charging the capacitor and the counter starts counting the clock pulses.
3. When $V_c = V_{in}$, the comparator changes state. This state change is detected by the logic circuit, and a number of actions occur:
   (a) The charging of the capacitor is stopped.
   (b) The counter is stopped.
   (c) The contents of the counter are transferred to the buffer to be displayed as the result.
   (d) The capacitor is discharged.
   (e) The counter is reset to zero.
4. The entire cycle starts over.

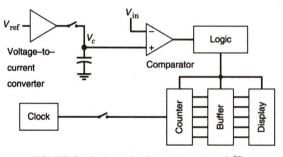

FIGURE 36-4 A single-slope integrating A/D.

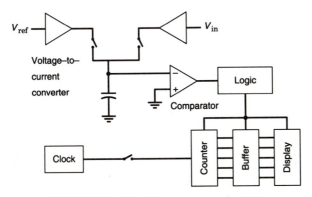

FIGURE 36-5 A dual-shape integrating A/D.

The accuracy of this type of A/D depends on the value and the stability of the value of the capacitor, the accuracy of the comparator, the accuracy of the reference voltage, and the accuracy of the time base. When the calibration is done correctly, the counter reads a number that is directly proportional to the applied voltage. This type of A/D is not very common because it is relatively hard to maintain good accuracy.

DUAL-SLOPE INTEGRATING A/Ds

A schematic of **dual-slope integrating A/D** is shown in Figure 36-5. This is similar to the single-slope integrating A/D except that there are two current sources. One is used to charge the capacitor and the other, to discharge the capacitor. The operation of this A/D is as follows:

1. The capacitor and the counter are both zeroed.
2. The capacitor is charged by a current proportional to V_{in} for a fixed time period that is some multiple of $\frac{1}{60}$ of a second. This helps suppress 60-Hz noise.
3. The capacitor is discharged by a current proportional to V_{ref}. The counter counts the time this discharge takes. With the appropriate calibration, this time is proportional to V_{in}.
4. The control logic detects when the capacitor voltage has reached zero, transfers the counter contents to the buffer, and starts the cycle over.

This type of A/D does not depend on the exact value of the capacitor or, of the offset of the comparator, since variations in these wash out because of the dual integration process. The accuracy depends on the time base, which can be made fairly stable without much trouble, and on the accuracy of the reference voltage. As a result, this type of A/D gives excellent results, and some variation on it is used in most DVMs.

Chapter 37

VOLTAGE-CONTROLLED OSCILLATORS AND FUNCTION GENERATORS

OBJECTIVES

1. Become familiar with one design of a voltage controlled oscillator.

2. Become familiar with one way of implementing a function generator.

INTRODUCTION

This chapter introduces the **voltage-controlled oscillator**, commonly abbreviated **VCO**, and the system called a function generator, which can be made from a VCO and a number of simple op-amp circuits. These circuits involve both analog and digital electronics. There are a number of different ways to implement both VCOs and function generators; only one of these is illustrated here, but this is enough to introduce the important features of these circuits.

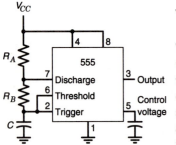

FIGURE 37-1 A fixed-frequency astable multivibrator based on the 555.

VOLTAGE-CONTROLLED OSCILLATORS

The voltage-controlled oscillator, VCO, is an oscillator whose frequency is controlled by an input control voltage. The circuit is sometimes called a **voltage-to-frequency converter**. The particular implementation discussed here is a digital oscillator; that is, its output is a square wave. However, as will be shown in the following section, there are ways to convert a square wave into a sine wave. Other designs for VCOs directly give rise to sine waves, but they are less common in this era of digital electronics.

Figure 37-1 shows the 555 timer used as an astable multivibrator; this figure is taken from Chapter 23. As described in that chapter, the voltage on the capacitor oscillates between $\frac{1}{3}V_{CC}$ and $\frac{2}{3}V_{CC}$, these limits being set by the internal voltage divider chain that feeds the inputs on the comparators. The frequency of the oscillation depends on C, R_A, and R_B and is quite stable once the components have been fixed.

318

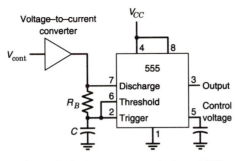

FIGURE 37-3 A variation on the basic VCO.

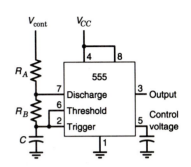

FIGURE 37-2 A voltage-controlled oscillator based on the 555.

By changing the configuration to that shown in Figure 37-2, the circuit has been converted into a voltage-controlled oscillator. The control voltage marked V_{cont} controls the rate at which the capacitor is charged; thus, varying V_{cont} varies the frequency of the oscillator. This arrangement provides a perfectly acceptable way for making a voltage-controlled oscillator.

The range over which the control voltage can be varied is limited by the fact that the comparators are connected to a voltage divider which is connected to V_{CC}. However, the input marked "control voltage" on the 555 (pin 5) provides a way by which this limit can be overcome. Use of this input allows the VCO above to be used over a much wider range of frequency—see the data sheets for the 555 in Appendix B.

There are many possible variations on this circuit. Both V_{cont} and R_A really are current sources. They can be replaced with voltage-to-current converters, as shown in Figure 37-3. For this circuit, to vary the frequency by a factor of 10 requires the control voltage to vary by a factor of 10. It is sometimes useful to have a VCO that can vary over a very wide range of frequency without the control voltage having to vary over an equally large range. A exponentiating circuit on the input means that a much smaller range of input voltage can cause a wide variation in frequency. A schematic of such a circuit is given in Figure 37-4.

Of course, at other times it is useful to have a VCO whose frequency changes very little while the input voltage changes over a wide range; this could be done with an input circuit that takes a logarithm of the input. Obviously, the list of such variations can go on almost forever.

There are many different VCOs commercially available. The 74124 shown in Figure 37-5 is one such circuit. This particular circuit has two control in-

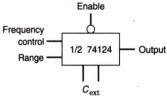

FIGURE 37-5 A commercial VCO.

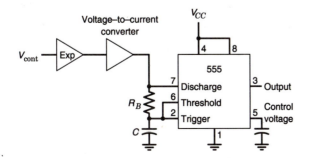

FIGURE 37-4 A VCO with a very wide range of frequency variation for a small-input voltage variation.

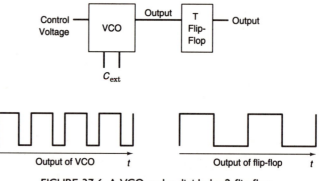

FIGURE 37-6 A VCO and a divide-by-2 flip-flop.

puts: one is indicated as frequency control, and the other as frequency range. The data sheets for this circuit give plots of the frequency versus the various control voltages as a function of the external capacitance. This circuit can also be used as a crystal-controlled oscillator (see Chapter 23).

The basic VCO described above may have unequal off and on times, and the ratio of these times changes as the frequency changes. By connecting the output to a type-T flip-flop, an output with equal off and on times is generated. However, the frequency is a factor of 2 lower than the output of the basic VCO oscillator. This is shown in Figure 37-6.

FUNCTION GENERATOR

The **function generator** is a common instrument in electronics shops. It is a signal generator that has a variable-frequency output. The output can be changed from a square wave to a triangular wave or to a sine wave. Usually, the amplitude of the output can be adjusted, sometimes it can be inverted and, on some models, a dc offset can be added to the output.

A function generator is based on a VCO and a number of op-amp circuits to shape the output. A block diagram of a function generator is given in Figure 37-7. Most of the circuits are very simple. A summing amplifier, an inverting amplifier, and a variable-gain amplifier make up the final shaping network. The voltage follower simply isolates the rest of the circuit from whatever is connected to the input.

Most function generators have a switch that controls the frequency range, changing it by factors of 10. Clearly, this switch changes the value of C in the VCO. Function generators also have a knob that changes the frequency; this is a potentiometer that changes the control voltage and hence, the frequency.

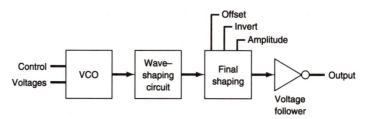

FIGURE 37-7 A block diagram of a function generator.

In general, a frequency calibration is given on the dial of the instrument. The wave-shaping circuits are more interesting and perhaps not as obvious.

The output of the VCO discussed above is already a symmetrical **square wave**. Thus, to make a square wave, nothing more is needed except the final shaping network and the voltage follower.

Integrating a square wave with an op-amp circuit gives a **triangular wave**. Figure 37-8 shows both the square wave and the triangular waveform created by integrating the square wave and giving it an offset so that its average is zero.

There are several ways to generate a **sine wave** from a square wave. None of these will generate a perfect sine wave—all will have some higher harmonics. Integrating the triangular wave gives a slightly fat sine wave. This really is a set of quadratics connected together. The graph in Figure 37-9 shows the result of integrating the triangular wave; a pure sine wave is shown as a dotted line for comparison. (Note that the scales have been expanded from the previous graphs.) The approximation is fairly good, and a simple low-pass filter would make it even better.

There are a number of wave-shaping networks that convert square waves into approximate sine waves. One of the most common networks uses seven diodes to approximate the sine wave by seven segments of exponential curves. An analysis of this circuit would be far beyond the scope of an introductory text.

The LM 566 made by National Semiconductors is a typical VCO, whereas the ICL 8038 made by Datel-Intersil is a typical voltage-controlled function generator.

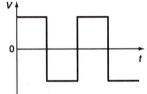

FIGURE 37-8 A triangular wave made by integrating a square wave.

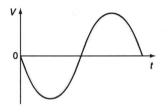

FIGURE 37-9 The integral of a triangular wave (solid line) and a true sine wave (dotted line).

SOME FINAL COMMENTS ON THE USE OF THESE CIRCUITS

The two circuits discussed above together with their many variations have many uses, some readily apparent and others less so. Obviously, the function generator circuit forms the basis of the common function generator instrument used in electronics laboratories. Almost as obvious is the fact that the VCO together with various wave-shaping circuits forms the basis for electronic organs and music synthesizers. Perhaps less obvious is the use of VCOs in electronic variable-speed motor drives or the use of VCOs and wave-shaping circuits in speech synthesis circuits. These really are very useful building-block circuits.

Chapter 38

POWER SUPPLIES

OBJECTIVES

1. Become familiar with the purpose and specifications of power supplies.

2. Become familiar with the circuits that make up a power supply, including rectifiers, filters, and regulators.

INTRODUCTION

Power supplies are probably the single most common functional unit in all of electronics; almost every piece of apparatus contains one or more of them. This chapter is a brief introduction to power supplies, their purpose and parameters, and the circuits from which they are made.

Generally, the easiest and best way of getting a power supply for a given application is to buy one. Many different manufacturers make many different types of power supplies. It is almost always possible to find one that is more or less correctly suited for the application at hand. If the power supply has been made by a reputable manufacturer and if the unit has been correctly chosen for the application, buying a supply is almost always easier, better, and often cheaper than building one. To aid in the process of selecting a power supply, the first section of this chapter discusses the purpose of power supplies and their most important specifications.

Since there are often situations when it is necessary either to build, repair, or modify a power supply, it is also necessary to understand the circuitry used in power supplies. The later sections of this chapter describe the various circuits that are used to make up power supplies. A simple, complete design is included at the end of the chapter for reference.

THE PURPOSE AND SPECIFICATIONS OF POWER SUPPLIES

The purpose of a dc power supply is to provide a unidirectional voltage and current. Ideally, the power supply should act like a perfect voltage source. That is, the output voltage should be constant, no matter how much current

322

is drawn by the load; the voltage should not change as the ac line voltage varies, as the temperature varies, or as time passes. Furthermore, the output should have no ac component; that is, there should be no ripple in the output.

The basic specifications for a power supply should be fairly clear: what voltage and how much current it puts out, how well it is regulated and so on. Unfortunately, there are no universally accepted definitions for these specifications. Each manufacturer of power supplies uses slightly different definitions, which makes it difficult to compare a power supply from one manufacturer with one from another manufacturer. The definitions of the various specifications given here are the most commonly used ones. In any case, the following list of specifications need to be considered in building or selecting any power supply.

Voltage, Current, and Power Ratings

These specifications are more or less self-evident. What voltages does the supply put out, and what are the maximum currents that can be drawn? For supplies with multiple voltage outputs, the power rating is often an overall power rating; that is, not all the voltages can supply the maximum currents at the same time, but rather some maximum amount of power can be drawn from the unit. For example, a unit might be rated at +5 V at 20 A, +15 V at 5 A, and −15 V at 1 A and have an overall power rating of 120 W. If 20 A of the +5 V are being used, there are only 20 additional watts of +15 and −15-V power available before the power rating is exceeded. This means about 1.4 A total can be drawn from the +15 and −15 V lines. (The supply may well put out more power than it is rated for, but this will result in reduced life expectancy of the supply.)

Percent Load Regulation

This is a measure of how much the output voltage of the supply varies as the current drawn from the load varies. The usual definition is

$$\% \text{ load regulation} = 100\% \times \frac{V_{\text{no load}} - V_{\text{full load}}}{V_{\text{no load}}} \qquad (38\text{-}1)$$

With modern techniques, load regulations of 0.1% or better may be easily obtained. However, such good regulation is not needed for all purposes—for example, TTL logic can tolerate voltage changes of almost 10% with no ill effects.

Percent Line Regulation

This parameter tells how much the output voltage will change as a result of a change in the input 60-Hz ac voltage. There is no universally accepted definition of this parameter, but one possibility is as follows:

$$\% \text{ line regulation} = \frac{\% \text{ voltage change}}{1 \text{ V line change}} \qquad (38\text{-}2)$$

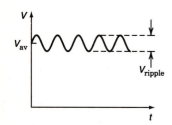

FIGURE 38-1 A dc voltage with a small ac ripple.

Ripple

An ideal dc power supply produces a pure dc output voltage with no ac component. Figure 38-1 shows a typical output with a small amount of ac **ripple** superimposed on the dc voltage.

Obviously, the less ripple, the better the power supply. The **ripple percentage** is often defined as

$$\% \text{ ripple} = 100\% \times \frac{V_{\text{ripple}}}{V_{\text{dc}}} \tag{38-3}$$

The ripple voltage can be specified as the peak-to-peak ripple voltage, or as the rms value. Since the rms value is always smaller than the peak-to-peak value, ripple specifications that refer to rms ripple values always look better than those that refer to peak-to-peak values. The ripple percentage is also a function of the load current. Generally, a supply will have a worse ripple percentage as more current is drawn from the supply. This gives the manufacturers another way to disguise this specification.

Drift

The output voltage of a power supply will change slowly with time. This is referred to as **drift**. Obviously, the less the drift, the better.

Thermal Regulation

The output voltage of a power supply will change with changes in the temperature of the circuit. Specifications giving the percentage change in the output voltage for a 1°C temperature rise are often given. An important fact to remember is that the temperature of the circuit may rise by 40°C or more as it warms up from the off temperature to normal operating temperature.

Thermal Derating

A power supply can put out more power when it is cool than when it is hot. In fact, power limitations on power supplies are usually set by the maximum temperature allowed before some component fails. Usually, some sort of specification or graph showing how the maximum power available for the supply decreases with temperature is available.

MTBF

This is the **mean time between failure**. When specified, these times are usually very long (typically years or more). But as any modern computer user knows, power supply failures are perhaps the most common cause of electronic malfunction today. Anyone who is manufacturing products for resale worries very much about the cost of repairs due to failure of components. For them, this is a very important specification.

Cost

The meaning of this specification is clear. Although it often appears as though buying a single power supply is more expensive than building one, experience shows that this is often false economy. Nevertheless, for many reasons, building power supplies is a very common activity in the laboratory.

RECTIFIER CIRCUITS

The remainder of this chapter is devoted to discussions of the detailed circuitry that is used in the typical power supply. The normal power supply converts 60 Hz ac power (50 Hz in some countries) into various dc voltages. There are three main sections to a typical power supply: (1) the transformer/rectifier section, which converts the ac into a pulsating dc voltage of approximately the right voltage; (2) the filter section, which smooths out the massive ripple that comes from the rectifier; and (3) perhaps a regulation section, which serves to keep the output voltage constant no matter what else changes. Each of these circuits is discussed below. This section deals with the transformer and rectifier portion of the circuit.

Transformers were discussed in Chapter 11. The transformer in a power supply converts the input ac voltage into suitable ac voltages for rectification and filtering. For low-power use in the United States, the input voltage is almost universally 60 Hz ac in the range of 110 to 120 V rms. The secondary winding of the transformer may or may not have a center tap. Furthermore, for power supplies that produced multiple voltages, there may be several independent secondary windings on the transformer. For the following, only a single-output power supply is discussed. Multiple-output supplies are made by simply repeating these circuits for each output voltage.

Diodes conduct current only one way. Thus diodes can be used to convert ac into unidirectional current—pulsating dc current. Diodes used in this mode are traditionally called **rectifiers** because they rectify the ac into dc. All the examples in this chapter are drawn for the production of positive voltages. To make a power supply to produce negative voltages, simply reverse all the diodes.

The schematic symbol for a solid state diode is given in Figure 38-2. The symbol is really an arrow indicating which way conventional current can flow through the diode. Thus current flows when the triangular end of the symbol is positive with respect to the bar, which is indicated in the figure. Diodes are discussed in more detail in Chapter 41.

The simplest rectifier circuit is the **half-wave rectifier** shown in Figure 38-3. The diode conducts during the half-cycle when the upper terminal of the transformer is positive with respect to the lower terminal and does not conduct during the other half-cycle. The circuit gets its name from the fact that only one half of the input sine wave survives to the output. For this circuit with ideal transformers and diodes, it is possible to show that

$$V_{peak} = 1.4V_{rms}$$

$$V_{av\ dc} = 0.45V_{rms}$$

where V_{peak} and $V_{av\ dc}$ are respectively the peak value and the average value of the output voltage, and V_{rms} is the rms value of the ac voltage at the secondary of the transformer. The advantage of this circuit is its simplicity. Its main disadvantage is that it is very hard to filter the output voltage because of the gaps in the voltage. In addition, the diode and the transformer are

Conventional current

FIGURE 38-2 The schematic symbol and allowed current direction for a diode.

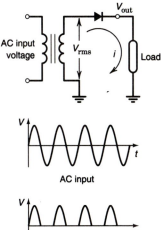

FIGURE 38-3 A half-wave rectifier circuit and the resultant waveforms.

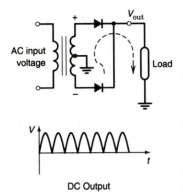

FIGURE 38-4 A full-wave rectifier and the output voltage waveform.

only conducting half the time, so the heating effects (which vary as I^2) are greater than they need to be for a given output current. This circuit is rarely used in modern equipment.

A more complex and useful rectifier circuit is the **full-wave rectifier** shown in Figure 38-4. This circuit requires two diodes and a center tap in the transformer. During one half-cycle, the upper half of the transformer and the upper diode are conducting; the polarities are indicated, and the current path during this half-cycle is shown as a dotted line. During the other half-cycle, the polarities are reversed, and the lower half of the transformer and the lower diode are conducting. Because of the way the diodes are arranged, the current always flows the same way in the load. The resultant voltage at the load is shown in the figure. The circuit gets its name from the fact that the full input cycle is present in the output—although half of it has been inverted.

Again, for an ideal transformer and ideal diodes, it can be shown that

$$V_{\text{peak}} = 0.7 V_{\text{rms}}$$

$$V_{\text{av dc}} = 0.45 V_{\text{rms}}$$

where V_{peak} is the voltage across the entire transformer secondary. This circuit produces an output that is easier to filter than that produced by the half-wave rectifier. However, it requires two diodes and a center tap on the transformer. Furthermore, only half the transformer is being used at any time. Thus the ratio of peak current to average current in the transformer is still unnecessarily high, and extra heating is taking place.

The **full-wave bridge rectifier** circuit is shown in Figure 38-5. This circuit requires four diodes but no center tap on the transformer. During any half-cycle, the entire transformer is conducting, but only two of the diodes are conducting. The current flow at the indicated polarities is shown for one half-cycle as a dotted line. During the other half-cycle, the polarities are reversed, and the other pair of diodes are conducting. The arrangement of the diodes is such that the current is always flowing in the same direction in the external circuit. The circuit gets its name from the fact that it uses the full sine wave and from the similarity of the circuit to the Wheatstone bridge.

For the ideal transformer and diodes, it can be shown that, for the full wave bridge,

$$V_{\text{peak}} = 1.4 V_{\text{rms}}$$

$$V_{\text{av dc}} = 0.9 V_{\text{rms}}$$

This circuit uses the full transformer on each half-cycle, so it is most efficient in the use of the transformer but it does use four diodes. Since reasonable

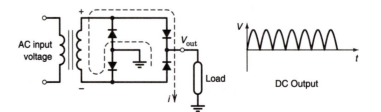

FIGURE 38-5 A full-wave bridge rectifier and the resultant output waveform.

rectifier diodes cost less than $1.00 today and transformers cost $10 and up, this is the preferred circuit at this time. In fact, so-called **bridge rectifiers** are commercially available; they are four diodes packaged as one unit, connected as shown above with only four external leads. They simplify the construction of full-wave bridge rectifiers to the point where there is no real excuse for not using them all the time.

FILTERS

The purpose of the **filter** is to smooth out the bumps in the voltage coming out of the rectifier circuit to make it into smooth dc voltage. The simplest possible filter and one that is part of every other filtering circuit is a single capacitor connected to the output of the rectifier circuit. The circuit and some voltage and current waveforms are shown in Figure 38-6. The operation of this filter can be understood as follows.

1. Because of the nature of the diode, current flows at point B only when the voltage across the transformer is greater than the voltage across the capacitor. During this time, the capacitor is charging. When the voltage across the capacitor is greater than the voltage across the transformer, the diodes prevent current flowing in the other direction.

2. While current is flowing, the capacitor stores charge. During this time, the voltage on the capacitor tends to follow the voltage across the transformer. Thus current flows only during the part of the ac cycle when the voltage across the transformer is increasing, that is, on the rising portions of the cycle. Once the voltage across the transformer starts to decrease, the diodes block the current from flowing at point B. This gives rise to the asymmetrical current waveform.

3. When the current is not flowing at point B, current is still flowing at point E and the voltage on the capacitor is decaying away like any other RC circuit. In other words, during the charging portions of the cycle the voltage on the capacitor rises like a sine wave, while during the non-charging portions of the cycle it decays away exponentially. This gives rise to the voltage waveform at the output.

4. The peak current at point B is limited by the resistance of the transformer and the diodes. As a result, the diodes must be able to handle large amounts of peak currents for short periods of time (a few milliseconds).

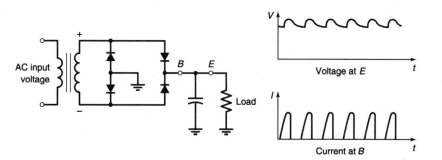

FIGURE 38-6 The simplest filter attached to a bridge rectifier and the key voltage and current waveforms.

The above gives all the details about the operation of this filter. However, the operation can be thought of in two conceptually elegant ways. First, the capacitor stores charge during parts of the cycle and delivers the charge to the load during other parts of the cycle. Second, the circuit acts as a low-pass filter or as a current divider for the ac ripple current—the ripple flows through the capacitor and the dc through the load.

The main problems with this filter are that it does not do a very good job of getting rid of the ripple—it has very poor load regulation and no line voltage regulation at all. Hence, it is not a very good filter/regulator. However, no matter what other kinds of filters and/or regulators are added, the power supply must begin with this circuit to provide the necessary charge storage during those parts of the cycle when the diode is not conducting. A method of picking the size of the capacitor is discussed in the last section of this chapter.

Finally, the portions of the power supply already discussed up to this point contain the parts of the circuit that are most likely to fail. Because of the heating effects and the high peak currents, transformers and rectifiers are prone to fail in otherwise apparently normal circumstances. Likewise owing to the heating effects of the ripple current flowing through the capacitor, the capacitor heats up and is also prone to fail.

A somewhat better filter is shown in Figure 38-7. In this and in all subsequent figures, the bridge rectifier is not shown, but only the circuit to the right of point B in Figure 38-6 is shown. This is simply to prevent unnecessary information from making the figures hard to understand.

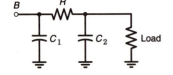

FIGURE 38-7 An RC Π section filter.

This circuit is called a π **section filter** because of the shape of the circuit diagram. Several of these sections can be stacked up, one after the other. The filter may be thought of as one low-pass filter followed by another. This type of filter gives better ripple suppression than the simplest filter. However, it still does not have good load regulation and no line voltage regulation. It was frequently used in the past, at least in vacuum tube days, and as a result there are many design curves for this type of regulator in the electronic handbooks.

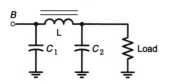

FIGURE 38-8 An LC Π section filter.

An even better filter is shown in Figure 38-8; this is the LC π **section filter.** This filter is made from two capacitors and an inductor rather than a resistor. Replacing the resistor with the inductor improves the load regulation (since the voltage drop across the resistor is no longer present) and improves the ripple suppression. But this filter has no line voltage regulation, and inductors are big, heavy, and expensive. However, for many years, this kind of filter was all that was used in high-fidelity circuits and in other high-quality electronics. Since the regulators to be described below are smaller, lighter, and cheaper and do a better job, this style of filter is no longer used in new equipment.

ZENER DIODE REGULATORS

A simple **Zener diode voltage regulator** circuit is shown in Figure 38-9 along with the voltage–current relationship for an ideal Zener diode. The Zener diode has a very sharp turn-on voltage in the reverse direction. This turn-on characteristic is used in the regulator. The regulator circuit is drawn for positive voltages. The Zener diode is installed so that it is conducting in the

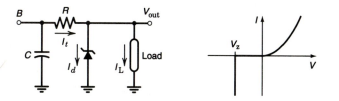

FIGURE 38-9 A Zener diode regulator and the *V–I* graph for an ideal Zener diode.

reverse direction. The operation of this filter/regulator circuit can be understood by considering the *V–I* graph:

1. The capacitor *C* stores charge during the dips in the ac cycle. Neither the bridge rectifier nor the Zener diode can provide power during these dips; it must be stored in the capacitor. If the storage capacity of the capacitor is exceeded, there will be dips in the output of the regulator.
2. For a perfect Zener diode and a fixed voltage at point *B*, as long as any current is flowing through the Zener diode, the output voltage across the load will be V_Z. Thus the load current i_L can vary from 0 to i_t without change in V_{out}, but i_d will change to compensate for these changes in the load current.
3. For a perfect Zener diode and a fixed value of load current i_L, as V_B varies, i_d changes so that the voltage drop across *R* changes to keep V_{out} constantly equal to V_Z.

The Zener regulator gives good ripple suppression and good line and load regulation. Zener diodes are available for voltages ranging from about 3 to 200 V, with power ratings ranging from $\frac{1}{2}$ to 1000 W or more. (Remember that the power dissipated in the Zener diode is the current through the diode times the voltage across the diode. Since the voltage may be large, the power can easily be a very large number.)

There are two major problems with Zener diode regulators. First, the Zener voltage drifts with temperature. Typical thermal drifts are a few parts per million per °C which, although small, are often large enough to be troublesome in precision circuits. It is possible to make a thermally stable reference diode that has essentially no thermal drift, but this can be done for 7 V only. Second, the Zener voltage does vary with the current through the diode. The ideal Zener has an effective dynamic resistance of 0 Ω once the reverse current has begun; real diodes have a dynamic resistance of a few ohms. Thus the load regulation is not perfect. The example below shows the magnitude of this effect.

A fairly complete design example will illustrate the problems and considerations in constructing a Zener diode voltage regulator. For this example, it is assumed that the voltage out of the bridge rectifier is 20 V, that the output of the power supply is to be 6.8 V, and that the load current will vary from 0 to 40 mA.

The 1N4736A Zener diode is a 1-W diode with a Zener voltage of 6.8 V; it is suitable for this example. The thermal drift of this diode is

$$T_C = 5 \times 10^{-4}/°C = 0.05\%/°C$$

and its dynamic resistance is

$$Z_d = 3.5 \ \Omega \text{ at } 37 \text{ mA}$$

Having picked the diode, the next step in designing the regulator is to pick the total current that will flow through the resistor. This current must be greater than the largest value the load current will ever have; for this case, the value of 45 mA is chosen. This means that when the load current is zero, there will be a total of 45 mA going through the Zener; when the load current is 40 mA, 5 mA will be going through the Zener. Remember that the Zener diode current can never drop to zero or else regulation will fail. (Note that there is nothing special about the 5-mA value of the minimum current through the diode other than it is clearly greater than zero and yet not excessively large.)

The value of the resistor can then be computed from Ohm's law:

$$R = \frac{V_A - V_{\text{out}}}{I_t} = \frac{20 - 6.8}{0.045} = 293 \ \Omega$$

The power dissipated in the resistor is

$$P_R = 13.2 \times 0.045 = 0.6 \text{ W}$$

Using a 2-W resistor would be a fairly conservative choice. The power dissipated in the diode is

$$P_d = 6.8 \times 0.045 = 0.306 \text{ W}$$

which is safely less than 1 W.

Finally, the variation in the output voltage as the load current changes from 0 to 40 mA can be calculated. During the change, the current through the diode changes by 40 mA, so that the change in the voltage is given by

$$\Delta V_Z = \Delta i_d \times Z_d = 0.040 \times 3.5 = 0.14 \text{ V}$$

and the load regulation is given by

$$\frac{\Delta V}{V} = \frac{0.14}{6.8} = 2\%$$

Thus this simple regulator has a 2% load regulation, good line regulation, good ripple suppression, and only a moderate thermal drift. While this would have been considered very good a few years ago, much better results can now be obtained with active regulators.

ACTIVE REGULATORS

A conceptual design of an **active regulator** is shown in Figure 38-10. The operation of this regulator is quite simple in principle. The capacitor stores the

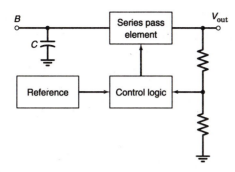

FIGURE 38-10 A schematic outline of an active regulator.

charge from pulse to pulse just as it does for any other regulator. The control logic compares V_{out} (or some fraction of it) to a reference voltage and adjusts the series pass element to keep V_{out} constant. Thus if V_{out} starts to decrease, the series pass element is adjusted to compensate for the decrease. This type of regulator gets its name from the fact that the series pass element is continually adjusted; thus, **active** efforts are made to keep the output voltage at the preset value.

A man or a woman, a variable resistor, and a meter can be combined to make a simple active regulator. The variable resistor is the series pass element, and the person reads the meter that is measuring V_{out}. If V_{out} starts to decrease, the person decreases the resistance of the series pass element to compensate; if V_{out} starts to increase, he or she increases the resistance of the series pass element. In this way, the series pass element is continually adjusted so that the output voltage remains constant.

An op-amp circuit can compare V_{out} to a thermally stable reference voltage. A transistor voltage follower circuit can be used as the series pass element (see Chapter 43). The resulting circuit of a simple active regulator is given in Figure 38-11. The analysis of this circuit is simple; its operation will be explained by examining the sequence of events that occurs if the output voltage starts to change from its correct value:

1. Assume that, for some reason, V_{out} starts to decrease.
2. This means that voltage at the inverting input of the op-amp decreases.
3. Because of this, the output of the op-amp increases and V_b increases.

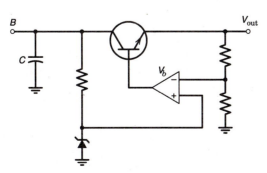

FIGURE 38-11 An electronic active regulator.

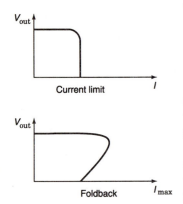

FIGURE 38-12 Current limiting and foldback current limiting.

4. This means that the forward bias on the transistor is increased, the current increases, and V_{out} must increase. The increase in V_{out} tends to cancel the original decrease in V_{out} which started this analysis series. Thus this circuit actively adjusts the output to keep V_{out} constant.

There are many refinements that can be easily added to this circuit. Four of these are discussed below. First, a thermally stable reference diode can be used in a circuit, which ensures that a constant current passes through it, no matter what either the output or input voltage of the circuit is.

Second, **current limiting** can be added so that the output voltage will be shut down if the output current exceeds some preset value. This will protect the series pass element and the rest of the circuit from damage if the output is shorted to ground. A graph of a current limited voltage output is given in Figure 38-12.

Third, a feature called **foldback** can be added to the circuit. This is essentially a more subtle version of current limiting. Not only must the maximum current through the circuit be limited but also the maximum power dissipation in the circuit must be limited. The maximum current depends on the input voltage; for a fixed input voltage, the maximum current will be less the lower the output voltage. The foldback circuit basically takes all this into consideration and computes the maximum output current permitted for each combination of input and output voltages. The shape of the typical graph of foldback current limiting given in Figure 38-12 shows where the name of the feature comes from.

Finally, a **thermal limiting** circuit can be added. This circuit monitors the temperature in the rest of the circuit and adjusts the maximum current allowed so that nothing burns out.

All of these functions can be combined into one 14-pin integrated circuit to give a voltage regulator circuit. The 723 is such a circuit; it is similar to that shown in Figure 38-11 with all of the features discussed above added to the circuit. This circuit gives load and line regulation values of the order of 0.01% with no further effort required.

The 723 regulator requires an external potentiometer to set the output voltage. This is extremely useful when a variable regulator is required. However, normal use is for regulators that give 5, 12, or 15 V. For these cases, a series of three-terminal regulators have been manufactured. These have part numbers of the form 78xx or 79xx. A 7812, for instance, is a +12-V voltage regulator; a 7905 is a −5-V regulator. These have all the components needed to make excellent voltage regulators. The simple resultant circuit is shown in Figure 38-13. The three-terminal regulators are made to handle large currents, up to 1 A. This means that they may dissipate a large amount of power, which may require the use of a heat sink to improve the cooling of the device.

Another important type of regulator is the so-called **tracking regulator**. This is a regulator that has both polarities of inputs and outputs. The outputs are V_+ and V_-, and the regulator ensures that

$$V_+ + V_- = 0$$

For instance, if the V_+ output is shorted to ground, then both outputs go to zero. Such a regulator is very useful for powering those op-amp circuits that require symmetrical positive and negative voltage supplies.

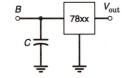

FIGURE 38-13 The use of a three-terminal regulator.

The main disadvantage of the regulators described here is the fact that substantial amounts of power may be dissipated in the series pass elements. This can be a problem in high-current supplies like those used in computers. A new type of regulator called the **switching regulator** has been devised to answer this problem. These regulators start with unregulated dc voltages, and use variable-frequency square wave oscillators, transformers, and simple capacitive filters to produce the output. The regulation takes place by varying the frequency to ensure that the output voltage remains a constant. The design of these regulators is very difficult, and errors can lead to outputs that are totally unusable because of the amount of high-frequency noise in the output. Presentation of their design and operation is beyond the scope of an introduction.

A COMPLETE EXAMPLE

Figure 38-14 shows a complete design for a simple power supply that will provide 250 mA at 5.0 V. A few brief comments are in order about the choice of each of the components that are used in this design. These comments will serve as a guide to most of the questions that must be answered in designing and building such supplies.

The transformer for this supply has a 110-V primary and a 12-V secondary, with a center tap and a secondary current rating of, at least, 1 A. Such transformers are readily available at all electronics parts suppliers.

A bridge rectifier is used. For this configuration, the peak voltage developed across the capacitor will be about 16.8 V. A 7805 is chosen as the regulator. To ensure that the 7805 regulator will operate correctly, there must be, at least, 8 V at its input. The filter capacitor must be chosen to ensure that this is true. The value of the capacitor must be sufficiently large to store enough charge to allow the regulator to operate during the portion of the ac cycle when the bridge rectifier is not conducting. The easiest way to ensure this is to assume that the capacitor must supply the load current for a whole half-cycle. At 60 Ha, this is 8 msec. Thus the charge needed is

$$Q = i_L \times 8 \text{ msec} = \Delta VC \tag{38-4}$$

where ΔV is the amount the voltage on C can decrease from its peak value and still be large enough for the regulator to operate. ΔV is roughly the dif-

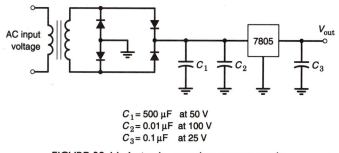

$C_1 = 500 \, \mu\text{F}$ at 50 V
$C_2 = 0.01 \, \mu\text{F}$ at 100 V
$C_3 = 0.1 \, \mu\text{F}$ at 25 V

FIGURE 38-14 A simple, complete power supply.

ference between the peak voltage from the rectifier and the minimum voltage required to operate the regulator for the desired output voltage.

For this example, the maximum voltage drop acceptable is about 8.8 V. To be conservative, a value of 6 V will be used in the calculation. The capacity needed is found by using Equation 38-4:

$$C = \frac{i_L \times 8 \text{ msec}}{\Delta V} = \frac{0.250 \times 8 \times 10^{-3}}{6}$$

$$= 3.3 \times 10^{-4}$$

$$= 330 \ \mu F$$

For a conservative design, a capacitor having a capacitance of 500 to 1000 μF would be reasonable. It must have a voltage rating of, at least, 20 V; 50 V would be more conservative.

The remaining components require only short comments. C_2 is a small ceramic disc capacitor that provides a low impedance route to ground for any high-frequency noise on the power line; it is required because the large capacitor C_1 does not have good high-frequency properties. C_3 is a 0.1-μF, 25-V capacitor that improves the high-frequency response of the 7805. If long leads are required between the power supply and the place where the power is used, additional capacitors should be located near the use of the power.

The 7805 will be dissipating about 3 W; as a result, it will need a heat sink. In addition, it will be very hot while in operation—allowance for adequate ventilation should be provided. The bridge rectifier should have a current rating of 1 A, a peak current rating of at least 10 A, and a peak inverse voltage rating of at least two times the peak voltage—50 V would be a conservative number.

Finally, as mentioned above, power supplies are one of the components that fail most often in electronic equipment; hence, conservative design easily justifies the slight extra costs.

Chapter 39

A COMPUTER INTERFACE FOR AN A-TO-D CONVERTER

OBJECTIVES

1. Become familiar with a simple computer interface.

2. Develop more insight into the function of the external data bus of a computer system.

3. Become more familiar with the use of many of the digital circuits already discussed.

INTRODUCTION

This chapter presents a fairly complete example of the design of an eight-bit analog-to-digital converter system interfaced to a computer. This example will provide further insight into the meaning and use of the various signal lines in the external data bus of the computer (Chapter 27), into the use of an analog-to-digital converter (Chapter 36), and into the use of many of the digital circuits discussed in Part Three of this book. In this chapter, many of the problems of digital design will be encountered and solved.

The design process for such a device proceeds in a number of steps. First, an overall description of what is to be done must be obtained, that is, what computer system, how accurate and fast an A-to-D, and so on. Second, the details of the external bus system for the computer must be investigated. Third, the details of the operation of the A-to-D must be investigated. Fourth, the design of the system must be worked out ensuring that all level, sign, and other conventions have been taken into account. Finally, the system must be built, tested, and made to work. This chapter will follow the outline sketched above ending when the design has been completed.

OVERALL DESIGN

Normally, the first step of the design process begins with a consideration of what is to be done. For a computerized A-to-D, this would involve considering what is to be measured by the A-to-D, what accuracy, speed, and other properties are needed, what computers are available, how much money is available, and so forth. In the case presented here, the design process starts after the A-to-D and the computer have been chosen.

An eight-bit A-to-D is to be interfaced to the eight-bit computer described below. This A-to-D system is to respond to three computer commands: a START CONVERSION command, which causes the converter to start converting its analog input signal; a TEST CONVERTER command, which tests if the converter has completed the conversion; and a READ CONVERTER command, which reads the results of the conversion into the computer.

THE EXTERNAL DATA BUS

For the computer being considered, there are 36 signals in the external data bus. Of these, 28 might be used here. They will be described below. Some further comments about features of these signals are as follows:

A0-A15. A 16-bit address bus. The address bus bits are active when High. The entire 16 bits are used when addressing memory, but only the low-order byte is used for addressing input/output (I/O) ports. With 16 bits, this computer can address 64k (65,536) bytes of memory. By using eight-bits to address I/O devices, 256 devices can be addressed.

D0-D7. An eight-bit data bus. This bus is active when High and tristate outputs are to be used to drive it.

$\overline{\text{MREQ}}$. The memory request signal. A Low on this line indicates that memory is being addressed.

$\overline{\text{IORQ}}$. Input/Output request. A Low on this line indicates that the input/output system is being addressed.

$\overline{\text{RD}}$. A Low on this line indicates that the CPU is reading data from the memory or I/O device being addressed.

$\overline{\text{WR}}$. A Low on this line indicates that data are being written from the CPU to the memory or I/O device being addressed.

The eight other signals on this external data bus have no function in the A-to-D system being designed here. In the address and data buses, the 0th bit is the least significant bit, LSB, whereas the highest-numbered line is the most significant bit. This and the fact that several of the lines are active when Low must be kept in mind during the design process.

As many computers do, this computer uses an addressing system for its input/output devices. This means that it has one set of I/O commands that is used for all I/O devices; which device responds to a particular command depends on the address used in the command. This is similar to the way the computer treats its memory. The computer has only a few commands that refer to memory; which particular location responds is determined by the address used in the command.

Some computers treat memory and I/O devices as parts of the same **address space**; sections of the address space are reserved for I/O devices, and sections are reserved for memory. This particular computer uses separate address spaces for its I/O devices and for its memory. Which space any particular address refers to is determined by the status of the $\overline{\text{MREQ}}$ and $\overline{\text{IORQ}}$ lines. The A-to-D system being designed here must identify its own address and also test that it is responding to an I/O request and not a memory request.

Since only eight bits of the address line are used for I/O requests, this computer can only address 256 different I/O devices. But, as will be seen, it can really address even fewer devices, since many devices will require several different addresses.

This computer has two instructions that can be used with I/O devices; these are "Read from I/O address *n*" and "Write to I/O address *n*" where *n* specifies a location in memory. Three different instructions are needed to operate the A-to-D system, which can be created from these two basic instructions by using different addresses. A Write I/O instruction with the appropriate address will be used to start the A-to-D conversion; this is the START CONVERSION instruction.

Two different Read I/O instructions with different addresses will be used. The first read instruction will transfer the output of the A-to-D to the CPU on the data lines. This is the READ CONVERTER instruction. The second Read I/O instruction will transfer the status of the converter to the CPU on the LSB of the data lines. The convention used will be that a result of 0 means that the A-to-D is not done converting, and a result of 1 means that the A-to-D is done converting and the result is ready to be read. In this way, this Read I/O instruction can be combined with a skip or jump instruction to provide the TEST CONVERTER instruction.

The information and conventions presented here are enough to allow the design of the interface. Before getting to the actual circuit, a little information about the A-to-D converter is needed. Finally, for anyone who is interested, the computer being described here is based on the Z80, one of the early, very popular eight-bit microprocessors.

THE ANALOG-TO-DIGITAL CONVERTER

For this particular application, the Analog Devices AD570 will be used. This is a low-cost, complete, integrated eight-bit A-to-D converter (see Appendix B for its data sheets). It is a successive approximations-type converter that takes 25 μsec for a conversion. For the purposes of this interface, there are 12 signal lines of interest.

Bit 1 to Bit 8. The output data bits. These are TTL-compatible outputs. The outputs are really tristate devices, but this will not be useful in this design. The only complexity is that bit 1 is the MSB, and bit 8 is the LSB.

B&$\overline{\text{C}}$. The blank and convert input. The trailing edge of a positive going pulse of greater than 2 μsec at this input turns off the output data bits and starts the conversion process.

$\overline{\text{DR}}$. The data ready line. This line is High when the output data lines are off, and Low when the output data are valid and ready and the output lines are turned on.

Analog In. The analog input line. This is where the input signal is connected.

Analog Common. The analog input signal can be referenced to a ground level which is not the same as the digital ground. This particular A-to-D can function with a difference as great as 200 mV.

Clearly, the output data bits will be the information transferred to the CPU by the READ CONVERTER instruction. The state of the $\overline{\text{DR}}$ line will be the information transferred by the TEST CONVERTER.

At some stage in the design and building process, provisions for power and ground for every IC must be made. No mention will be made here of the power supply connections for the A-to-D or any of the other chips used in this design. This A-to-D requires both a positive and a negative power supply voltages, as well as two grounds: digital and analog.

This particular A-to-D can handle either unipolar voltages in the range of 0 to +10 V, or bipolar voltages in the range of −5 V to +5 V. In this design, it is assumed that the analog input voltage has the correct sign and range to work with this A-to-D. In addition, it is assumed that any necessary signal conditioning has already been done before the signal is sent to this A-to-D. Computers are very noisy environments; the high-speed digital switching taking place radiates a great deal of noise to any nearby circuit—try taking a cheap AM radio near an operating computer. As a result, it is good practice to do any low-level analog signal amplification and conditioning far away from the computer and then bring only a high-level analog signal to the converter.

ADDRESS DECODING

This computer can address 256 different I/O devices. Each device attached to the computer must have a unique address; otherwise two or more devices might respond to the same I/O command and nonsense results would occur. Since there is no way of telling in advance which of the I/O addresses are already in use in a particular computer system, it is usual to build interfaces with some way of adjusting the address to which it will respond. This will be done here.

Two consecutive addresses will be used: an even address and the next consecutive odd address above this. The even address will be used for the START CONVERSION and for the TEST CONVERTER instructions. The odd address will be used for the READ CONVERTER instruction. Address bit A0 determines if the address is even or odd. To allow the rest of the address to be changed, seven switches are used to set the required values of the address bits A1 to A7. The values set by these switches will be compared with values on the external data bus lines by eight-bit magnitude comparator. Figure 39-1 shows this portion of the circuit. Some detailed comments about the circuit follow.

The 74LS682 is an 8-bit magnitude comparator. It compares two 8-bit numbers, P0 to P7, and Q0 to Q7. The output $\overline{\text{P} = \text{Q}}$ is Low if the two numbers are equal. To compare two 7-bit numbers requires connecting two of the inputs either to a High or to a Low; a Low is used here. The output of this unit is inverted and the resulting signal is called **equal**. This will occur whenever the lower 8 bits of the address lines are specifying the address to which this

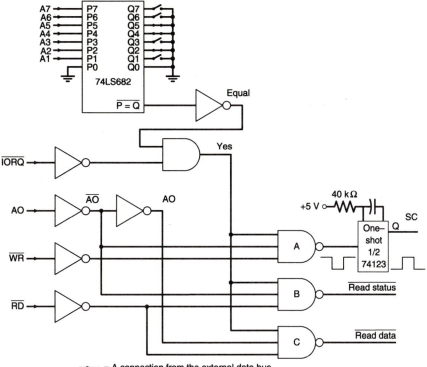

FIGURE 39-1 The address decoding logic for the computer interface. The switches are shown set so that the device will respond to Octal addresses of 312 and 313.

device might respond. The address to which this particular interface will respond is set by the switches shown in Figure 39-1.

The operation of the switches to set the address requires a few words of explanation. Each of the Q inputs of the 74LS682 has an internal 20-kΩ pull-up resistor. Figure 39-2 shows the effective configuration of this pull-up resistor and switch on the input. When the switch is open, the input is High; when it is closed, the input is Low and 0.05 mA is flowing to ground. This provides reliable noise-free operation of the gate. Arrays of small switches are available in DIP packages for mounting on boards that are designed for IC use. Thus the array of seven switches required on this circuit would all be mounted in one DIP package and would be similar to the **digi-switches**, which are found on almost all circuit boards for PCs and on most printers designed for use in PCs. In Figure 39-1, the switches are shown set so that the interface will respond to the addresses of 312 and 313 in octal, or 0CA and 0CB in Hex.

The next portion of the address decoding logic combines the Equal signal and the $\overline{\text{IORQ}}$ signal to see if this is an I/O instruction. If it is, the **Yes** line is True, indicating that an I/O instruction with the address of this unit is occurring.

The Yes line is combined with the status of the LSB address line, A0, and the $\overline{\text{RD}}$ and $\overline{\text{WR}}$ lines to determine which command is being executed by the CPU. The three-input NAND indicated as A in Figure 39-1 will have a Low output for a Write I/O instruction with an even address intended for this device. This is the START CONVERSION command. The output of the

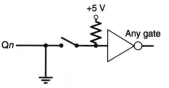

FIGURE 39-2 The equivalent circuit of a Q input of the 74LS682 as it is being used in this interface.

NAND will be a Low for this case; this is connected to a one-shot so that the leading edge of the output starts the one-shot. The timing components are chosen to generate a pulse 3 μsec long. This positive going pulse, called SC, is connected to the B & $\overline{C}$ input of the A-to-D and starts the conversion process.

The three-input NAND indicated as B in Figure 39-1 will have a Low output for a Read I/O instruction with an even address that is intended for this unit. This is the TEST CONVERTER instruction. This signal is given the name $\overline{\text{Read Status}}$. A NAND and, hence, a Low going output is chosen to match the bus drivers used to gate the data onto the data lines of the external data bus (see below).

The three-input NAND indicated as C in Figure 39-1 will have a Low output for a Read I/O instruction with the odd address that is intended for this unit. This is the READ CONVERTER instruction. The signal is given the name $\overline{\text{Read Data}}$. Again, a negative going signal is chosen because of the gating conventions of the bus drivers discussed below.

Finally, an extra inverter is used on the A0 input line so that the interface will present only one load unit to the external data bus.

OUTPUT SECTION

Little needs to be said about the A-to-D converter itself. The converter is entirely self-contained and requires no external components other than power supplies. The one-shot in the address decoding section generates the pulse required to start the conversion cycle.

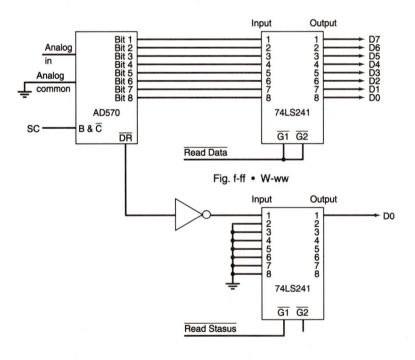

→ = A connection to the external data bus

FIGURE 39-3 The output section of the computer interface.

The data lines from the A-to-D are gated onto the lines D0 to D7 of the external data bus by the Read Converter instruction. Because B8 is the LSB of output data from the A-to-D, it must be connected to line D0, and so on. This is shown in Figure 39-3. A 74LS241 Octal Buffer/Line Driver with three-state outputs is used as the bus driver. This unit has two enable lines, $\overline{G1}$ and $\overline{G2}$, each of which enables four of the outputs. The outputs are enabled when the enable signals are Low. The $\overline{\text{Read Data}}$ line is used as the gate for this unit.

The $\overline{\text{DR}}$ line is inverted and gated onto line D0 of the external data bus by the $\overline{\text{Read Status}}$ signal. One-half of another 74LS241 is used for this; the unused inputs are tied to ground to ensure that their input states are well defined.

CONCLUDING COMMENTS

The entire interface consists of the two parts shown in Figures 39-1 and 39-3. The next step in the design would be to pick the appropriate gates to provide the buffers, inverters, and NANDs needed to build the circuit, lay out the circuit on a suitable circuit board, put pin numbers on the drawings, and then wire the circuit. In the process, it would be necessary to remember to include power and ground lines to each IC and to provide suitable decoupling capacitors on the power lines to prevent noise pickup.

PART SIX

SOLID STATE ELECTRONICS

The current electronic revolution is largely due to the availability of tiny integrated circuits containing circuits of almost unimaginable complexity. This situation is a direct outgrowth of the invention of the transistor at Bell Telephone Laboratories in 1948 and the subsequent understanding of the physics of solid state electronic devices. This part of this book is an introduction to the basic physics of solid state devices, to a few of the most common devices including diodes and transistors, and to the most rudimentary elements of transistor circuit design and analysis.

With the exception of the chapter about power supplies, the entire book thus far has been almost totally devoid of specific references to the details of circuits within the integrated circuit packages. It is possible to do a great deal of electronics using integrated circuits, both digital and linear, without having any idea of what is happening inside them. However, there are at least three reasons for including a short introduction to the details of solid state electronics. First, for some purposes, diodes, transistors, and other more complex devices are essential; these areas include high-power, high-voltage, and high-frequency circuits. Second, the data books for integrated circuits often have sections devoted to explaining the operations of the circuits in terms of transistors and other components. A mastery of this material is often necessary to get the fullest use out of the integrated circuits. Finally, curiosity ought to make a person working with electronics want to have at least an understanding of what is happening inside the circuits.

The first chapter of this part, Chapter 40, is an introduction to the basic physics of semiconductors and the behavior of the basic *pn* junction—the basis of all semiconductor devices. Chapter 41 is an introduction to real solid state diodes, including a discussion of their uses and the terminology used to describe them.

Chapter 42 is an introduction to the transistor, and Chapter 43 is a catalog of simple transistor circuits. This is a small sample of the material that formerly made up the bulk of introductory electronics courses. However, with the ever-growing versatility of integrated circuits, the need for this material in an introduction is rapidly shrinking. Chapter 44 presents a procedure for gaining an understanding of an existing transistor circuit. This is a skill that is still necessary because, even though few transistors are being used in new equipment, a great deal of existing equipment still contains transistors. As a result, understanding the circuit is a necessary part of much repair and modification work. Finally, the last two chapters of this part present a brief introduction to FETs and to the techniques used to manufacture transistors and integrated circuits.

The chapters in Part Six of this book constitute only a very brief introduction to a vast amount of information. Their main purpose is as a preparation for reading more detailed and more advanced material. For this reason, with the exception of Chapter 44, there are no problems included in any of these chapters.

PREREQUISITES

For all chapters, a familiarity with the basic concepts of electrical circuits and basic circuit analysis is needed—Chapters 2 and 3.

CHAPTER 40 AN INTRODUCTION TO THE PHYSICS OF SOLID STATE DEVICES
1. A background in elementary physics and chemistry is useful.

CHAPTER 41 SOLID STATE DIODES
1. An understanding of the basic properties of a *pn* junction—Chapter 40.
2. The description of the uses of diodes makes reference to essentially all parts of electronics.

CHAPTER 42 AN INTRODUCTION TO TRANSISTORS
1. A familiarity with the basic properties of *pn* junctions and diodes—Chapters 40 and 41.
2. A familiarity with the concept of dynamic impedance—Chapter 14.

CHAPTER 43 A CATALOG OF TRANSISTOR CIRCUITS
1. A familiarity with the basics of transistor circuits—Chapter 42.
2. A familiarity with op-amp circuits will help understand some of the results and terminology—Chapter 29.
3. A knowledge of the logical NAND is needed to understand the example of the NAND—Chapter 19.

CHAPTER 44 HOW TO UNDERSTAND TRANSISTOR CIRCUITS
1. A familiarity with transistor basics—Chapter 42.

CHAPTER 45 FETs
1. A familiarity with the elementary properties of semiconductors—Chapter 40.

CHAPTER 46 TRANSISTOR FABRICATION
1. The basics of semiconductors—Chapters 40 through 42.

AN INTRODUCTION TO THE PHYSICS OF SOLID STATE DEVICES

OBJECTIVES

1. Gain familiarity with the physical processes that are taking place in intrinsic and extrinsic semiconductors and in *pn* junctions as a background for understanding the operation of the diode and the transistor.

2. Understand how a *pn* junction acts as a diode.

INTRODUCTION

This chapter is a very informal introduction to the basic physics of solid state devices as well as the phenomena that take place at junctions between different types of semiconductor material. Although there are passing references to many topics of physics, the chapter is primarily descriptive and should be understood by anyone who has had some exposure to physics or chemistry.

WHAT IS A SEMICONDUCTOR?

Elsewhere in this book, only two classes of material have been discussed: conductors and insulators. We have assumed that all conductors are perfect conductors and that all insulators are also perfect. These are such good assumptions that no further consideration needs be given to the subject in an introduction to electronics. Only in very specialized applications do more accurate representations of normal conductors and insulators have to be made.

Semiconductors differ from both insulators and conductors in two important ways. The first of these ways is the magnitude of the resistivity of the semiconducting material. As discussed in Chapter 2, the resistance of an object

345

depends on the material of which the object is made and the shape of the object. The resistance of a wire or similar object is

$$R = \frac{L\rho}{A} \tag{40-1}$$

where

R = resistance of a wire or other regular shape
L = length of the wire
A = cross-sectional area of the wire
ρ = resistivity of the material

Only the resistivity is a property of the material out of which the object is made; the other items are geometrical terms describing the shape of the object. As was mentioned in Chapter 2, conductors have resistivities of the order of 1×10^{-8} ohm-meter, insulators have resistivities of the order of 1×10^{11} ohm-meters. On the other hand, semiconductors have resistivities in the range of 10^{-6} to 10^4 ohm-meter. Thus semiconductors are neither good conductors nor good insulators hence, their name.

The second main difference between semiconductors and the other two classes of material is the fact that the resistivity of a semiconductor is a very strong function of temperature. For normal circuit temperatures, 0 to 80°C, the resistivities of both conductors and insulators change very little with changes of temperature. However, the resistivity of a semiconductor is a very strong function of the temperature in this range; in fact, the conductivity (the reciprocal of resistivity) of a semiconductor increases exponentially with the temperature.

INTRINSIC SEMICONDUCTORS

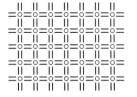

FIGURE 40-1 A schematic representation of a germanium or silicon crystal.

The most common semiconductors are germanium and silicon. These two materials both lie in the same column of the periodic table as does carbon, and they share many properties with carbon. Both of these materials have four electrons in their outer shell, and both form regular crystals in which each atom shares one of these outer electrons with each of four adjacent atoms. In this way, each atom is surrounded by eight outer electrons, a particularly stable arrangement. A schematic diagram of this situation is shown in Figure 40-1. In this diagram, the atomic cores are shown as small circles, while the outer valence electrons are shown as lines. The diagram illustrates the sharing of the valence electrons between the atomic cores. Although it shows this arrangement as a two-dimensional grid, for semiconductor devices it is a three-dimensional structure based on a tetrahedral arrangement. It should be noted that carbon also forms this same crystalline structure. The three-dimensional arrangement is called diamond; the two-dimensional arrangement of the bonds is called graphite.

To the extent that all the electrons are firmly bound to the atoms, they cannot move about in the material, and there is nothing that can carry an electric current in the material. Thus if the electrons were very tightly bound to a particular atom pair, these materials would not conduct electricity and would be insulators. This is true of diamond.

However, because of the size of the germanium and silicon atoms, the binding energy of each electron to its parent atom is not much greater than the average vibrational energy of each of the valence electrons at room temperature. As a result of the statistical way in which the vibrational energy is shared among all of the elements of the crystal, a few of the electrons are freed from their parent atom by thermal excitation. Such a situation is shown in Figure 40-2.

The freed electron is essentially free to move about the crystal; unless it happens to encounter an atom that is lacking an electron, it will simply bounce about. When an external electric field is applied to the crystal, the free electron will drift through the crystal under the influence of this field, and an electric current will result.

The parent atom which lost an electron due to thermal excitation is missing a negative charge and, hence, looks like a positive charge in the crystal. The atom itself cannot move, since it is still firmly bound into the crystal, but it can borrow an eighth electron from some neighboring atom, thus causing the apparent positive charge to move. Since all of the atoms in the crystal are identical, the concept of parent atom is not meaningful, and all that matters is which atom is missing an electron. To simplify the terminology, this lack of electron will be called a **hole**.

By the argument above, the hole will move about the crystal, and until an electron happens to fall into it, it will also continue to bounce about. Again, if an external electric field is applied to the crystal, the holes will drift through the crystal acting like positive particles. Thus in a pure germanium or silicon crystal, it will appear as if the current is being carried by both positive and negative particles.

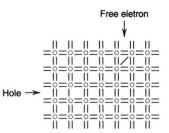

Free eletron

Hole →

FIGURE 40-2 A schematic crystal in which one valence electron has broken free because of thermal excitation.

A correct description of the behavior of electrons in matter requires the use of quantum mechanics. Although a complete quantum mechanical treatment of this problem shows that a literal interpretation of the pictures drawn above is incorrect, this same quantum mechanical treatment shows that all the results obtained with these arguments and pictures are essentially correct. In a semiconductor, it appears as if current is carried by both positive and negative particles, each having the same magnitude of charge as an electron and each having masses approximately the same as an electron. The details of a quantum mechanical treatment of solid state physics are a good topic for a graduate course in physics but not for an introduction to electronics. Some of the results that come from a quantum mechanical treatment are particularly useful. These are discussed below.

First, for a pure semiconductor, the number of holes is exactly the same as the number of free electrons. Thus for an intrinsic semiconductor,

$$n_e = n_h \tag{40-2}$$

When a hole and a free electron collide, they may recombine, that is, simply go away leaving a little extra heat in the crystal at that location. This means that the electron becomes bound to the atom that is lacking an electron, and some extra vibrational energy is left in the crystal as the electron becomes bound. The rate at which free electrons and holes collide and recombine is proportional to the density of both the free electrons and the holes. If n_e designates the free electron density (number per cubic centimeter), then the recombination rate will be proportional to the product of the two densities, $n_e n_h$.

At room temperature, only a tiny fraction of the valence electrons is freed by thermal excitation; at most, 1 part in 10^9 parts is free. Since the production of the free electron–hole pairs is a statistical process and since the density of these pairs never becomes more than a tiny fraction of the atomic density of the crystal, the production rate of these pairs is a constant and depends only on the temperature. From statistical mechanics, it is known that the production rate of any process that depends on the statistical fluctuations of thermal excitations goes as an exponential function of the temperature. Thus

$$\text{Rate of production} = f(T) = A \exp(bT) \tag{40-3}$$

where A and b are constants and T is the temperature.

An equilibrium situation occurs when the rate of recombinations equals the rate of production of electron–hole pairs. Since the recombination rate is proportional to the product $n_e n_h$, Equation 40-3 becomes

$$n_e n_h = f(T) = C \exp(bT) \tag{40-4}$$

where C is the constant of proportionality. Hence the density of holes and free electrons depends only on the temperature of the crystal. Since the conductivity of a material depends on the density of the free charges that are carrying the current, the conductivity of a semiconductor will increase exponentially as the temperature increases.

This argument has been able to explain why germanium and silicon are semiconductors while diamond is an insulator and why the conductivity of a semiconductor depends so strongly on the temperature of the crystal. This same type of argument will be used to explain the behavior of extrinsic semiconductors and semiconductor junctions in subsequent sections of this chapter.

Finally, the name: Since the processes discussed thus far depend only on the inherent properties of germanium and silicon and nothing else, this kind of semiconductor is called an **intrinsic semiconductor**. Just as there would be no electronics if there were only resistors and insulators, there would be no solid state electronics if there were only intrinsic semiconductors. The properties of a germanium or silicon semiconducting crystal can be dramatically changed by the addition of a small amount of impurities. Since these semiconductors depend on the properties of things other than the silicon or germanium, they are called **extrinsic semiconductors**.

EXTRINSIC SEMICONDUCTORS

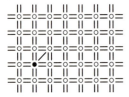

FIGURE 40-3 A donor atom in a crystal.

If a tiny amount (a few parts per million or less) of a **donor** material having five outer electrons is added to the germanium or silicon before forming the crystal, the situation shown in Figure 40-3 results. In this figure, the impurity is shown as a solid dot and its extra, or fifth electron is still attached to it. Typical materials used are arsenic (As), lead (Pb), Tin (Sb), and others.

Although these extra electrons are originally attached to the impurity atoms in the crystalline lattice, they are only very weakly bound to the impurity; hence, they are virtually free to move about the crystal. This situation is shown in Figure 40-4. At room temperature, essentially all of the donor atoms have been ionized; that is, they have lost their fifth valence electron. Furthermore, the impurity atoms have very little tendency to acquire a fifth electron. Although the donor atom is a residual positive charge left behind when the electron drifts off, this positive charge remains fixed in the crystalline lattice and does not move.

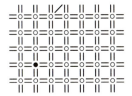

FIGURE 40-4 An ionized donor atom in a crystal.

Hence, the addition of a donor atom has created a free electron but not a hole. The result is that the number of holes does not equal the number of free electrons in an extrinsic semiconductor (see below). The addition of one part per million impurity increases the conductivity of the semiconductor by a factor of about 1000 at room temperature. In this type of semiconductor, the current is predominantly carried by the **majority carrier**, the free electron; hence, the name, *n*-type semiconductor for this type of material. The **minority carrier** in an *n*-type semiconductor is the hole.

If an **acceptor** impurity is added to the crystal, a material having three valence electrons such as alumimum (Al), the situation shown in Figure 40-5 results. As the crystal originally forms, the acceptor atom is sharing only seven bound electrons. However, it has a very strong tendency to borrow an eighth electron from a neighboring atom, thus creating a hole. The hole moves freely about the crystal, leaving behind a negative charge bound into the matrix of the crystal.

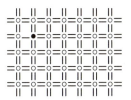

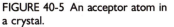

FIGURE 40-5 An acceptor atom in a crystal.

Most of the comments about *n*-type semiconductors also hold true for this type with suitable changes of sign. At room temperature, essentially all the acceptor atoms are ionized, producing free holes. The addition of one part per million of impurities increases the conductivity by a factor of about 1000 compared with the intrinsic situation. Each acceptor atom results in a hole but not a free electron, so that the majority carriers in this type of semiconductor are the holes. Since the current is carried by positive particles, this is called *p*-type semiconductor.

Again, there are a number of additional features about these systems that are needed to understand their behavior in actual electronic devices. First, under normal circumstances, and this includes all the situations discussed thus far, the crystal remains microscopically neutral. This means that in any small piece of the crystal the number of free charges is the same as the number of bound charges.

Second, a quantum mechanical treatment of the situation involving donors and acceptors shows that the product of the hole concentration and the free electron concentration remains constant as impurities are added to the crystal. Thus Equation 40-4 continues to hold:

$$n_e n_h = f(T) = C \exp(bT) \tag{40-4}$$

where $f(T)$ is the same function of temperature as it was for an intrinsic semiconductor. Thus as donors are added, the electron density n_e increases and n_h decreases so that the product remains the same. If enough donors are added to the material to substantially increase the conductivity, then the number of holes must have decreased almost to zero so that this product remains constant. (Note that this is very similar to the concentration of OH⁻ and H⁺ ions in water as acids or bases are added).

Finally, if a number of minority carriers are somehow injected into a semiconductor, they will diffuse away from the source and gradually be lost because of recombination with the majority carriers. Thus if a few holes are dropped into an *n*-type semiconductor, the holes will diffuse away from the injection point and will gradually recombine with the negative majority carriers. On the average, the minority carriers will go several hundred microns before they recombine.

THE HALL EFFECT

This section is a digression that will make sense to those who have studied enough physics to understand the forces on charged particles moving in a magnetic field. The material in this section is not required to understand the rest of this chapter or any of the others that follow.

The Hall effect permits the verification of the existence of two kinds of semiconductor, having different signs of majority carriers. Remember that the magnetic force on a charged particle moving in a magnetic field is given by the following expression:

$$\mathbf{F} = q\mathbf{v} \times \mathbf{B} \qquad (40\text{-}5)$$

where q is the charge of the particle, $\mathbf{v}$ is the vector velocity of the particle, and $\mathbf{B}$ is the vector magnetic field.

A simple experimental situation is shown in Figure 40-6. Two pieces of semiconductor, one *n*-type and one *p*-type, are put into a magnetic field and connected to an external circuit so that (conventional) current flows through the pieces. In both cases, the conventional current flows from left to right. This means that the holes in the *p*-type semiconductor are drifting from left to right while the free electrons in the *n*-type semiconductor are drifting from right to left.

Since the magnetic field is into the paper, the holes will experience a magnetic force deflecting them upward, and they will tend to follow a path similar to the one shown as a dotted line in the figure. Thus the upper edge of the piece of *p*-type semiconductor will be positive with respect to the lower edge.

The electrons will also experience an upward force (both their charge and their velocity has the opposite sign), so they will follow a similar path. Thus the upper edge of the *n*-type semiconductor will be negative with respect to the lower edge.

This difference can be measured and was originally used as proof that positive particles were carrying the current in *p*-type semiconductors.

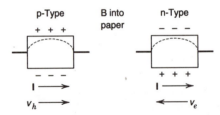

FIGURE 40-6 A simple Hall effect experiment.

SUMMARY OF SEMICONDUCTOR PROPERTIES

Before we consider the behavior of a junction between a piece of *n*-type semiconductor and a piece of *p*-type semiconductor, here is a summary of the important results obtained thus far:

1. The product of the hole density and free electron density in a semiconductor depends only on the temperature of the crystal, not on the amount of doping:

$$n_e n_h = f(T) = C \exp(bT) \qquad (40\text{-}6)$$

2. Each type of semiconductor is neutral under normal conditions.

3. For *n*-type semiconductors, the current is carried by free electrons, the majority carrier. The minority carrier (holes) density is almost zero, and there are positive donor atoms fixed throughout the crystal.

4. For *p*-type semiconductors, current is carried by holes, the majority carrier. The minority carrier (free electrons) density is almost zero, and there are negative acceptor atoms fixed throughout the crystal.

pn JUNCTIONS

This section examines the phenomenon that takes place at the junction between a piece of *n*-type semiconductor and a piece of *p*-type semiconductor. The situation described and all the conclusions reached in this section are shown in the various parts of Figure 40-7. The step-by-step argument will be presented first. Refer to the figure while studying the step-by-step argument.

The situation being considered consists of two blocks of semiconductor material, one *n*-type, the other *p*-type. As is shown in Figure 40-7a, the *p*-type is on the left. (Note that there is nothing important about the left/right positions; but by making it explicit, the language of the arguments can be made simpler. All that really counts is which is *p*-type and which is *n*-type.) At the start, both of the pieces of semiconductor are neutral as shown. The two pieces of semiconductor are bonded together so that the crystalline structure is continuous across the junction. (How this kind of junction can be made is one of the technological wonders of the microelectronics industry.)

When the junction is formed, the free charged particles in each of the pieces of semiconductor can diffuse or drift across the junction into the other type of semiconductor. In particular, holes can drift from the *p*-type into the *n*-type semiconductor. Once they enter the *n*-type semiconductor, they are minority carriers; they diffuse about until they collide with a free electron and recombine—vanish. At the same time, electrons are drifting from the *n*-type semiconductor into the *p*-type, drifting around, colliding with holes, and recombining. This situation is indicated in Figure 40-7b.

There are several consequences of this diffusion process. Although these consequences are stated as if they were separate, they really are all different aspects of the same process. First, there is a net diffusion of holes from the left to the right, that is, from the *p*-type to the *n*-type semiconductor. This represents a net conventional current from left to right. Likewise, there is a

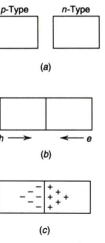

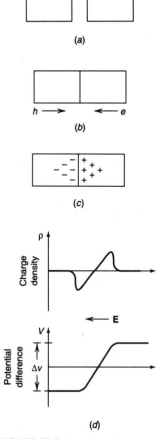

FIGURE 40-7 A *pn* junction.
(a) The two pieces of semiconductor before the junction is formed. (b) the Two pieces of semiconductor bonded together showing the direction of the net hole and free electron diffusion. (c) The two pieces of semiconductor after the formation of the dipole layers. The plus and minus signs are schematic representation of the fixed charge centers in the crystal. (d) Plots of the charge density and potential difference across the *pn* junction. Also shown is the direction of the resultant electric field, indicated as **E**.

net diffusion of electrons from right to left; this also represents a net conventional current from left to right. As a result, the left side becomes less positive and the right side more positive, and there is an electric field across the junction.

Second, when a hole leaves the *p*-type material, it leaves behind a negative ion fixed in place in the crystal. Likewise, when an electron leaves the *n*-type material, it leaves behind a positive ion fixed in the crystal. Since only holes or electrons near the junction cross the junction because of the nature of random motion, these excess positive and negative charges are located near the junction. They form a **dipole layer** at the junction. The effect of this dipole layer is to create an electric field across the junction that tends to oppose the diffusion of more holes and electrons across the junction.

Finally, as a result of the electric field in the region of the junction, holes in the *p*-type material are pushed to the left while free electrons in the *n*-type material are pushed to the right. Thus the number of free charges in the region of the junction is decreased, or the region is depleted of free charges. As a result, the region on each side of the junction is often called the **depletion region**, or **depletion layer**.

This process started with two neutral pieces of semiconductor being bonded together. As a result of diffusion, free carriers started drifting across the junction, colliding with carriers on the other side and recombining. This represents a net current that creates a dipole layer that tends to reduce the rate at which future diffusion takes place. Eventually, an equilibrium point is reached at which no net charge diffuses across the junction, that is, there is no net current flowing. These processes are all indicated in the last portions of Figure 40-7.

The end result of this process is that a potential difference forms between the two types of semiconductor material. The magnitude of this voltage depends on the bulk semiconductor material and on the junction temperature but not on the doping materials. For room temperature, this voltage Δv is roughly

$$\Delta v \approx 0.3 \text{ V for germanium}$$

$$\Delta v \approx 0.7 \text{ V for silicon}$$

This potential difference is created by the fixed charges in the dipole layer. An electric field exists in this dipole layer, and its direction is such that it sweeps the majority carriers on each side away from the junction. Thus a region is left that is rich in fixed charges but depleted of free (movable) charges. The depletion layer is typically a few microns thick, so that the electric fields in the depletion region are of the order of 10^4 V/cm; these are very high fields.

Finally, although this has been described in terms of the processes that start with the joining of two neutral blocks of semiconductor, the equilibrium conditions hold true no matter how the junction is formed.

THE *pn* JUNCTION IN AN ELECTRICAL CIRCUIT

The usefulness of the *pn* junction only becomes apparent when its behavior is considered in an electrical circuit. If two wires are connected to the ends of the pieces of semiconductor, then the junction can be put into an external circuit as shown in Figure 40-8. If an external voltage is applied to the *pn* junction such that the *p*-type semiconductor is made negative with respect to the *n*-type, as shown in the figure, this only makes the potential barrier between the two types of semiconductor higher and increases the width of the depletion layer. Since a region devoid (depleted) of free charges cannot conduct current, no current flows across the junction and, hence, no current flows in the external circuits. This arrangement is called **reverse bias**, or **backward bias**.

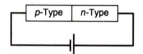

FIGURE 40-8 A reverse biased diode.

If the *p*-type semiconductor is made positive with respect to the *n*-type semiconductor, as shown in Figure 40-9, then the potential barrier between the two types is decreased, and charge is forced into the depletion region. As a result, current flows. This arrangement is called **forward bias**.

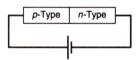

FIGURE 40-9 A forward biased diode.

As can be understood from the discussion above, the *pn* junction conducts electricity in one direction but not in the other. A device that conducts electricity in one direction and not in the other is called a **diode**. Thus a *pn* junction is a diode. The properties of real diodes are discussed in the next chapter.

Like the two preceding sections, this section ends with a listing of some of the more important features of a *pn* junction. In general, these features have not been demonstrated here, but are simply stated for future reference. First, when the junction is biased forward, somehow holes are being created at the wire *p*-type semiconductor junction. These holes are drifting through the neutral part of the *p*-type semiconductor until they are forced (injected) into the *n*-type material by the external voltage. Once the holes are in the *n*-type material, they diffuse away from the junction until they recombine with an electron. Similarly, electrons are being created at the wire *n*-type semiconductor junction and diffusing toward the *pn* junction.

Second, when the *pn* junction is biased backward, there really is a small current, called the **reverse saturation current** I_s. There are two sources of this current: (1) thermally created electron–hole pairs created in the depletion region, which cannot be avoided; and (2) impurities of the wrong type in the materials (that is, donors in *p*-type semiconductor, etc.). This reverse saturation current can be reduced by using ultrapure semiconductor and by being very careful in all the processing.

It can be shown that for a *pn* junction, the current is given by

$$I = I_s\left(\exp\left\{\frac{qV}{kT}\right\} - 1\right) \tag{40-7}$$

where

q = charge of the electron
V = voltage applied
T = junction temperature
k = Boltzmann's constant
I_s = reverse saturation current

This expression is plotted in the next chapter.

Finally, it should be noted that the depletion layer has some of the properties of a capacitor. However, its width, and hence its capacity, is a function of the bias voltage. Thus the diode may have unusual and potentially useful high-frequency characteristics.

For anyone who wants a more detailed introduction to the physics of semi-conductor devices (and who does not want to spend years studying physics), the books published by the Semiconductor Electronics Education Committee are excellent references. Particularly recommended are Volume, 1, Chapters 1 and 2, *Introduction to Semiconductor Physics*, by R. B. Adler, A.C. Smith, and R. L. Longini; and Volume 2, Chapters 1, 2, and 3, *Physical Electronics and Circuit Models of Transistors*, by P. E. Gray, D. DeWitt, A. R. Boothroyd, and J. F. Gibbons. Both were published by John Wiley & Sons in 1964.

Chapter 41

SOLID STATE DIODES

OBJECTIVES

1. Become familiar with the major types of diodes including rectifiers, switching, zener, LEDs, and tunnel diodes.

2. Become familiar with the main uses of each of these types of diodes.

3. Become conversant with some of the terminology used to describe diodes.

INTRODUCTION

This chapter is an introduction to the variety of types and uses of solid state diodes. Included is a brief glossary of the terms used to describe or to specify diode performance. The purpose of this chapter is to present enough information about the properties and language of diodes to provide an understanding of the discussions in more advanced sources.

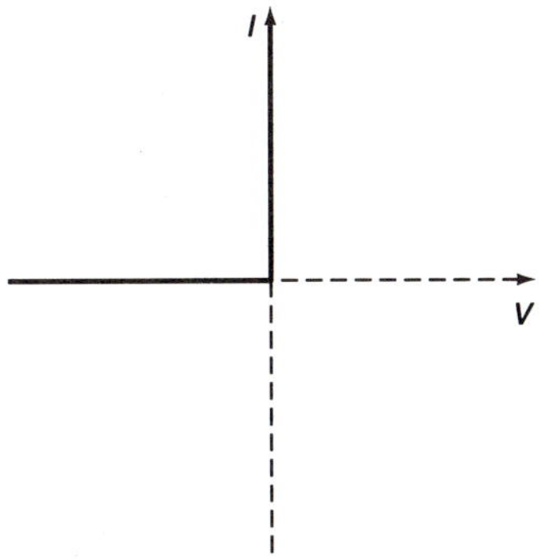

IDEAL DIODES

An ideal diode has the voltage–current graph shown in Figure 41-1. The ideal diode has zero resistance for current flowing one way through it and infinite resistance for current flowing the other way. From another point of view, it allows current to flow through itself in only one direction. The direction in which the current may flow is called the **forward** direction, whereas the way in which current may not flow is called the **backward** direction.

The name **diode** itself dates back to the early days of vacuum tubes. A diode was a vacuum tube with two electrodes; similarly, a triode was a tube with three electrodes, and so on. The first vacuum tube diodes were made by Thomas Edison as part of his effort to make the filaments in light bulbs last longer. The addition of another electrode did not help, so he did not follow up the interesting properties of the diode. This was left for others who are much less famous.

FIGURE 41-1 The voltage–current graph for an ideal diode.

THE SOLID STATE DIODE

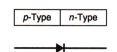

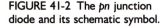

FIGURE 41-2 The *pn* junction diode and its schematic symbol.

As was shown in the preceding chapter, a *pn* junction allows current to flow relatively easily in one direction while strongly resisting the flow of current in the other direction. Thus, the *pn* junction is a reasonable approximation of an ideal diode. The basic *pn* junction and the schematic symbol for this device are shown in Figure 41-2. The arrow head in the symbol designates the *p*-type semiconductor material, and the bar in the symbol designates the *n*-type material. **The arrow points in the forward direction.** Thus conventional current will flow through the diode in the direction the arrow points; but current will not flow in the opposite direction.

As was shown in the preceding chapter, the *pn* junction is forward biased when the *p*-type material is positive with respect to the *n*-type material. This gives rise to a simple mnemonic for remembering the biasing direction, *p—*positive, *n—*negative. **Memorize this.**

The current flowing in a *pn* junction is given by the following expression:

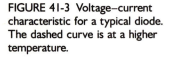

$$I = I_s \left[\exp\left(\frac{qV}{kT}\right) - 1 \right] \tag{41-1}$$

where

I_s = reverse saturation current of the diode

T = temperature of the junction

q = electron charge

k = Boltzmann's constant

FIGURE 41-3 Voltage–current characteristic for a typical diode. The dashed curve is at a higher temperature.

This equation is plotted for two different temperatures in Figure 41-3. At room temperature, kT/q is about 26 mV; thus, the current increases by a factor of *e* (about 2.72) for each 26 mV of increase in forward bias. The plot of the *V–I* graph on semilog graph paper for a typical diode is a straight line for many decades.

As can be seen in Figure 41-3, when plotted on a linear scale, the *V–I* graph for a diode shows a fairly sharp turn-on voltage in the forward direction. This voltage depends on the bulk semiconductor material from which the diode is made. For germanium, this turn-on is at about 0.3 V; for silicon, it is at about 0.7 V. This graph should be compared with the graph for an ideal diode (see Figure 41-1).

DIODE TYPES

Diodes are made for many different purposes. Their parameters can be optimized for some particular use, or variations on the basic *pn* junction structure can give rise to even more complex *V–I* graphs than the curve given above. Some of these different types of diodes are introduced and briefly discussed in this section. Many of the terms used here are defined in the following section.

Switching diodes are used in many low-power situations. They are used as detectors to convert low-power high-frequency ac into slowly varying dc in radio and TV receivers. Diodes can be combined with transistors to form logic gates (DTL, or diode transistor logic). Diodes can be used anywhere in a circuit where it is desired to limit the current flow to one direction only. All of these applications are places where low-power, high-speed switching diodes would be used. These diodes are designed to have high switching speeds and small reverse leakage currents, but they are not intended to carry high currents or withstand large reverse voltages. The diode is generally very small—the size is mainly limited by the fact that human beings have to be able to see them, pick them up, and wire them into circuits. Switching diodes are also very cheap: in units of thousands, one popular switching diode costs only a few cents each.

Rectifiers are diodes used to convert ac voltages into dc voltages in a power supply—see Chapter 38 for the details of this process. A diode intended for this purpose is designed to carry large currents and to withstand high reverse voltages. Since normal ac voltages operate on a time scale of milliseconds, switching speed is not very important for this purpose. The diode is generally called a **power diode**, or a **rectifier**. Rectifiers that can handle hundreds of amperes and thousands of volts exist; any electronics catalog will have a reasonable selection. Rectifiers for low-power circuits—a few volts and a few amperes—are small and cheap, often costing 25 cents or less. Rectifiers for high-power circuits can be huge and expensive and may include provisions for water cooling.

Some diodes that are made out of more exotic materials than simple germanium or silicon give out light when they are conducting. These are called **light emitting diodes**, or **LED**s. LEDs have become ubiquitous as the pilot lights on all sorts of electronic equipment and as the LED displays on calculators, radios, stereos, and microwave ovens. The typical small red LED requires about 1.6 V forward bias to cause the 10-mA current needed to develop full brightness. Of course, a wide variety of colors, sizes, and shapes now exist. Some use much less power and produce only tiny weak colored dots; others use substantial power and generate very intense light beams.

In a certain sense, the **photodiode** is the complement of the LED. This is a diode made of special material that can be used to detect light. The diode is biased backward, resulting in a large depletion region that is also the photosensitive volume. Light absorbed in the depletion region causes electron–hole pairs to be formed. The very strong electric fields in the depletion region sweep the electrons one way and the holes the other way, resulting in a current flow across the junction while the hole–electron pair is being swept out. The current flowing through the reverse biased junction is proportional to the amount of light striking the depleted region. This kind of detector can be made to be very fast and sensitive.

Some cameras use diodes of this nature as part of the light meter circuit. This kind of diode, when operated at very low temperatures to eliminate the possibility of thermal excitation of electron–hole pairs, can be used as the detector in telescopes and other very sensitive optical instruments. Variations on this diode can be used to detect light beyond the range that the human eye can detect; for instance, an array of these diodes can be used to detect infrared light, giving rise to the night vision devices used in the military and the heat leak detectors used in energy conservation.

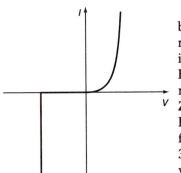

FIGURE 41-4 The voltage–current graph for a Zener diode.

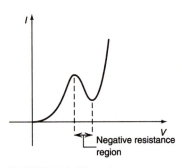
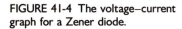

FIGURE 41-5 The voltage–current graph for a tunnel diode.

The **Zener diode** was briefly discussed in Chapter 38. A normal diode will break down if a sufficiently high reverse bias is applied. Breakdown in a normal diode is usually accompanied by the destruction of the *pn* junction. Thus, in normal operation, breakdown is the same as burnout and is a bad thing. However, it is possible to construct a diode that will have a nondestructive reverse bias breakdown at a well-defined voltage. Such a diode is called a **Zener** diode. The voltage–current plot for a typical Zener diode is shown in Figure 41-4. The reverse breakdown voltage can be set by adjusting the manufacturing parameters. Zener diodes with breakdown voltages in the range of 3.6 to 50 V exist—see any electronics catalog. Zener diodes can be used as voltage references and as voltage regulator elements in power supplies.

The **tunnel diode** has a more complex junction structure, namely, a multilayer structure that takes advantage of the quantum mechanical phenomena called **tunneling**. This is a phenomenon that has no classical analog. The basic phenomenon is that an electron or hole suddenly goes from one side of a potential barrier to the other side, passing through a region where it cannot be; that is, it tunnels through the potential barrier. Diodes can be constructed to have the voltage–current relationship shown in Figure 41-5. As the graph shows, tunnel diodes have a region of negative resistance in their *V–I* graphs, the region where the current decreases with increasing voltage. Tunnel diodes are very specialized elements that can be used to make very fast switching circuits.

DIODE PROPERTIES

Many different specifications and parameters are used to describe diodes. Obviously, not all of these are of use for all types of diodes. The following sections briefly list most of the common parameters and/or specifications relevant to the selection of a diode. For those cases where it is not immediately obvious, the meaning of the parameter is defined. The parameters are given roughly in the order in which they need to be specified when making a selection.

Purpose

Of course, the most important specification needed to select a diode is what its purpose is. It makes no sense to use an LED as a rectifier, even though it might be colorful.

Material

For some uses, it is important to know the material out of which the diode is made. In particular, since the forward turn-on voltage depends on the bulk material, uses that require a specific turn-on voltage require a specific material. For general use, silicon is the preferred material for two reasons. First, the higher melting temperature of silicon allows higher currents than are allowed in similar-sized germanium devices. Second, at a fixed temperature, silicon devices have lower reverse leakage currents than do similar germanium devices.

Forward Voltage Drop

The forward voltage drop is the voltage across a diode at a specified current. Frequently, signal diodes are specified at 20 mA in the forward direction. Rectifiers may be rated at much higher currents.

Current Limits

The maximum current in the forward direction is frequently specified for a diode. There are two ways in which this current is specified: one is the average forward current; the other is the instantaneous forward current. Diodes in rectifier currents are often required to handle instantaneous peak currents (for a few milliseconds) that are many times larger than the average currents.

Power Limitation

A certain amount of power is dissipated in the diode. This amount is given by the voltage drop across the diode times the current through the diode. For a typical rectifier, there is essentially no power dissipated when the diode is biased backward because the reverse leakage current is so small. Hence, for a typical rectifier, the power limitation should be consistent with the product of the forward voltage drop times the peak forward current. For a Zener diode, on the other hand, the power limitation will occur in the reverse direction.

Peak Reverse Voltage

A typical diode can withstand a certain peak backward voltage before it breaks down in a destructive mode. This voltage is specified as the **peak reverse voltage**. Note that for a rectifier circuit, this voltage may be almost three times as large as the dc output of the circuit.

Reverse Leakage Current

This specifies how much current flows when the diode is biased backward. A complete specification includes the temperature of the measurement and the value of the backward bias. This parameter is of interest in switching diodes and of little interest in a rectifier.

Switching Speed

This is a measure of how fast the diode can begin to conduct or stop conducting when the bias is suddenly switched. A modern switching diode has a switching time of a few nanoseconds or less. This is another parameter that is only of interest for switching circuits.

Dynamic Forward Resistance

One last parameter that is sometimes given for signal diodes is the dynamic forward resistance of the diode. This is the slope of the voltage–current graph at some particular forward current. It is of interest when creating small-signal circuit models. A dynamic reverse resistance value is often given for Zener diodes.

Size and Shape

These parameters are primarily of interest in the case of rectifiers where space, ventilation, and cooling are problems. Fortunately, most diodes come in a limited number of sizes and shapes.

Cost

The meaning of this parameter is evident. For most common uses, the cost of the diodes for a circuit is almost irrelevantly small.

Chapter 42

AN INTRODUCTION TO TRANSISTORS

OBJECTIVES

1. Become familiar with the basic bipolar transistor structure and operation.

2. Be able to analyze a transistor's operation by using the characteristic curves.

3. Become familiar with several transistor models and see at least one model used.

4. Become familiar with the terminology used to describe transistors.

INTRODUCTION

This chapter is an introduction to transistors. It includes a discussion of what they are, how they work, their characteristics, and the simplest amplifier and circuit models and concludes with a short list of some of the most common terms used to describe transistors. The next chapter catalogues a few of the most common circuits.

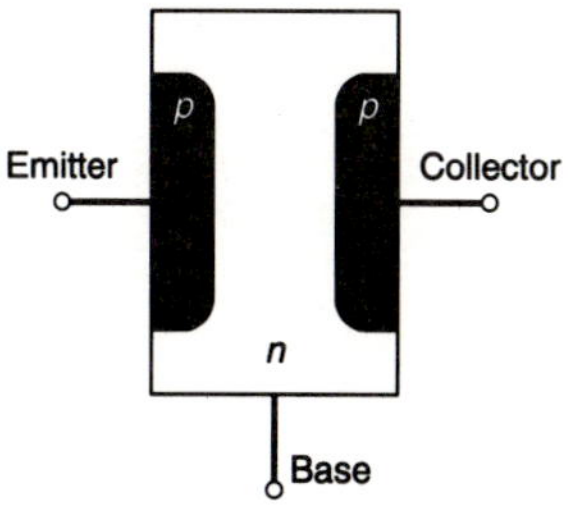

FIGURE 42-1 A simple *pnp* transistor.

BASIC STRUCTURE

The simplest transistor structure is shown in Figure 42-1. This transistor is built out of a block of *n*-type semiconductor, which is called the **base** of the transistor. In the faces of this block, there are two thin regions of *p*-type semiconductor; these are called the **emitter** and the **collector**. (The reasons for these names will become clear as the discussion progresses.) Metallic (wire) connections are made to the emitter, base, and collector. The drawing is a considerable distortion of the actual shape of a typical transistor. In normal situations, the base is a few thousandths of an inch thick, and it is a few tenths of an inch in the other two directions. If there is any difference in size between the emitter and the collector, the collector is larger.

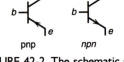

FIGURE 42-2 The schematic symbols for transistors.

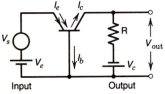

FIGURE 42-3 A common base amplifier.

The transistor shown in Figure 42-1 is a *pnp* **transistor**, where the name refers to the semiconductor type of the emitter, base, and collector, respectively. An *npn* **transistor** has an *n*-type emitter, a *p*-type base, and an *n*-type collector. Otherwise, it is exactly the same as the *pnp* transistor. Since the currents in these transistors are carried by both polarities of carriers (holes and electrons), they are often called **bipolar** transistors, in contrast to FET transistors, which are discussed in Chapter 45.

The schematic symbols for a *pnp* and an *npn* transistor are shown in Figure 42-2. In the schematic symbols and in most discussions, the emitter is abbreviated as *e*, the base as *b*, and the collector as *c*.

THE OPERATION OF A TRANSISTOR

Figure 42-3 shows the simplest possible circuit for a transistor. This particular configuration is called the **common base configuration**, because the base is common to both the input and the output circuit. The operation of the transistor in this configuration is as follows:

1. The collector–base junction is biased backward. The collector–base depletion region may reach the emitter–base junction. If there were only the collector–base junction, there would be no current flowing in the collector circuit.

2. If the emitter–base junction is biased backward, no current flows in this junction either and, as a result, no current flows anywhere in the transistor circuit. This is a relatively uninteresting situation.

3. In Figure 42-3, the emitter–base junction is biased forward. As a result, holes are injected from the emitter into the base region—this is the source of the name **emitter**.

4. The holes injected into the base region diffuse away from the base–emitter junction; but because of the geometry (it is hundreds of times farther to the base connection than it is to the collector) and because of the reverse bias on the base–collector junction, most of the holes injected into the base are swept into the collector by the base–collector dipole field; only a few are lost to the base. Thus the collector current is slightly less than, but almost equal to, the emitter current. Remember that holes in the *n*-type base are minority carriers, and the fields in the depletion region push them toward the collector. Clearly, the name **collector** comes from the collection of the minority carriers in the base.

KCL applied to the transistor gives

$$I_e = I_c + I_b \tag{42-1}$$

The argument above about the relationship between the collector and emitter current indicates that

$$I_e \approx I_c \quad \text{but that} \quad I_e > I_c \tag{42-2}$$

Two parameters are commonly used to describe transistors. These are the two current ratios

$$\alpha = \frac{I_c}{I_e} \tag{42-3}$$

$$\beta = \frac{I_c}{I_b} = \frac{\alpha}{1 - \alpha} \tag{42-4}$$

For a typical transistor, α is in the range of 0.9 to 0.999 and β is in the range of 10 to 1000. (Note that $\alpha = 0.99$ means $\beta = 100$.) Another name for β sometimes used in the literature is h_{fe}.

THE COMMON BASE CONFIGURATION AS AN AMPLIFIER

To understand how the transistor acts as an amplifier in the situation shown in Figure 42-3, consider the following events. A small emitter voltage variation produced by the input signal source v_s causes a variation in the current flowing in the base–emitter junction. This variation is roughly approximated by

$$i_e = \frac{v_s}{r_{in}} \tag{42-5}$$

where r_{in} is the dynamic impedance of the emitter–base junction. Typically, this is of the order of 100 Ω or less.

For a normal transistor, $i_e \approx i_c$ and the variation in the collector current is essentially the same as the variation in the emitter current. The output voltage is given by

$$v_{out} = i_c R_L \tag{42-6}$$

in all of these equations, lowercase symbols refer to changes.

The voltage gain G is given by

$$G = \frac{v_{out}}{v_s}$$

$$= \frac{i_c R_L}{v_s}$$

$$= \frac{\alpha i_e R_L}{v_s}$$

$$= \frac{\alpha v_s R_L}{r_{in} v_s}$$

$$= \frac{\alpha R_L}{r_{in}}$$

$$\approx \frac{R_L}{r_{in}} \tag{42-7}$$

Since typical values of R_L may be 10,000 Ω whereas r_{in} is 100 Ω or less, this configuration may easily have a voltage gain of 100.

Although many other amplifier circuits are possible, this example illustrates most of the usual features. In general, the base–emitter junction will be biased forward while the base–collector junction is biased backward. Small voltage changes in the emitter–base voltage cause current changes in the emitter current which causes changes in the collector current; in general, the emitter current controls the collector current. The emitter circuit is a low-power circuit, whereas the collector is a higher-power circuit. Thus the low-power input signals control the higher-power signals in the output or collector circuit.

The simple arguments used above can be used to roughly analyze most transistor circuits. However, for many purposes a better analysis scheme is needed. Both a graphic scheme and a mathematical scheme are presented below.

CHARACTERISTIC CURVES

Two graphs are presented as the characteristic curves for a transistor. One of these is a voltage–current plot for the base–emitter junction. Since this graph is largely independent of the collector voltage, only one curve is given. The other characteristic curve is the collector characteristic plot, a plot of I_c versus the collector–base voltage. Since this curve depends explicitly on I_e, a family of plots is usually given, each plot being for one particular emitter current. A typical set of curves is given in Figure 42-4.

There are several points to be made about these curves. First the emitter characteristic curve looks like the voltage–current graph for a diode, which it is; however, this plot is usually rotated 90° from the orientation used elsewhere in this book. The turn-on voltage V_f for a germanium transistor is about 0.3 V; for a silicon transistor, it is about 0.6 V.

The collector characteristic graph shows that for the base–collector voltage above some value, typically about $\frac{1}{2}$ V, the collector current depends only weakly on V_{cb} but that it depends strongly on the emitter current. (The collector characteristics would be horizontal if there were no dependence on the base–collector voltage.)

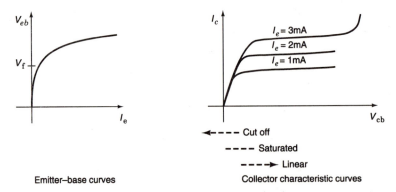

FIGURE 42-4 Characteristic curves for a typical transistor; the three operating regions are indicated below the V axis.

The dotted portion of the collector characteristic, Figure 42-4, is a thermal effect. At some point, as the power being dissipated in the transistor increases, it begins to heat up substantially. Then, the reverse leakage current starts to increase and the power being dissipated in the transistor increases even more. This situation called thermal run-away is usually followed by burnout.

There are several different operating regions on a transistor's characteristic curves. These regions all have names and distinct properties. They are indicated in Figure 42-4 and are briefly defined here:

1. If no collector current flows, either because no emitter current flows or because there is no collector bias, the transistor is said to be **cut off**.
2. If collector current flows, the transistor is said to be **active**.
3. If the collector current varies very rapidly with collector voltage, the transistor is **saturated**.
4. If the collector current does not vary rapidly with collector voltage but does depend on emitter current, the transistor is operating in its **linear operating region**.

A GRAPHIC ANALYSIS OF THE COMMON BASE AMPLIFIER

It is possible to analyze the operation of the common base amplifier shown in Figure 42-3 by means of the characteristic curves. The battery V_e in the emitter circuit provides a forward bias for the base–emitter junction and sets the **operating point**, or **quiescent point**. The emitter current at the operating point is I_{eo}. This is indicated on Figure 42-5; I_{eo} determines the operating point collector current I_{co}.

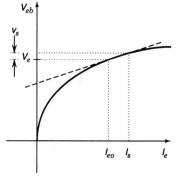

FIGURE 42-5 The effects of the input signal on the emitter current.

A small input signal v_s is added to the battery voltage. The goal of the analysis is to find out how V_{out} varies with v_s. In quantitative terms, if v_s is positive, then I_e will increase and, because of this, I_c will also increase. Because of the increase in I_c, the voltage drop across R will increase. This means that $|V_{out}|$ decreases and V_{out} becomes less negative. Similarly, if v_s becomes negative, V_{out} becomes more negative.

To do a graphic analysis, a small positive change in the base–emitter voltage is indicated in Figure 42-5. The resulting change in the emitter current is also indicated; the new emitter current is called I_s. To the extent that the base–emitter voltage–current curve may be approximated by a straight line in the region of interest—the dashed line shown in Figure 42-5—the variation in the emitter current is given by

$$\Delta i_e = \frac{v_s}{r_{in}} \tag{42-8}$$

where r_{in} is the dynamic resistance corresponding to the slope of the dashed line.

For any given emitter current, the output voltage can be determined from the collector characteristic curve by use of a **load line**. The voltage drop across the load resistor R is given by $I_c R$. The voltage indicated as V_{out} in Figure 42-3 is given by

$$V_{out} = V_c - I_c R \tag{42-9}$$

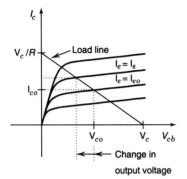

FIGURE 42-6 The load line drawn on a graph of the collector characteristic curve.

The graph of this expression is a straight line. When there is no current, the voltage is its maximum, or V_c; when the voltage is zero, the current has the value of

$$I_c = \frac{V_c}{R} \tag{42-10}$$

These two points determine a straight line, which is indicated on the collector characteristic curve in Figure 42-6; the line is marked **load line**. The output voltage must always lie on this load line.

The curves due to the values of the emitter current that are caused by the variations in the input voltage can be drawn on the graph by interpolation. The output voltage change can then be determined by the fact that it must lie on both the load line and on the collector characteristic curves. The two values of the output voltage are indicated on the graph along with the change in the output voltage. (Note that there is a sign convention hidden in this graph; the collector is really negative with respect to the base, but by convention the graphs are always drawn as if the voltages were all positive.) In this manner, the output voltage corresponding to any input voltage can be determined. Since the nonlinearities in the transistor are represented in the graphs, there are no approximations in this analysis. However, this analysis does not lend itself to mathematical representation.

TRANSISTOR CIRCUIT MODELS

As mentioned previously, the goal of circuit analysis is to predict the properties of a circuit without having to build and measure the circuit. In general, a mathematical model of a circuit is a much easier and faster way to explore variations in the circuit than is actual measurement. To create a mathematical model of a circuit involving a transistor, it is necessary to replace the transistor with a suitable mathematical model. Ideally, this would be a linear model so that all the powerful techniques for dealing with linear equations could be used. However, as the characteristic curves show, a transistor is inherently nonlinear. This is a problem.

The usual process is to separate transistor circuit analysis into two steps. The first step deals with the inherent nonlinearities, whereas the second step can be approximated in a linear manner. The biasing of the transistor, the establishment of the **operating point**, inherently involves dealing with the nonlinear nature of the transistor. The biasing of a transistor circuit is done or analyzed by means of graphic techniques or by means of rules-of-thumb. Once the operating points of the transistors in the circuit have been determined, variations about these operating points due to **small signals** can be treated by a linear mode. For small enough variations, a linear approximation to the transistor operating curves is highly accurate. For intermediate to large signals, the models are much less accurate. Obviously, determining the scope of applicability for a model is a very difficult problem which goes far beyond the scope of an introduction. For this chapter, several small-signal models will be introduced with no further comments about their range of applicability.

Figure 42-7 shows one possible circuit for a transistor. Typical values for the resistors in this model are

$$r_e \approx 10\ \Omega$$

$$r_b \approx 800\ \Omega$$

$$r_c \approx 1\ M\Omega$$

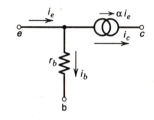

FIGURE 42-7 The *T* network model of a transistor.

This model reflects the structure of the transistor. The resistor across the current generator models the effects of the collector–base voltage on the output. Also, the model is fairly easy to use and lends itself to approximations while in use. However, for some reason it is not used very much in the literature.

The simplest model for the transistor is called the **reduced T model**; it consists of one resistor and one current generator as is shown in Figure 42-8. The current generator produces a current that depends on the emitter or base current. Remember that

$$I_c = \alpha I_e = \beta I_b \tag{42-11}$$

For a typical transistor, r_b is slightly less than 1000 Ω. None of the imperfections in the transistor is included in this model. This model is roughly equivalent to that analyzed in the section on "Common Base Configuration" above. Although this is the simplest and least accurate model described here, it is the only one that will be used in this book.

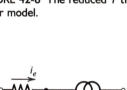

FIGURE 42-8 The reduced *T* transistor model.

The **Ebers–Moll model** is somewhat similar to the reduced T model; however, it explicitly incorporates some of the nonlinear effects of the transistor (see Figure 42-9). For this model, the collector current is given by

$$I_c = I_s\left[\exp\left(\frac{V_{be}}{V_T}\right) - 1\right] \tag{42-12}$$

where

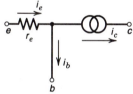

FIGURE 42-9 The Ebers–Moll model.

$$V_T = \frac{kT}{q} = 25.3\ \text{mV} \quad \text{at} \quad 68°\text{F} \tag{42-13}$$

and

$$r_e = \frac{V_T}{I_c} = \frac{25}{I_c}\ \Omega \tag{42-14}$$

when I_c is measured in mA.

As the Ebers–Moll model makes explicit, the parameters in the various models depend on the operating point of the transistor. The parameters also depend on temperature and frequency; in some cases, they may even be complex numbers. For some of the most common transistors, graphs of the parameters versus voltages and currents are available in the data books.

The parameters depend on the individual transistor. For a particular transistor type, some of the parameters may vary by a factor of 10 or more from one transistor to another. In calculations, generally typical values are used.

For high-frequency calculations, even more complex models are needed. This requires no changes in principle, but the calculations become sufficiently difficult that they are suitable only for computers. From physical arguments, expressions that relate the parameters in the equivalent circuits to physics properties of the circuit, such as size, materials, and so on, can be derived. These relationships are very useful for transistor designers.

AN EXAMPLE

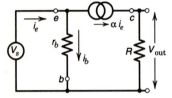

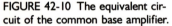

FIGURE 42-10 The equivalent circuit of the common base amplifier.

As an example of the use of a transistor model, the common base amplifier shown in Figure 42-3 will be analyzed by using the reduced T model. This analysis concerns only the **small-signal** behavior; all the effects of the biasing batteries are involved in determining the operating point. All the voltages and currents involved in the small-signal analysis are simply small changes in the operating point values.

The circuit in Figure 42-3 is replaced by the equivalent transistor model; the batteries that have to do only with the determination of the operating point are omitted. The resulting circuit is shown in Figure 42-10. From this figure, using KCL and KVL and Equations 42-3 and 42-4, it follows that

$$v_s = i_b r_b \tag{42-15}$$

$$v_{\text{out}} = i_c R = \alpha i_e R = \beta i_b R \tag{42-16}$$

and

$$i_b = (1 - \alpha)i_e \tag{42-17}$$

The voltage gain is given by

$$G_v = \frac{v_{\text{out}}}{v_s} = \frac{\beta i_b R}{i_b r_b}$$

$$= \frac{\beta R}{r_b} \tag{42-18}$$

which is just what was obtained before in Equation 42-7 with r_{in} replaced by r_b/β.

Also, for this circuit the current gain can easily be calculated:

$$G_i = \frac{i_{\text{out}}}{i_{\text{in}}} = \frac{I_c}{I_e} = \alpha \tag{42-19}$$

From these two results, the power gain is given by

$$G_p = G_i G_v = \alpha \beta \frac{R}{r_b} \tag{42-20}$$

For a typical transistor circuit with $R = 10{,}000 \ \Omega$, $r_b = 800 \ \Omega$, and $\beta = 100$, the voltage gain is 1250, the current gain is 0.9988, and the power gain is

1248. Furthermore, the output impedance of the amplifier is 10,000 Ω and the input impedance r_{in} is 8 Ω.

TRANSISTOR TERMINOLOGY

As with all electronic components, a wide variety of parameters may be used to describe the capabilities of a transistor. With a knowledge of the terminology used to describe IC amplifiers and solid state diodes, most of the transistor terminology should be self-explanatory. However, just in case it is not, a brief glossary of some of these terms is given in this section. They are organized in more or less the order in which they are usually given in data sheets.

Type

The transistor **type** is usually described in one or two sentences. This description indicates the use for which the transistor is intended (high frequency, high power, small signal, switching, etc.); its structure, *npn* or *pnp*, or some kind of FET structure; perhaps some brief indication of the fabrication technique; and usually some indication of the size and package shape.

Maximum Ratings

Various maximum ratings are often given. These include the **maximum collector–emitter voltage**, the **maximum collector–base voltage**, and the **maximum emitter–base voltage**. These are the maximum voltages that the transistor can withstand without breakdown.

The **maximum collector current** the transistor can handle is often specified. Only rarely will a maximum emitter or base current specification be given.

The **maximum power dissipation** for the transistor is generally specified; frequently, this specification will be given at two or more temperatures, and there may even be a graph of maximum power versus temperature. For high-powered devices, this may depend on the nature of the heat sink and the cooling techniques used.

Thermal Characteristics

The **operating and storage junction temperature range** gives the temperature range over which the transistor can be used and over which it can be stored. There may also be some sort of specification about how hot the leads can be for how many seconds without damaging the junction; this is of interest if the transistor is to be soldered into a circuit.

For transistors that are intended to handle substantial power (low-power transistors can handle less than 100 mW), data about thermal resistance are often given. The most common parameter is the **thermal resistance, junction to ambient.** This parameter has units of °C/mW or °C/W and is used to calculate the temperature rise as a function of power dissipated. For high-power transistors, these data may be given in terms of the heat loss to various kinds of heat sink hardware.

Off-Characteristics

Several parameters describing the transistor in its cutoff state are generally given. These include the **collector–emitter breakdown voltage**, the **collector–base breakdown voltage**, and the **emitter–base breakdown voltage**. These are the voltages at which the various junctions will break down. These breakdown voltages are generally the same as those given in the maximum ratings.

Frequently, the **collector cutoff current** and the **emitter cutoff current** will be specified. These are the currents that flow in the reverse biased junctions; their specifications will include temperature and biasing information.

On-Characteristics

A somewhat larger number of parameters are given to describe the transistor when it is in its active region. These include the **dc current gain** (this is β or h_{fe}). As the name implies, it is measured at dc. A complete specification will include the voltages and currents at which the measurement was made. The **collector–emitter saturation voltage** and the **base–emitter saturation voltage** are the voltages between the various elements when the transistor is driven into saturation. A saturated transistor acts somewhat like a three-way connection with fixed voltage drops between the terminals.

Small-Signal Characteristics

Depending on the use for which the transistor is intended, there may be only a few or many parameters given under this heading. The **small-signal current gain** is another value of β. It is the value that refers to small signals. The value and test specifications may or may not be the same as for the value of dc current gain. The **high-frequency current gain** is a value for β with specifications about the voltages, currents, and frequency at which the measurement was made. This can be used as an indication of the performance of the transistor at high frequencies. The **current gain–bandwidth product** f_T is another way of specifying the high-frequency capabilities of the transistor. Careful study of the test setup is needed to understand what this parameter means.

The **output capacity** and **input capacity** are the values of the parasitic capacitances across the base–emitter junction and the base–collector junction. These are also needed for high-frequency work.

Finally, a **noise figure** of some nature may be given for transistors that are intended for low-noise operations.

Switching Characteristics

Several parameters are often given that describe how the transistor will operate when it is used in situations where it is switched between cutoff and saturated states. These include **delay time**, which is a measure of the delay after a change takes place at the input of the transistor circuit before the output begins to change. This delay is related to how long it takes the charges to drift across the various elements of the transistor. The **rise time** measures how rapidly the collector current can increase.

If a transistor is driven into saturation and then suddenly switched off, the collector current does not start to decrease for a period of time that is determined by how long the charges stored in the emitter–base structure take to diffuse out. The **storage time** is a measure of this time. Transistors are rarely symmetrical—the collector current can often rise faster than it can fall, or vice versa. The **fall time** is a measure of how fast the collector current can drop from its initial to its final value.

Chapter 43

A CATALOG OF TRANSISTOR CIRCUITS

OBJECTIVES

1. Become familiar with the important features of the most common transistor configurations, including the common base amplifier, the common emitter amplifier, the split load amplifier, the emitter follower, the Darlington pair, the difference amplifier, and the NAND gate.

2. Know the most common ways of biasing simple transistor circuits.

INTRODUCTION

This chapter is a short list of transistor amplifier and biasing circuits. The first sections deal with the most common amplifier configurations whereas the last section deals with biasing techniques. The most complex circuits discussed are a simple NAND, which illustrates the basic circuit used in TTL, and a simple difference amplifier, which forms the basis of most op-amps.

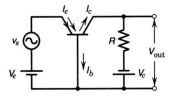

FIGURE 43-1 The common base configuration.

COMMON BASE AMPLIFIER

The common base configuration shown in Figure 43-1 was discussed in Chapter 42. Only the results of that analysis are given here. Typical values for the circuit components and transistor paramters are

$$R = 10{,}000 \ \Omega$$

$$r_b = 800 \ \Omega$$

$$\beta = 100$$

From the circuit analysis and with these values, the voltage gain is

$$G_v = \beta \frac{R}{r_b} = 1250 \qquad (43\text{-}1)$$

the current gain is

$$G_i = \alpha = 0.990 \qquad \text{(43-2)}$$

and the power gain is

$$G_p = G_v G_i = 1238 \qquad \text{(43-3)}$$

The input resistance is

$$r_{\text{in}} = \frac{r_b}{\beta} = 8 \ \Omega \qquad \text{(43-4)}$$

and the output resistance is

$$r_{\text{out}} = R_L = 10{,}000 \ \Omega \qquad \text{(43-5)}$$

This circuit is useful when a low input impedance and a high output impedance is needed. In such situations, it serves as an impedance transformer while still producing some voltage and power gain. It is a noninverting amplifier, which means that the output has the same sign as the input.

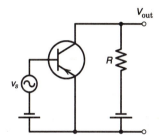

FIGURE 43-2 The common emitter configuration.

COMMON EMITTER AMPLIFIER

The **common emitter amplifier** shown in Figure 43-2 is made with a *pnp* transistor, as are all the examples in this chapter. If an *npn* transistor were used, all the biasing batteries would have to be reversed, but none of the conclusions reached in the analysis would change. As with all the normal amplifier configurations, the emitter–base junction is biased forward whereas the collector–base junction is biased backward. In this example, the bias is provided by the batteries; in the last section of this chapter, other schemes of biasing the circuits are discussed.

If the input signal v_s causes the base to become slightly more positive, the base current will decrease slightly. The collector current will decrease β times as much; as a result, the voltage drop across R will decrease, and v_{out} will become more negative. Thus a positive going input causes a negative going output, and this configuration has inverting voltage gain. A mathematical expression for the gain may be obtained by use of the modified T equivalent circuit for the transistor. Using the model and omitting the batteries, which are not related to the small-signal analysis, gives the circuit shown in Figure 43-3. The only tricky point is getting the signs correct. As can be seen

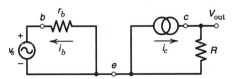

FIGURE 43-3 The equivalent circuit for the common emitter amplifier.

from the diagram, a positive v_s causes a negative i_b, as was stated above. The analysis is easy; by using the same values for $R, r_b,$ and β as above,

$$i_b = -\frac{v_s}{r_b} \tag{43-6}$$

$$v_{\text{out}} = \beta i_b R = -\beta \frac{v_s R}{r_b} \tag{43-7}$$

The voltage gain is given by

$$G_v = -\beta \frac{R}{r_b} = -1250 \tag{43-8}$$

where the minus sign means inverting. The current gain is

$$G_i = \beta \tag{43-9}$$

and the power gain is

$$G_p = G_v G_i = 125{,}000 \tag{43-10}$$

The input resistance of the amplifier is

$$r_{\text{in}} = r_b = 800 \ \Omega \tag{43-11}$$

and the output resistance is

$$r_{\text{out}} = R = 10{,}000 \ \Omega \tag{43-12}$$

The common emitter configuration is the standard configuration where high gain is needed; it has both voltage and current gain and the highest possible power gain. However, it has two deficiencies. First, it has a moderately low input resistance so that it tends to load down the signal source. But the main problem with this configuration is that the gain directly depends on the values of β and r_b. Since these parameters vary greatly from transistor to transistor, this means that each circuit has to be individually tweaked to obtain the desired results. This is not good for mass production, and it means that repairs that require changing a transistor can be very difficult. The solution to these problems is to use some negative feedback. This is discussed in the next section.

SPLIT LOAD AMPLIFIER

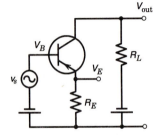

FIGURE 43-4 The split load amplifier.

The addition of one resistor, R_E, to the common emitter circuit dramatically changes the nature of the amplifier; this is called the **split load amplifier** (Figure 43-4). This resistor introduces negative feedback into the circuit and makes the gain much less dependent on the transistor's β and r_b. To understand the operation of the amplifier, consider the following sequence of events, each of which is caused by the preceding event:

Start by assuming that V_s goes slightly positive; this means that V_B increases. This means that the forward bias on the base–emitter junction is decreased.

Therefore, i_b decreases.

Therefore, i_c decreases β times as much.

But i_c flows through R_E. As a result of the decrease in i_c, the voltage drop across R_E decreases and V_E becomes more positive—less negative.

But the bias on the junction is the difference between V_B and V_E. As a result of these changes, the decrease in the forward bias is not as great as it would have been without R_E being present.

This sequence began with an increase in V_B due to v_s. As a result, V_E increases. Thus the forward bias is not increased as much as it would have been if R_E were not there. Thus the change in i_c and the change in v_{out} is less, and the gain is less than if R_E were not there. R_E has fed back some of the output into the input circuit to cancel the changes in the input. This is negative feedback.

In Figure 43-5, the transistor has been replaced by its equivalent circuit. Using KCL and KVL (watching the signs) gives

$$i_e = i_b + i_c = i_b + \beta i_b$$
$$= (\beta + 1)i_b \tag{43-13}$$

$$v_s = i_b r_b - i_e R_E$$
$$= -i_b r_b - (\beta + 1)i_b R_E$$
$$= -i_b[r_b + (\beta + 1)R_E] \tag{43-14}$$

$$v_{out} = i_c R_L = \beta i_b R_L$$
$$= -\frac{\beta R_L v_s}{r_b + (\beta + 1)R_E} \tag{43-15}$$

The voltage gain is

$$G_v = \frac{v_{out}}{v_s} = -\frac{\beta R_L}{r_b + (\beta + 1)R_E} \tag{43-16}$$

The minus sign means that the amplifier inverts. Using the same typical values as above plus a value of $R_E = 1000\ \Omega$ for the emitter resistor gives us

$$G_v = -9.823$$

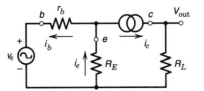

FIGURE 43-5 The equivalent circuit for the split load amplifier.

Notice that this is almost the same as

$$\frac{R_L}{R_E} = 10.0 \approx -G_v \tag{43-17}$$

The most important feature of Equation 43-16 is that it is almost independent of the values of β and r_b. This can be seen both numerically and by approximating the equation. For typical values, r_b can be ignored compared with $(\beta + 1)R_E$, so the first step in the approximation is

$$G_v = -\frac{\beta R_L}{r_b + (\beta + 1)R_E} \approx -\frac{\beta R_L}{(\beta + 1)R_E}$$

Again, for normal values, $\beta/(\beta + 1)$ is almost 1, yielding

$$G_v \approx -\frac{R_L}{R_E} \tag{43-18}$$

This approximate result is clearly independent of the transistor parameters. The numerical values above show that, for this particular case, the approximation differs from the exact result by less than 2%, which is generally far more accurate than the values of r_b, β, and perhaps even R_L and R_E. The similarity between this result and the basic inverting op-amp circuit should be obvious.

The analysis above also shows that the input resistance is given by

$$r_{in} = r_b + (\beta + 1)R_E >> r_b$$
$$= 101{,}800 \ \Omega \tag{43-19}$$

and that the output impedance is still R_L.

The split load amplifier is the workhorse of transistor circuits. It has a high input impedance so that it does not load down the signal source. It has a reasonable voltage gain that is almost independent of the transistor parameters. As is discussed in the last section of this chapter, the negative feedback also gives the amplifier thermal stability and prevents thermal runaway.

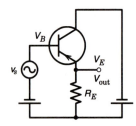

FIGURE 43-6 The emitter follower configuration.

EMITTER FOLLOWER

The **emitter follower,** or **voltage follower,** is simply a split load amplifier with no collector resistor (Figure 43-6). Since the base–emitter junction has a typical diode characteristic, the base voltage V_B is essentially one diode drop more than the emitter voltage. Thus for a silicon transistor, V_{out} will be about 0.6 V less than the base voltage V_B. (To be perfectly precise, all the signs have to be correct; since V_B and V_E are both negative, V_E will really be 0.6 V more positive than V_B. But the statement above is correct for the absolute values of the voltages.) As a result, the emitter voltage v_E will follow the base voltage V_B, and the emitter current will be $\beta + 1$ times the base current.

The quantitative analysis has already been done in the previous section. The output voltage can be obtained from the results of the previous section (again watching the signs). Using Equation (43-13),

$$V_E = -i_e R_E$$

$$= -(\beta + 1)i_b R_E$$

$$= \frac{v_s(\beta + 1)R_E}{r_b + (\beta + 1)R_E} \tag{43-20}$$

and the voltage gain is

$$G_v = \frac{V_E}{v_s} = \frac{(\beta + 1)R_E}{r_b + (\beta + 1)R_E} = 0.992 \tag{43-21}$$

The approximations used in the previous section give a gain of 1. As the numerical value shows, this approximate value differs by less than 1% from the actual value. The current gain, measured for the emitter current, is

$$G_i = \beta + 1 = 101 \tag{43-22}$$

The input resistance is the same as for the split load amplifier:

$$r_{in} = r_b = (\beta + 1)R_E = 101,800 \ \Omega \tag{43-23}$$

and the output resistance is

$$r_{out} = R_E = 1000 \ \Omega \tag{43-24}$$

The emitter follower is a noninverting unity-voltage-gain current amplifier. As a sample use, it may be used to allow a low-power op-amp output to control a high-power, high-current device. By including the transistor in the feedback loop as shown in Figure 43-7, even the 0.6-V drop can be compensated. Thus in the circuit shown in the figure, the output voltage tracks the voltage input to the op-amp almost exactly, but the possible current available in the output circuit is β times as large as that available from the op-amp. Obviously, this can require a macho power transistor.

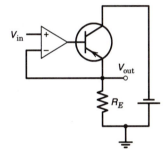

FIGURE 43-7 An emitter follower current booster for an op-amp voltage follower.

DARLINGTON PAIR

The configuration shown in Figure 43-8 is called a **Darlington pair**. An analysis of this circuit is shown on the figure. The configuration is equivalent to a single transistor having an equivalent gain of $(\beta + 1)^2$. Thus two transistors having a β of 100 connected together in this configuration have an equivalent β of 10,200. But this gain is not obtained for nothing. The input and output capacitances of the circuit are larger by a factor of β than those for a single transistor. The delay times, storage times, and rise and fall times are all much longer than for a single transistor. For this reason, the configuration can only be used at low frequencies.

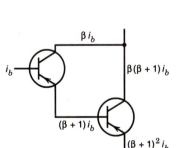

FIGURE 43-8 A Darlington pair with the analysis results indicated.

Darlington pairs can be purchased already made up in a case with three leads. They are very useful in power supplies, emitter followers, switching circuits, and difference amplifiers—see the next section.

DIFFERENCE AMPLIFIERS

The circuit shown in Figure 43-9 is a slightly simplified version of a **difference amplifier.** A real difference amplifier would have some sort of network to provide the biasing voltages but, since these add nothing but complexity, they are merely indicated as bias in the diagram. This amplifier consists of two split load amplifiers sharing a common feedback (emitter) resistor. Thus a signal input to either input will be fed back into both sides of the amplifier. The operation of the circuit depends on the fact that the total current through R_E remains more or less constant; the inputs simply change the way in which this current is shared by the two transistors. The operation of this circuit is moderately complex. A step-by-step quantitative analysis with comments follows.

With no input signal, the bias supplies are adjusted so that the voltage drops across R_{L1} and R_{L2} are the same and $V_{out} = 0$. This adjustment corresponds to the offset adjustment on most op-amps.

If V_{in1} increases slightly, this decreases the forward bias on Q_1; hence I_{Q1}, the current through Q_1, decreases. This means that the voltage drop across R_{L1} decreases and the voltage at point A is closer to the negative power supply, or V_A is more negative.

Also, because of the reduced current in Q_1, V_E increases. This increases the forward bias on Q_2, which in turn causes I_{Q2} to increase, which causes V_B to become more positive. This increase in the current in Q_2 tends to cancel the increase in V_E. Because of the changes in the currents in the two transistors, there is a V_{out}. If everything is well matched; an increase in V_{in1} has caused current to switch from Q_1 to Q_2, leaving the total current unchanged. Because of this, the arrangement is sometimes called a **current mirror,** or a **current sharing** circuit.

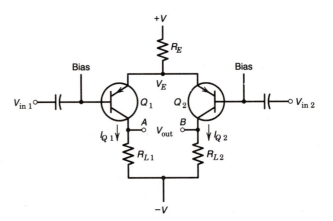

FIGURE 43-9 A simple difference amplifier.

Remember, an increase in V_{in1} caused V_A to become more negative and V_B to become more positive. A similar analysis shows that if V_{in2} increases slightly, V_A becomes more positive and V_B more negative. Thus again, there is a V_{out} but of opposite sign than before. Depending on the sign conventions used, one of the two inputs will be inverting and the other, noninverting.

Finally, if both the input voltages increase by the same amount, the ratio of the current sharing is not changed and V_{out} remains 0. Hence, this is a difference amplifier, that is, an amplifier whose output is proportional to $V_{in1} - V_{in2}$.

A complete quantitative analysis of this circuit would take many pages of algebra. But, here, the solutions for two special cases will simply be stated. For these two special cases, it is assumed that the two transistors are identical. This means that

$$r_{b1} = r_{b2}$$

$$\beta_1 = \beta_2$$

The differential mode can be investigated by assuming that only one input is nonzero. Because of the symmetry of the amplifier, there will be no difference other than a minus sign, depending on which input is chosen. Here, the case where $v_{in1} = v$ and $v_{in2} = 0$ will be investigated. With this input and dropping small terms, the following results are obtained:

$$i_{b1} \approx \frac{v}{2r_b} \tag{43-25}$$

and

$$i_{b2} \approx -\frac{v}{2r_b} \tag{43-26}$$

Note that, to the accuracy of the approximations used here, the current sharing is exact. More exact solutions can be used to investigate the accuracy of the current sharing if desired.

If the two load resistors are matched, that is, if

$$R_{L1} = R_{L2} = R_L$$

then the differential voltage gain is given by

$$A_{diff} = \frac{V_{out}}{v_{in}} = \beta \frac{R_E}{r_b} \tag{43-27}$$

For a normal transistor circuit, with $\beta \approx 100$, $R_L = 10,000$, and $r_b = 1000$, the differential voltage gain $A_{diff} \approx 10^3$. If the circuit were made with a Darlington pair with $\beta = 10^4$, then the differential gain would be $A_{diff} \approx 10^5$.

The response of an amplifier to the same input at both inputs is called the **common mode** response of the amplifier. To show that this really is a difference amplifier, it is necessary to show that the common mode response is nearly zero. To do this, the response of the amplifier is calculated for an input of $v_{in1} = v_{in2} = v$. When this is done, the following result is obtained:

$$A_{com} = \frac{V_{out}}{v} = \frac{R_{L1} - R_{L2}}{2R_{E\pi}}$$
(43-28)

This is small in any case, and it is zero if $R_{L1} = R_{L2}$.

The difference amplifier forms the basis of essentially all op-amps, either integrated versions or amplifiers made with discrete components. As has been shown, with well-matched transistors and resistors, the amplifier can have very high differential gain and essentially zero common mode gain. Modern laser trimming technology makes it possible to build amplifiers that have very accurately matched components so that very low common mode gains can be obtained.

All that really is missing from this difference amplifier is some means of converting the existing output into a single-ended signal that is referenced to ground. For an amplifier that is to work only in ac circuits, either transformer coupling or capacitive coupling can be used to do this. For amplifiers that are to work at dc, a more complex level shifting circuit is needed.

Integrated op-amps are built with Darlington pairs in the difference amplifier. Furthermore, since it is easier to build a transistor than it is to make a high resistance on an IC, transistors are used as the resistors R_L and R_E. With this introduction to difference amplifiers, most of the schematic for a normal op-amp such as the 741 should be understandable (see Appendix B).

A NAND CIRCUIT

The circuit shown in Figure 43-10 is a logic NOT, or a logic inverter. The transistors are operated as saturated switches; this means that the transistors are either turned off (base–emitter junction biased backward), or they are turned on so hard that they go into saturation and the base, emitter, and collector voltages are all essentially the same. The analysis is relatively simple.

If the input is Low, transistor T_1 is turned on. Current is flowing through the transistor and out the emitter. The transistor is driven into saturation. As a result, the base and the collector of T_1 drop almost to zero. This means that the emitter of transistor T_2 (point A) cannot be much above zero either. Therefore, transistor T_2 is turned off and the output voltage is essentially V_{cc} (+5 volts). Thus a Low input causes a High output.

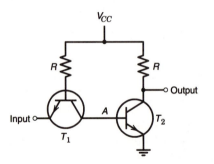

FIGURE 43-10 A logic NOT made from two *npn* transistors.

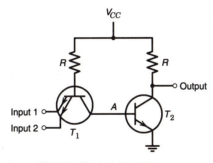

FIGURE 43-11 A NAND circuit.

When the input is High, the base–emitter junction of transistor T_1 is not biased forward, so the transistor is not saturated. This means that the base of T_1 is at a voltage not too far below V_{cc}. In addition, point A cannot be more than about 0.6 V positive, since it is only one diode drop above ground. Because the collector of T_1 is connected to point A and the base is positive, the T_1 collector–base junction is biased forward. Hence, current flows through R, then through the base–collector junction of T_1, and then through the base–emitter junction of T_2. This also means that transistor T_2 is turned on and is saturated; and, as a result, the collector voltage of T_2, V_{out}, is approximately 0 V (less than 0.3 V). Thus a High input causes a Low output, and the fact that this circuit is a NOT has been verified.

The configuration shown in Figure 43-11 is a logic NAND. Its operation is exactly the same as that of the NOT, except that the input transistor T_1 has multiple emitters (two in the case shown). If either of the emitters is Low, then transistor T_1 is turned on and the output is High. Only if both the inputs are High will the output be Low. Converting this into a truth table gives the following:

Inputs		Output
H	H	L
H	L	H
L	H	H
L	L	H

This is the truth table of a NAND gate. Finally, if the concept of a multiemitter transistor seems strange, a transistor having three emitters is shown in Figure 43-12.

With the technology to make multiemitter transistors, NANDs with any number of inputs can easily be manufactured. As was pointed out in Chapter 19, DeMorgan's laws say that all other logic functions can be constructed from NOTs and NANDs. Thus in principle all the integrated logic functions can be constructed from the two circuits shown in Figure 43-12. In actual fact, there are other ways to construct ANDs and ORs other than by simply combining NOTs and NANDs; but these are outside the scope of an introduction.

The circuits shown in Figure 43-12 are the basic circuits used in TTL logic. The analysis shows why this is current sinking logic. For an input to be pulled Low, the current flowing through the base–emitter junction must flow out of the circuit through the input. This current is set by the value of V_{cc}/R. Likewise, when an input is High, only the leakage current needs to be sup-

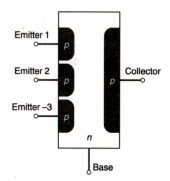

FIGURE 43-12 A three-emitter transistor.

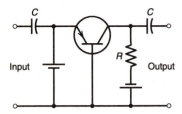

FIGURE 43-13 A common base amplifier biased with two batteries.

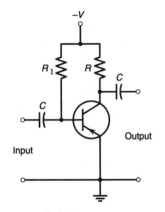

FIGURE 43-14 A very poor way to bias a common emitter circuit.

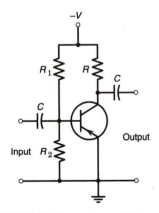

FIGURE 43-15 A voltage divider biasing scheme.

plied. These are the reasons for the widely asymmetrical values for the input currents when the input is held High and Low (see Chapter 22).

BIASING TECHNIQUES

The last practical problem to be discussed is how to bias transistor circuits. This is a major problem in the design of multitransistor circuits, since without considerable care, many different power supplies or batteries are needed to bias each of the transistors so that its output is at a suitable voltage for the next stage.

The general problem is that the first few tenths of a volt applied to the emitter–base junction do nothing. A voltage of about 0.3 V for a germanium transistor or 0.6 V for a silicon transistor is needed before the transistor starts working. In practice, the operating point of a transistor being used as an amplifier should be chosen at some convenient point on the load line—generally in the middle, not at either end. In this way, a small input of either sign will cause changes in the collector current with no threshold effects.

The simplest way to bias a transistor circuit is to use separate batteries or power supplies on the emitter and the collector circuits while coupling signals from one stage to the next by means of capacitors (Figure 43-13). This, of course, prevents the circuit from operating at dc or low frequencies. It also quickly gets very expensive when multiple stages are used.

The common emitter circuit is easier to bias because it needs only one polarity of power supply. The circuit shown in Figure 43-14 is very simple, but it has a number of problems. The operating point is not well defined because the currents flowing will depend on the β of the transistor. Also, the operation is not thermally stable—this is discussed below. As a result, this biasing method should be used only with the greatest caution.

The configuration in Figure 43-15 is much better. The operating point is determined by the voltage divider, which is independent of the transistor parameters. Thus the operating point can be put exactly where it is wanted. However, the circuit is still thermally unstable. Remember that the current through a forward biased diode depends on the temperature of the *pn* junction in the diode, and the current increases as the junction gets warmer. In this circuit, as the transistor heats up, the current at the fixed bias voltage increases, the power dissipated in the transistor increases, the junction heats up more, more current flows, and trouble may result. Circuits built this way tend to operate satisfactorily up to some given room temperature and then suddenly run away, perhaps even resulting in burnt-out transistors.

The values of the resistors in the voltage divider are a compromise between being low enough so that the standing current in the divider is much larger than the base current in the transistor and high enough so that the parallel combination of the resistors does not load down the signal source. Fortunately, it is generally easy to satisfy both goals. Remember that since the power supply voltage does not change, it is essentially signal ground.

In Figure 43-16, a split load amplifier is biased with a voltage divider from a single power supply. The operating point is determined by the voltage divider and is stable as a result. The circuit is protected from thermal runaway by the negative feedback due to the emitter resistor R_E. For a fixed base voltage, if the emitter current starts to increase because of thermal effects, the voltage drop across the emitter resistor increases, the emitter voltage increases, and the forward bias on the base—emitter junction is decreased. As a result, this circuit is thermally stable. If the reduction in gain of the circuit is a problem, then the emitter resistor can be bypassed with a capacitor, as is shown in Figure 43-17.

Finally, if operation at dc is needed, the coupling capacitors between circuits must be removed. By using two power supplies and the split load amplifier shown in Figure 43-17, the input can be biased to zero volts, and at the same time the load line can be adjusted so that the output with no input is also at zero volts. Then, dc changes in the input will result in dc changes in the output. Obviously, for dc operation, it is impossible to use a bypass capacitor on the emitter resistor.

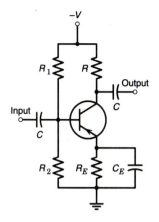

FIGURE 43-16 A voltage divider biasing circuit with emitter degeneration for dc. The emitter resistor can be bypassed with a capacitor, as is shown in the figure for greater ac gain.

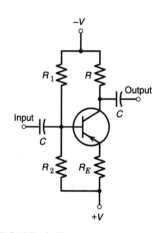

FIGURE 43-17 A dc amplifier.

Chapter 44

HOW TO UNDERSTAND TRANSISTOR CIRCUITS

OBJECTIVES

1. To become familiar with the techniques for determining what a transistor circuit is doing.

INTRODUCTION

Probably the most discouraging problem encountered by anyone working in electronics is to attempt to fix a nonworking circuit he or she did not design. Under ideal circumstances, there is complete documentation for the circuit, spare parts are readily available, and the repair goes fairly easily. In the more normal situation, only a schematic exists (or can be made from the circuit). This is a difficult problem. Success is not guaranteed, but there are some techniques that can be used to understand what the various parts of a circuit are doing. This chapter is a quick introduction to these techniques for bipolar transistor circuits. Obviously, only long experience will give confidence and facility in this process.

THE GOLDEN RULES

In general, a transistor can be operating in only one of three states: linear, saturated, or cut off. For each of these states there is a set of relationships between the voltages and currents at the three terminals of the transistor. These rules were called the **golden rules** by people working with transistors. These rules are outlined in this section, and a strategy for working out a circuit is given in the next section. Several examples are included at the end.

A transistor that is active—not cut off—and operating as a amplifier—linear and not saturated—is said to be in the **linear state**. In this case, the collector current is not zero and is not a strong function of the collector–base junction voltage. For a transistor operating in the linear range, the base current is small and the base–emitter voltage drop is small and relatively constant. The base–

emitter voltage drop is about 0.6 V for a silicon transistor and for many purposes can be approximated as zero so that the base and emitter act as if they were connected together. Finally, the collector current is approximately equal to the emitter current. Mathematically, these statements are

$$i_b \approx 0 \quad \text{(for } \beta \text{ very large)}$$

$$V_e \approx V_b$$

$$i_c \approx i_e$$

These statements hold for both *npn* and *pnp* transistors. By using the directions defined in the diagrams in Figure 44-1, the rules can be extended as given below. For active linear transistors,

pnp

npn

FIGURE 44-1 Current direction conventions in transistors.

pnp for Finite β	*npn* for Finite β
$i_e > 0$	$i_e > 0$
$i_c > 0$	$i_c > 0$
$i_b > 0$	$i_b > 0$
$V_c < V_b$	$V_c > V_b$
$V_c < V_e$	$V_c > V_e$

A **saturated** transistor acts as if the base, emitter, and collector are all connected together. Thus

$$V_e \approx V_b \approx V_c$$

For the current directions, defined in Figure 44-1, all of the currents are positive and are determined by Kirchoff's laws and the external components.

For a transistor that is **cut off**, there is no current flowing in any of the leads. The voltages of the various terminals are unrelated and are determined by the external components in the circuit, but they must satisfy the inequalities below or else the transistor will not be cut off:

$$i_b \approx i_c \approx i_e \approx 0$$

For an *npn*: $\qquad V_b < V_c \quad$ and $\quad V_b < V_e$

For a *pnp*: $\qquad V_b > V_c \quad$ and $\quad V_b > V_e$

A STRATEGY

The process of working out what an unknown circuit is doing involves a systematic trial-and-error approach. In general terms, a guess about how the circuit is operating is made. This guess (set of assumptions) allows calculating the voltages and currents in each transistor. These values can be compared with those required by the golden rules and the assumptions. If the values are consistent, then the assumptions are probably correct. If they are not consistent, then one or more of the initial assumptions have to be changed and the whole process must be repeated. In slightly more detail, these steps are as follows:

1. First, determine what is definitely known about the circuit. There usually are some voltages that can be calculated, no matter what the transistors are doing. Calculate all these values and indicate them on the diagram.
2. Start with one transistor, guess if it is linear, saturated, or cut off. Determine the voltages and currents using the golden rules and Kirchhoff's laws.
3. Work through the circuit transistor by transistor. Check the voltages on each transistor. If any contradictions are found, change an assumption and try again.
4. Usually only a single solution will be found. If more than one possible solution is found, there are real problems.

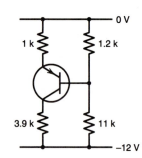

FIGURE 44-2 A simple transistor circuit. What is its state.

A SAMPLE ANALYSIS

The problem is to determine the state of the transistor shown in Figure 44-2. The author's strategy was to proceed as follows. First, try assuming that the transistor is cut off. If this does not work, try linear. If the linear solution works but $V_c \approx V_e$, then the transistor is saturated. At any step, an inconsistency means that the assumptions are wrong.

First, assume that the transistor is cut off. This means that

$$V_e = 0$$

$$V_c = -12 \text{ V}$$

$$V_b = -1.18 \text{ V}$$

where V_b follows from the fact that the base is connected to a voltage divider and no base current is flowing. But these results are not consistent with the transistor being cut off because $V_e > V_b$, which means that the base–emitter junction is biased forward.

Next, assume that the transistor is linear and that β is very large. This means that $i_b \approx 0$ and $V_b \approx -1.2$ V as above.

Therefore, $V_e \approx V_b \approx -1.2$ V (ignoring the V_{be} of 0.6 V at this time).

If V_e is 1.2 V, then I_e is 1.2 mA by Ohm's law and $I_c \approx 1.2$ mA, since $I_c \approx I_e$. With this collector current, the drop across the 3.9-kΩ resistor is 4.7 V and $V_c = -7.3$ V.

Finally, collecting these voltages gives

$$V_c < V_e \quad \text{and} \quad V_c < V_b$$

All this is consistent with the transistor being active.

Next, some refinements in the analysis are needed to check for any errors. It is necessary to ask if the base current is loading down the voltage divider connected to the base. Assume that β is about 100. Then,

$$i_b = \frac{I_c}{\beta} = \frac{1.2 \text{ mA}}{100} = 0.012 \text{ mA}$$

The standing current in the voltage divider is about 1 mA, so that the base current changes the operation of the voltage divider by about 1%, which can be ignored.

The last refinement of the analysis above takes into account the nominal voltage drop between the base and emitter of 0.6 V. This gives the following results:

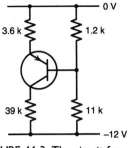

FIGURE 44-3 The circuit for Problem 44-1.

$$V_e = -0.6 \text{ V}$$

$$I_e = 0.6 \text{ mA}$$

$$I_c = 0.6 \text{ mA}$$

$$V_c = -9.6 \text{ V}$$

$$i_b = 0.005 \text{ mA}$$

and everything is still consistent with the transistor being linear.

PROBLEMS

44-1 Determine the state of the transistor in Figure 44-3.

44-2 Determine the state of the transistor in Figure 44-4.

44-3 Determine the state of the transistor in Figure 44-5.

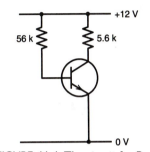

FIGURE 44-4 The circuit for Problem 44-2.

ANSWERS TO ODD-NUMBERED PROBLEMS

44-1 linear
44-3 cut off

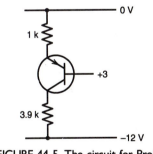

FIGURE 44-5 The circuit for Problem 44-3.

Chapter 45

FETs

OBJECTIVES

1. Become familiar with the basic structure and operation of FETs.

INTRODUCTION

Besides the bipolar transistors discussed in the previous chapters, other types of transistors and related solid state devices have been developed in recent years. Two of the more important are the JFET and the MOSFET. Because these two devices are planar structures manufactured by the same photo-optical processing technology that is used to make integrated circuits, these transistor types only become available at the same time that ICs began to replace complex discrete component circuits. Thus these transistor types have never become as common in discrete component circuits as are bipolar transistors. However, FETs are extremely important in IC design.

As a result, little if any information about FETs is needed in an introduction to electronics other than that needed to make some of the descriptions and terminology about ICs meaningful. This is what this chapter provides. It is a brief introduction to FET-type transistors. Its main purpose is to provide the background necessary to understand more fully some of the discussions of ICs and related circuits elsewhere in this book and in the literature as a whole. The actual design of FET circuits will require more advanced study.

JUNCTION FIELD-EFFECT TRANSISTORS

In the bipolar transistors discussed in the previous chapters, the collector current is controlled by the current injected by the emitter, that is, they are current-controlled devices. For the transistors discussed in this chapter, the controlling element is an electrostatic field; these are called **field-effect transistors**, or **FETs**. FETs are made from *p*- and *n*-type semiconductors, just as are bipolar transistors. However, the arrangement of the regions, the characteristics of the device, and the way in which they are used are different. As a

388

result, there are different schematic symbols and an entirely different terminology for FETs.

The controlling device in an FET is called the **gate**. There are two basic variations on the FET. In one, the gate forms a bipolar junction with the rest of the transistor structure; this is called the **junction field-effect transistor**, or **JFET**. In the other form, the gate is insulated from the rest of the transistor by a layer of oxide; this is the **metal-oxide-semiconductor field-effect transistor**, or **MOSFET**. The JFET is discussed in this section; the MOSFET is discussed in the next one.

The simplest JFET, the names of its elements, and its schematic symbol are shown in Figure 45-1. The conduction path from the source to the drain is through one kind of semiconductor material; in the case shown in the figure, this is an *n*-type semiconductor. This means that the source-drain voltage–current characteristic is resistive; it has no threshold effects, and it is the same for either polarity voltage. The resistance of the channel depends on its length, its cross-sectional area, and its resistivity; the resistivity in turn depends on the doping of the semiconductor material and the temperature. The source and the drain are interchangeable; that is, the device is symmetric.

The gate is of the opposite type of semiconductor material, in this case *p*-type. In the normal operation of a JFET, the gate is **always** biased backward. This creates a depletion region in the *n*-type channel near the *p*-type gate. The depletion region decreases the cross-sectional area of the conduction channel. Figure 45-2 illustrates this reduction of the area of the conduction channel in a schematic way. Thus the gate voltage controls the resistance of the source–drain conduction channel by controlling the cross-sectional area of the conduction channel. It is because the control of the source–drain resistance is due to the electrical field of the gate that this device is called a field-effect transistor.

The detailed characteristics of a JFET transistor depend on the details of the size and geometry of the structure and the levels of doping of the gate and the conduction channel. For some JFETs, conduction can essentially be cut off totally, or **pinched off**, by a suitable voltage on the gate. This kind of transistor can be used as a switch; it is either conducting with some nominal resistance, or it is totally cut off. Slightly more complex FET structures are used as **solid state relays** in power circuits; these have resistances of only tenths of a ohm, yet are capable of carrying hundreds of amps or holding off hundreds of volts ac.

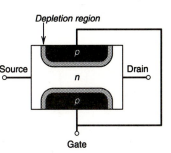

FIGURE 45-2 A JFET with the depletion region reducing the area of the conduction channel.

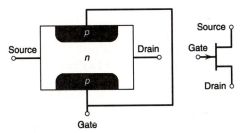

FIGURE 45-1 An *n*-channel JFET and its schematic symbol. The symbol for a *p*-channel JFET has the arrow pointing the other way.

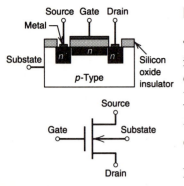

FIGURE 45-3 An *n*-channel depletion MOSFET transistor and its schematic symbol.

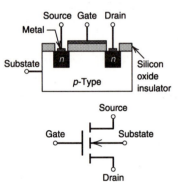

FIGURE 45-4 An *n*-channel enhancement MOSFET and its schematic symbol.

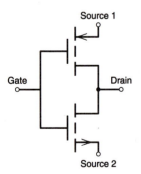

FIGURE 45-5 A simple CMOS configuration.

MOSFETS

The MOSFET has a structure similar to that of a JFET, except that the gate is insulated from the conduction channel by a layer of insulation—glass or silicon dioxide. A typical MOSFET is shown in Figure 45-3. It is this structure that gives rise to the name metal-oxide-semiconductor field effect transistor. Whereas the gate of a JFET is always biased backward, MOSFETs come in two varieties, depending on the nature of the applied bias. These are called enhancement and depletion types. They are each discussed briefly below.

Figure 45-3 shows the structure and symbol for a **depletion-type *n*-channel MOSFET**. Since the gate is insulated from the rest of the structure, no gate current will flow, no matter which polarity gate voltage is applied (aside from capacitive effects). Thus the gate has a very high input impedance. For the device shown, a positive voltage on the gate will attract more electrons into the conduction channel and will decrease the resistance of the channel. A negative bias on the gate tends to repel the negative carriers from the channel region; hence, the resistance increases. At some bias voltage, there are no carriers left in the channel, resulting in a very high channel resistance.

The structure and symbol for an **enhancement-type MOSFET** are shown in Figure 45-4. The operation of this device is not obvious from the previous discussions of semiconductor physics. With no bias on the gate, the *pn* diodes prevent current from flowing in either direction. This is indicated by the breaks in the line of schematic symbol. Positive charge on the gate material induces a negative charge just below the insulator, thus providing a narrow *n*-channel between the source and the drain. Thus if the gate is made positive, there are free negative charges in the channel and current can flow. If the gate is made negative, there is no free charge and no current flows.

CMOS DEVICES

Finally, this chapter ends with a very brief description of the most primitive CMOS-type device. A combination of *p*-channel enhancement and an *n*-channel enhancement device with a common substrate is schematically shown in Figure 45-5. This is called a **CMOS**, or complementary MOS, circuit because both types of FETs are present. In normal operation, one or the other of the two MOSFETs is turned on and the other is off. Thus the output of the circuit, the drain, is connected either to source 1 or to source 2 via a resistive channel. For the example shown, a positive voltage turns on the upper FET and the output is at the voltage of source 1. Likewise, if the input were a negative voltage, the lower FET is turned on and the output is at the voltage of source 2. If source 1 is negative and source 2 is positive, this is a logic inverter, since a high input causes a low output, and vice versa. In neither case is any current following through both FETs because one or the other is always off. Only when the output is switching from one state to the other is current drawn. As a result, this circuit uses very little power. This is the simplified basis of all CMOS logic.

Chapter 46

TRANSISTOR FABRICATION

OBJECTIVES

1. Become familiar with some of the ways in which transistors are made and some of the terminology that goes with these techniques.

2. Become familiar with the technique used to make an integrated circuit.

INTRODUCTION

The purpose of this chapter is to present a brief description of the techniques used to make transistors and integrated circuits. The intent of these descriptions is to give a flavor of the actual techniques. Most of the details and subtleties of the processes have been omitted. Many different techniques are used in making transistors and integrated circuits. New techniques are being continuously developed; this is an area of intense interest to the manufacturers, since an improvement in processing technology may mean profits of hundreds of millions of dollars. As a result, the details of the processes are a matter of continual change and are often highly secret. Finally, although the techniques are discussed here as if they were all independent, in actuality, combinations are often used in manufacturing a single device.

POINT CONTACT TRANSISTORS

The earliest transistors were called **point contact transistors**, since they were constructed by the contact points of wires on a block of bulk semiconductor. The nature of the point contact transistor is shown in Figure 46-1. To construct one of these transistors, two wires made of an appropriate metal are pressed against a block of a suitable type of semiconductor material. For instance, aluminum wires might be used with n-type semiconductor. Once the

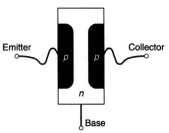

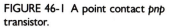

FIGURE 46-1 A point contact *pnp* transistor.

391

wires are in place, carefully controlled discharges through the contact points melt the ends of the wires and a little of the semiconductor just below the contact point. While the semiconductor is molten, a little of the wire material, aluminum in this example, diffuses into the melted material. Thus when the melted part of the semiconductor solidifies again, there are tiny regions of *p*-type semiconductor in the block of *n*-type semiconductor. In this manner, a *pnp* transistor has been formed.

To get good transistor action, the emitter and collector need to be very close together. Obviously, placing the wires on a semiconductor crystal and then producing just exactly the correct amount of melting to make a good transistor is a very difficult process to control. It was very difficult to manufacture consistent transistors by this technique. As a result, the parameters of early transistors tended to vary greatly from transistor to transistor. Furthermore, since the transistors were essentially made one at a time by this process, they were quite expensive. Although this was a suitable technique for making the first transistors in the laboratory, transistors were not very useful until better and cheaper methods of fabrication were developed. This construction technique is essentially no longer used.

GROWN TRANSISTORS

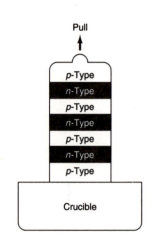

FIGURE 46-2 An alternating layer ingot being grown.

Large single crystals of semiconductor material are grown by slowly drawing a crystal out of a crucible of molten material. By starting with a tiny seed crystal and using very pure semiconductor material, single crystals of ultrapure silicon or germanium five or more inches in diameter and over a foot long can be produced. These are commonly called ingots. The crystals grow very slowly (millimeters per hour), and the conditions in the crucible and in the surrounding gases have to be extremely carefully controlled. The production of ultrapure ingots is an important step in the IC industry.

If a tiny amount of impurity is put into the crucible, then either *n*-type or *p*-type semiconductor material can be grown. If the type and amount of impurities in the crucible are changed periodically, it is possible to grow long crystals of semiconductor material that consist of alternate layers of *p*-type and *n*-type semiconductor. Such a process is shown schematically in Figure 46-2.

Once one of these ingots has been grown, it can be sliced and diced into little pieces. If the sections contain only two layers, diodes are constructed; if the sections contain three layers, transistors result. Since the crystals grow very slowly, it is fairly easy to grow thin regions of one type of semiconductor material. Thus fairly thick layers of *p*-type material might alternate with thin layers of *n*-type. When sliced, these will make *pnp* transistors as shown in Figure 46-3.

The result of the slicing and dicing processes is a large number of chips which become individual transistors. Wires must be bonded onto the chips in a way that does not form additional *pn* junctions; these are called ohmic or metallic contacts. Finally, the transistor must be mounted in something which protects the silicon from the atmosphere. This is a fairly reliable technique which produces many transistors or diodes at a time. As a result, it is also fairly economical. Many transistors and diodes are made by this technique.

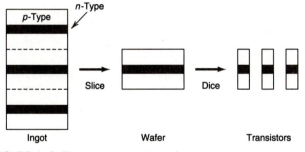

FIGURE 46-3 The slicing and dicing of an ingot to make transistors.

DIFFUSION TECHNIQUES

A thin wafer of semiconductor material cut from an ingot is loaded into an oven and heated in an inert atmosphere to just below the melting point of the semiconductor material (Figure 46-4). A small amount of a suitable impurity gas is introduced into the oven—a gas containing aluminum, arsenic, or tin or a similar dopant. Impurity atoms diffuse into the nearly molten semiconductor material. The rate of diffusion is relatively slow and depends on the temperature and the type of impurity. By controlling the time and other parameters of the exposure, layers of semiconductor material of either type can be diffused into the basic wafer. In this way, *npn* sandwiches with the middle layer only a few microns thick can be produced relatively easily.

Once the semiconductor wafers have gone through the diffusion process, they are ready to be diced into chips, have ohmic connections made to the chips, and to be packaged as transistors or diodes. The ease of making very thin layers and the ease of control makes this one of the dominant processes used today to make transistors, and the basic process is used in the fabrication of ICs.

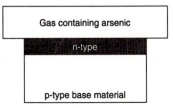

FIGURE 46-4 A schematic diffusion process.

ION IMPLANTATION

If a block of semiconductor material is exposed to a beam of high-energy, high-velocity ions of the appropriate type (aluminum, tin, etc.), the ions simply smash their way into the crystal (Figure 46-5). The depth to which they penetrate depends on their initial velocity; and as a result, it is possible to

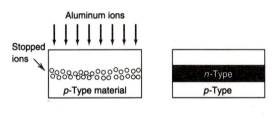

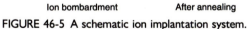

FIGURE 46-5 A schematic ion implantation system.

arrange it so that they are totally buried below the surface of the semiconductor. (Note that diffusion processes take place on the surface.)

Of course, the ions break up the nice crystalline structure of the semiconductor as they smash their way into the material. However, if the block of semiconductor is then **annealed** at a temperature just below the melting point for a suitable length of time, the damage to the crystal structure is repaired, and a region of the appropriate type of semiconductor material is formed.

PLANAR OR MESA TRANSISTORS

Planar or mesa transistors are made by a long sequence of steps which include the diffusion process described above. Although this process is described as if a transistor is the goal, essentially the same process can be used to make integrated circuits. (See the next section for some comments about this.) For the purposes of this discussion, all of the steps take place on one side of the wafer; the other side is assumed to be protected by some sort of mounting fixture.

All of the steps in this process take place either in a vacuum or in an inert gas to which carefully controlled amounts of other material have been added. Some of the steps take place at high temperatures, others at low temperatures. However, the environment for each step must be very carefully controlled. The machines that carry out these processes cost millions of dollars, and the machines and the techniques used to operate them represent some of the most important controlled technology in the cold and trade wars.

The example considered here will start with a wafer of p-type semiconductor, but there is no importance attached to this choice. Either type of material or ultrapure semiconductor can be the starting point; it is the goal of the processing that determines the beginning point. A partial listing of the steps in the process follows.

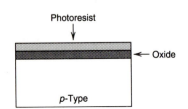

FIGURE 46-6 A wafer covered with layers of oxide and photoresist.

1. First, the surface of the wafer is cleaned and a thin layer of silicon dioxide (glass) is formed on the surface of the wafer.
2. A layer of **photoresist** is painted onto the wafer. When the photoresist is developed, the parts that were exposed to light remain, but the unexposed parts dissolve away. This is indicated in Figure 46-6.
3. The appropriate pattern of light and dark is printed onto the wafer by photo-optical means. Although the figures show only a single transistor being formed, it should be clear that hundreds to millions of different regions can be created during one processing cycle. The photo mask determines what happens all over the surface of the wafer, which may be 5 or 6 inches in diameter, while the region for a single transistor may be as small as a few microns square.
4. After the exposure, the photoresist is developed, and then the wafer is passed through an etching process. The result of this etching is that those areas of the underlying semiconductor that were not covered by developed photoresist are exposed, while the other regions are still protected by the developed photoresist and silicon dioxide. This state is shown in Figure 46-7.

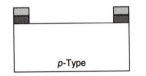

FIGURE 46-7 The wafer in Figure 46-6 after exposure to light, developing, and etching.

5. The wafer is next subjected to a diffusion process. For the example here, a fairly large *n*-type region is diffused into the bulk *p*-type material, as shown in Figure 46-8.

6. The wafer is now cleaned. All the photoresist and silicon dioxide are removed, and the whole photoetching process is repeated with a different optical mask pattern (steps 1 through 4).

7. Another diffusion process takes place. For the example shown here in Figure 46-9, this one puts a smaller *p*-type region in the middle of the *n*-type region that was created in step 5. As a result, there is now a very nice *pnp* structure, with the *n* region being fairly thin.

8. Another cleaning and photoetching process takes place. This time after the etching, small metallic regions are plated onto the various parts of the transistor, as shown in Figure 46-10.

9. Finally, all of the remaining photoresist is removed from the transistor, leaving the surface covered (protected) by either silicon dioxide (glass) or metal.

10. At this point, the wafer can be diced, yielding small chips each of which is a planar *pnp* transistor. The final steps involve welding wires to the metallic regions and mounting the transistor in its final package. The result is shown in Figure 46-11.

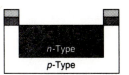

FIGURE 46-8 The wafer in Figure 46-6 after a diffusion process.

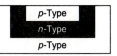

FIGURE 46-9 The wafer in Figure 46-6 after the second diffusion process.

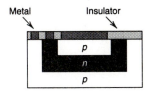

FIGURE 46-10 The wafer in Figure 46-6 after the metallic pads have been deposited.

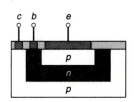

FIGURE 46-11 The final transistor structure on the wafer in Figure 46-6.

The process described above involves some 20 to 30 individual steps, requiring a vast amount of technology, skill, and capital. However, in one processing cycle, hundreds of wafers may be processed, each wafer producing thousands of transistors all essentially identical. Because of this, once the techniques have been mastered, transistors (and ICs) can be made in vast numbers for relatively low prices per item. Since making a different transistor or integrated circuit requires a new set of photo masks and mastering a different processing sequence, the reason for the desire of the manufacturers to produce more copies of fewer different devices is obvious. As was mentioned earlier, this has driven the design of ICs toward more complex, more powerful, and more flexible universal circuits.

Clearly, there is almost an infinite number of variations on the processes discussed above. Ion implantation can be used in the processing so that buried structures can be created. Layers of fresh semiconductor can be deposited on the surface of the wafer providing another way in which three-dimensional structures can be created.

INTEGRATED CIRCUITS

By using the planar technology described above, transistors, resistors, and capacitors can be constructed. Resistors are simply long, thin traces of lightly doped semiconductor, while capacitors can be formed by putting a metal layer on top of a glass layer on top of an area of heavily doped semiconductor. Furthermore, metallic traces can be created that connect one transistor with another. Thus entire circuits can be created by the same process.

Because it is relatively easy to make transistors that are a few microns square and to connect them on a wafer by metallic traces having a width of

2 microns or less, it is possible to create a circuit consisting of several transistors and resistors, such as a twofold NAND, in a space that is no larger than 40 microns square.

On a certain level, the only difference between making a planar transistor and an integrated circuit is the nature of the photo masks being used. For the transistor, the same simple pattern is repeated almost endlessly across the surface of the wafer. For the integrated circuit, a more complex pattern is repeated hundreds or thousands of times across the wafer. On a more detailed level, the number and sequence of steps will vary depending on what kind of device is being made.

This is the basis of the integrated circuit industry. Improving materials, technology, and understanding have enabled the progression from the simple initial circuits to the complete microcomputers of today which involve several hundred thousand transistors that fit into a circuit about a centimeter square.

PART SEVEN

COMPONENTS

In an introduction to electronics, resistors and other passive components are often treated as if there are no choices other than the value. In an instructional setting, resistors are simply something taken out of a drawer and used, but when it comes time to buy a resistor, a surprising number of choices must be made. The simple components, resistors, capacitors, and inductors are not as simple as has been indicated thus far. In this part of the book some of the complexities and the resulting terminology relating to these simple circuit elements will be introduced. Chapter 47 concerns resistors, Chapter 48 treats capacitors, and Chapter 49 describes inductors. Chapter 50 is slightly different, it is a brief introduction to transducers, those components that provide the input and output of electronic systems.

PREREQUISITES

All of the chapters in this section have essentially the same prerequisites. First, some familiarity with both ac and dc circuit analysis is required (refer to Chapters 3 and 8).

Second, the discussions in the chapters touch on essentially all the portions of electronics, at least briefly.

Chapter 47

RESISTORS

OBJECTIVES

1. Become familiar with the major types of resistors and their advantages and disadvantages.

2. Become familiar with the imperfections of resistors and the terminology used to describe them.

3. Comments in this chapter are related to all other portions of this book.

INTRODUCTION

This chapter contains a wealth of practical information about real resistors, including a discussion of resistor specifications, parasitic elements, and imperfections in resistors, a discussion of color codes and other marking schemes, and a discussion of newer packaging schemes. The chapter ends with a brief discussion of the properties, characteristics, and uses of a number of different types of resistors, both fixed and variable.

RESISTOR SPECIFICATIONS

Thus far, resistors have been specified only by their resistance value. This is the most important and most obvious parameter to be specified for a resistor, but there are many others. The most important of these are discussed below.

Tolerance

Thus far, it has been assumed that a resistor really has the value it is supposed to have; that is, a 47-Ω resistor really has a resistance of 47.00 Ω. Of course, this is not true. Resistors are manufactured with well-defined **tolerances**. Typical tolerances range from 10% to 0.1%. The meaning of the tolerance should be fairly clear; a 5% 47-Ω resistor will have a value that departs by no more than $\pm5\%$ from 47 Ω.

Precision resistors having tolerances of 1% or better cost much more than normal resistors, which have tolerances of 2% to 10% or sometimes more. Although there are some applications where very precise resistance values are needed, such as the voltage divider in a precise digital voltmeter, most applications do not require very precise values. In fact, it is a mark of a good design not to use precision components except in key locations.

The normal, wide variations in resistor values also set a limit on the accuracy needed for most calculations. After all, if a pair of 10% biasing resistors are to be used, it makes no sense to calculate their value to more than two significant digits. A related point is that in most cases where it is required to set something to a fairly precise value, it will be impossible to calculate the resistor values needed because they will depend on the exact values of other resistors, transistor gains, and so on. The normal process is to combine a fixed resistor in series with a variable resistor and adjust the variable resistor to obtain the desired operating point. In this situation, tolerance is not the main concern; rather, stability (see below) is of more interest.

Power Rating

Power is dissipated in resistors, and this means they may get hot. All resistors have **power dissipation ratings**, which tell how much power they can dissipate before their values change by more than their rated tolerance or before they fail. The power handling capability of commercially available resistors range from $\frac{1}{8}$ W to hundreds of watts. Lower-power resistors, $\frac{1}{2}$, $\frac{1}{4}$, and $\frac{1}{8}$ W, are generally satisfactory for op-amp, transistor, and logic circuits, whereas high-power resistors may be needed in power supplies. It never hurts to check the power dissipation in designing a circuit. It is considered good practice to use resistors well below their rated power limits.

Stability

The value of a resistor will change because of the passage of time, changes in temperature, and perhaps other parameters such as humidity. Resistor values also change slowly when subject to very high voltage, that is, voltages in the kV range. In general, all these changes are small and can be ignored. Obviously, in precision circuits, these changes may be important, and the stability data from the manufacturers will have to be consulted.

RESISTOR IMPERFECTIONS

Thus far, a resistor has been considered as a circuit element that always obeys Ohm's law exactly. In particular, when an ac voltage is applied to an ideal resistor, the voltage and current are always in phase; that is, the phase shift between the voltage and current is 0°. If this phase shift is measured for any real resistor, at some frequency it is found that the phase shift ceases to be 0°, and thus the resistor is acting as if it has an inductive or capacitive component. This should be no surprise, since even a plain wire has a small self-inductance and the two ends of a resistor form a capacitor.

The inductive and capacitive components of a resistor are called **parasitic** components. The parasitic components depend on the construction techniques that are used in making the resistor and may also depend on the frequency. When working at high frequencies, it may be necessary to take these parasitic elements into consideration when analyzing the circuits. Figure 47-1 presents a possible schematic diagram for a real resistor showing these parasitic elements.

Noise

Current flowing through a resistor causes electronic **noise**. The amount and nature of this noise depend on the current, the resistance, and the type of resistor. Any further comment about this far exceeds the scope of an introduction, except that certain types of resistors discussed below are noted as being low noise, that is, they produce less noise than other types.

RESISTOR MARKING SCHEMES

There are several schemes for marking the values of the resistance on a resistor. The three most common schemes are discussed below. Most precision resistors have their values and tolerances directly printed on the resistor. Not much needs to be said about this kind of identification, since a marking such as 47 kΩ, 1%, $\frac{1}{4}$ W is fairly self-explanatory.

Another scheme that is being used on many resistor networks and also on modern capacitors is some variation of the following. Two digits of the resistance (or capacitance) value plus a power-of-10 multiplier are given. Thus 473 would mean $47 \times 10^3 \ \Omega$; 105 means $10 \times 10^5 \ \Omega$, or 1 MΩ. Tolerances in this scheme are often indicated by a letter code. Unfortunately, there are no accepted conventions for these letter codes, so that the only way of interpreting these codes is to figure out what manufacturer made the resistor and then look up any information that this manufacturer makes available.

Most resistors are small, colored cylinders with wires sticking out of each end of the cylinder. The value, tolerance, and sometimes other information about each resistor can be determined from the colored bands on the resistor. A typical resistor is sketched in Figure 47-2, and the meaning of the various colors are given in Table 47-1.

Ideal

Representation of a real resistor

FIGURE 47-1 A real resistor and a representation of this resistor and its parasitic elements.

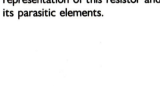

1st Digit
2nd Digit
Multiplier
Tolerance

FIGURE 47-2 A typical resistor showing the color band markings.

TABLE 47-1 Resistor Color Code

Color	First Digit	Second Digit	Multiplier	Tolerance
Black	0	0	1	—
Brown	1	1	10	—
Red	2	2	100	—
Orange	3	3	1000	—
Yellow	4	4	10,000	—
Green	5	5	100,000	—
Blue	6	6	1,000,000	—
Violet	7	7	10,000,000	—
Gray	8	8	100,000,000	—
White	9	9	—	—
Gold	—	—	0.1	± 5%
Silver	—	—	0.01	±10%

Thus a resistor that has color bands of red–red–red–silver is a 2.2-kΩ 10% resistor. Yellow–violet–green–gold is 4.7 MΩ 5%. For these resistors, the power dissipation value is indicated by the size. To determine the power dissipation rating of an unknown resistor, it is necessary to compare the unknown to resistors of known power ratings.

RESISTOR NETWORKS

For many years, all resistors were cylinders as sketched in Figure 47-2. Higher-power resistors might look somewhat different, but all low-power resistors were small cylinders. However, the advent of integrated circuits, the miniaturization of circuits, and the creation of an environment in which many similar resistors might be wanted in a small space have given rise to many different configurations of resistor networks. Many different resistor networks can be obtained today for use in IC circuits. These are multiple resistors mounted in either DIP (dual-in-line) packages or SIP (single-in-line) packages with 0.1-inch lead spacing, which can be used on the same circuit boards that are used for IC with no further modifications. These resistor networks are available with many different values and configurations of resistors. There is no universal packaging scheme or configuration scheme. But any time it is found that a number of identical resistors are needed in an IC circuit, it is worth looking through a catalog to see if the resistors needed are available as a prepackaged network. Using such a network will save much time in mounting resistors in IC circuits. Two sample networks are shown in Figure 47-3.

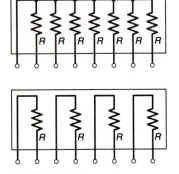

FIGURE 47-3 Two resistor networks.

FIXED RESISTORS—TYPES AND CHARACTERISTICS

The major types of fixed resistors are listed in this section. The characteristics and properties of each type are briefly discussed.

Carbon Composition

For many years, carbon composition resistors were the only readily available low-cost, low-power resistors. These resistors continue to be made from carbon powder held together with fillers and binders and are molded into small cylindrical shapes. The resistors are held together by an inert cylinder that has the color-coded value printed on it. The leads of the resistor are embedded into the ends of the carbon cylinder and bonded to the external cylinder.

These resistors were low in cost and readily available. They were available in low-accuracy values (20, 10, and 5%), in a resistance range running from 10 Ω to 22 MΩ, and in power ratings running from $\frac{1}{4}$ to 5 W in the standard ranges and to much higher-power values in special models. Composition resistors have relatively poor thermal stability and change value with changes in humidity, and the like. However, despite their deficiencies, they were the workhorse of the electronics industry.

Carbon Film

Carbon film resistors are made by depositing a pure carbon film on a ceramic or glass rod. A spiral is then engraved into the film; in this way, a long, thin carbon wire is constructed on a nonconducting core. By adjusting the width of the spiral and the total length and the thickness of the film, the resistance value can be adjusted. Finally, wires are bonded to the film and the whole structure is sealed with a coating of epoxy or special paint. The epoxy or lacquer coatings are generally moisture-proof. Figure 47-4 shows some of the steps in the construction of one of these resistors.

These resistors are the workhorses of modern equipment. They have better performance characteristics in all ways than do carbon composition resistors. They are more stable and have somewhat better noise characteristics. The protective covering is moisture proof and, as a result, humidity does not affect them as much as it does carbon resistors. These resistors have good stability under high voltage, but only fair thermal stability. They are available in the same resistance and power ranges as composition resistors, with 5% being the normal tolerance.

Finally, since more modern technology is used in their manufacture, they are surprisingly lower in cost than are composition resistors. Hence, they are the preferred resistor for noncritical applications in any new construction.

Metal Film or Metal Oxide

A metal film (usually nichrome) or a metal oxide is deposited on a ceramic or glass rod. The rest of the process is similar to that for carbon film resistors. These resistors have much better characteristics than the types described above. They have much better thermal stability, lower noise characteristics, and are generally preferred for critical circuitry. They are available only in lower-power values; a few types are available in 1- or 2-W sizes, but most are

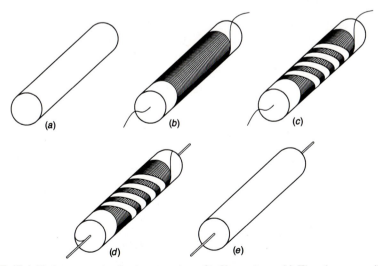

FIGURE 47-4 Various steps in the construction of a film resistor: (a) The glass core; (b) the glass core with resistive film coated on it; (c) a spiral cut into the resistive film coating; (d) wire leads bonded to the resistor structure; (e) the film resistor after the outer coating has been applied.

only available in $\frac{1}{4}$ W values or less. These film resistors are generally available only in semiprecision (1% or 2%) or high-precision (0.5% to 0.01%) tolerances. Of course, they are more expensive than the types above.

Wire-Wound

A high-resistance wire of some nature (frequently nichrome) is wound onto a cylinder. The wire and cylindrical core are covered with epoxy or vitreous enamel. If no special precautions are taken, wire-wound resistors have high inductance; but, by using special winding patterns, the inductance can be minimized. Wire-wound resistors are generally used in high-power equipment where high wattage resistors are needed. Wire-wound resistors are available in a resistance range of 0.1 to 100 kΩ with a few special models being available as high as 1 MΩ. Power ratings run from 1 to 1500 W. Wire-wound resistors can be made with close tolerances but generally are available in the 5% to 10% range. Their noise and stability characteristics are quite good; they are those of a metal wire and, hence, are easily predicted. Wire-wound resistors are generally quite expensive and large. As the power levels in electronics have dropped, the need for wire-wound resistors has decreased; however, they still are used in power equipment such as speed controls on subway cars.

VARIABLE RESISTORS—TYPES AND CHARACTERISTICS

The problems of specifying and/or selecting a variable resistor are more complex than those for a fixed resistor because there are more variables. In this section, each of the major choices is discussed.

Number of Terminals

There are two-terminal variable resistors, which are called **rheostats**, and there are three-terminal variable resistors, which are called **potentiometers**. These are both discussed in Chapter 3. Of course, if one end of a potentiometer is not connected, a rheostat results. For this reason, the rest of this chapter will be written in terms of potentiometers without any loss of generality.

Linear Versus Rotary

Conceptually, the simplest variable resistor is a very long wire with some sort of sliding contact that can be moved along the wire. The two ends of the wire form the fixed ends of the potentiometer, and the sliding contact is the variable connection of the potentiometer. The sliding contact is called the **wiper**, or the **slider**. This conceptual variable resistor has the disadvantage that it might be very long and, hence, quite awkward to use.

However, there are two solutions to the length problem. One is to use a higher-resistance material than simple copper wire so that large enough resistances can be obtained in a short length. The second solution is to roll up the element into a circle so that the wiper is moved in an arc along the resistance element by turning a shaft. Both of these are commonly done. The first type is called **linear potentiometers** and the second is called **rotary potentiome-**

ters. The rotary models are so common that any unqualified reference to a **pot**, or a potentiometer, is almost always a reference to a rotary potentiometer.

Linear potentiometers are most often used as sensors to measure the location of something that is attached to the wiper. Thus a linear resistor can be used to measure the compression of a spring in a load-bearing strut and, hence, to measure the load in the strut. Rotary pots are more often used as control elements, although they can also be used as sensor elements to measure the rotation of a shaft. The rest of this chapter deals with rotary potentiometers, although most of the material applies equally to either type.

Single Versus Multiturn

The resistive element in a rotary potentiometer can be bent into a simple circular shape or into a multiturn spiral. If the element is a simple circle, the slider can move no more than 300° at most; whereas if the element is bent into a multiturn spiral, then the shaft and the attached slider can be rotated by multiple turns. Obviously, multiturn potentiometers can have a much higher precision in their settings.

Very precise linear potentiometers are manufactured with a five- or ten-turn range. Fancy multiturn knob mechanisms are available to use with these potentiometers so that the settings of the shaft can be read out accurately. Multiturn potentiometers were never used for consumer equipment because of their cost; however, they were frequently used in laboratory equipment where precision settings were needed. In modern equipment, they are being replaced by digit switches, resistance ladders, and DACs, which give better performance without all of the related mechanical problems of the multiturn hardware.

Taper

The resistance between one end of the potentiometer and the slider may vary linearly as the shaft is rotated, or it may vary either more-or-less rapidly than linearly. This variation is called the **taper** of the potentiometer. Typical tapers are linear, logarithmic, or exponential. Each of these tapers is useful for some application. For instance, because of the nonlinear response of the human ear, a taper approximating an exponential makes a better volume control for audio equipment than does a linear taper. The catalogs from the manufacturers contain very cryptic comments about the tapers of their potentiometers. Fortunately, most often what is available is what is needed, and ignorance is bliss. However, it doesn't hurt to pay attention to the taper when buying a potentiometer.

Enclosure

The major problem with all potentiometers is that the slider must both move and make good contact with the underlying resistive material. In a new potentiometer, this is true; the potentiometer can be adjusted without any undesired effects—such as occasional open circuits or sudden jumps in the resistance value. However, as the potentiometer ages, dirt or oxide on the surfaces or corrosion of either the wiper or the resistive material, which may be caused by electrochemical processes, can cause the contact quality to decrease. The noise, crackle, snaps, and pops made by the volume control of an older radio are examples of the deterioration of this contact.

Inexpensive potentiometers are made with the resistive element and the wiper open and exposed to the air. Obviously, this type will age most rapidly. Very expensive potentiometers are made with totally enclosed elements, with gas-tight seals around the shaft so that the slider and element are permanently protected from airborne damage. Obviously, the desired lifetime and the desired cost are two things to consider when buying a potentiometer.

Material

Finally, the last thing to consider when selecting a potentiometer is the material used in the resistive element. The list of choices is essentially the same as for fixed resistors, but the advantages are somewhat different, since the materials have different resistance to wear caused by the slider. The common choices are described in the following paragraphs.

A **wire-wound** resistor can be bent into a circle, one side of it left unprotected and with a slider mounted to make contact along this side. The main problem with this type of variable resistor is that the wiper is making contact with one turn or another; thus, as the wiper is pushed along, the resistance increases in steps, each step corresponding to the resistance of one turn of the wire around the cylinder. If this is no problem, then the arrangement makes a perfectly acceptable variable resistor. Beyond the stepwise nature of the wire-wound potentiometer, all the comments above about wire-wound resistors apply. A single wire can be arranged in a multiturn spiral and is the basis of most multiturn precision potentiometers. Since these are made with a single wire and a slider, there is no stepwise resistance changes.

A **carbon composition** resistor can be made with a circular shape and a slider mounted on it, producing a rotary potentiometer. This type of variable resistor is quite cheap but has all of the problems of the carbon composition resistor discussed above. Most cheap controls are made in this way; they do not age well, but in a "throw away" society this may be a virtue.

Cermet resistors are made from a mixture of ceramic and metal. They can be formed just as carbon composition potentiometers can. They have better stability, better temperature coefficients, and better resistance to the effects of atmosphere on the surfaces than do composition potentiometers. Although they cost more than composition potentiometers, they are becoming the dominant form used in most equipment because of their better properties.

Chapter 48

CAPACITORS

OBJECTIVES

1. Become familiar with the major types of capacitors and their advantages and disadvantages.

2. Become familiar with the imperfections of capacitors and some of the terminology used to describe them.

3. Comments in this chapter are related to almost every other portion of this book.

INTRODUCTION

This chapter is a quick introduction to practical capacitors. Its purpose is similar to the previous chapter on resistors. Because the properties of capacitors vary so much and because the departures from ideal behavior are often quite large, capacitor selection is generally more difficult than resistor selection. This chapter presents an introduction to most of the topics needed to make an intelligent selection of a capacitor.

CAPACITOR USAGE

Capacitors are used for many different purposes in electronic circuits. Each of the different uses typically places different demands on the properties of the capacitor and, hence, perhaps requires a different choice of capacitor type. In this section, a number of different uses are described by one or two sentences, and the main characteristics needed are mentioned.

Large capacitors are used in power supply circuits to **store charge**, or **energy**, for delivery at a latter time or to **filter** out 60-Hz ripple. (These are two different points of view of the same functions.) For a filter capacitor, the main parameters are the capacitance (values to 20,000 μF and more are used in low-voltage, high-current supplies) and the ac current-carrying ability of the capacitor—multiampere ripple currents may flow through the filter capacitor. The various types of electrolytic capacitors are intended for these purposes.

407

The function of a **bypass** capacitor is to ensure that the dc component of a signal appears on some circuit element but that the ac component is shorted out or bypassed around the element. Bypass capacitors are used in modern digital electronics to prevent high-frequency noise on the dc power line from entering into the logic via the power leads. The high-frequency properties of the bypass capacitor are of greatest interest. Various small ceramic, plastic, or mica capacitors are used for this purpose.

A **blocking capacitor** blocks the dc component of a signal from propagating to another section of a circuit while allowing the ac signals to get through. Blocking capacitors are often called **coupling capacitors**. The most important parameters are the capacitance (to calculate the cutoff frequency) and the dc leakage current.

Capacitors are often used in **frequency discrimination**, or **timing**, circuits. *RC* and *LC* networks are used to create shaped frequency responses. *RC* networks are often used to define time constants that determine how fast a circuit operates. In general, these are low-power applications where the main parameters are the value, the tolerance, and the stability.

Capacitors are used in **integrating** circuits for measuring charge or in analog-to-digital converters. For these applications, the important properties are the value, the tolerance, the stability, and the leakage current.

CAPACITOR PROPERTIES

As described in Chapter 8, the simplest capacitor design is two flat conductors separated by an insulator. The capacitance is proportional to the area of the conductors, proportional to the dielectric constant, and inversely proportional to the spacing between the two conductors. Thus to get larger capacitance, there are three choices: larger area, larger dielectric constant, or smaller gap spacing. All three of these are used in real capacitors.

Table 48-1 shows the dielectric constants for the most common types of material used in capacitors. Other properties of some of these insulators are described in the discussion that follows.

TABLE 48-1 Capacitor Dielectrics

Material	Dielectric Constant
Vacuum	1.000 (exactly, be definition)
Air	1.0001
Teflon	2.0
Polystyrene	2.5
Polycarbonate	2.7
Mylar	3.0
Polysulfone	3.06
Kapton	3.2
Polyethylene	3.3
Kraft paper	2.0–6.0 (depends on impregnating material)
Mica	6.8
Aluminum oxide	7.0
Tantalum oxide	11.0
Ceramics	35.0–6000 and up

One important property of any dielectric is how much voltage can be applied across it before it breaks down in some way and starts to conduct. This dielectric strength is an important parameter of the dielectrics used in making a capacitor. Some materials, such as glass and mica, have very high dielectric strengths whereas others, such as air, have much lower dielectric strengths. For all dielectrics, the breakdown voltage increases as the thickness increases. Thus to make a capacitor that can withstand a very high voltage requires a fairly thick insulating layer. A thick insulator, in turn, makes it more difficult to get very high capacitances. Thus getting a large-capacitance, high-voltage capacitor is much harder (and more expensive) than getting either a large low-voltage capacitor or a small high-voltage capacitor.

A second important parameter about the dielectric is how it responds after it breaks down. Some dielectrics are **healing**; this means that after they break down and the discharge is terminated, the dielectric reforms and is essentially unchanged. Vacuum and air are the two best examples of healing dielectrics, but there are others. In contrast, some dielectrics, such as glass or mica, are essentially ruined once they have broken down. This behavior is an important consideration in situations in which there may be occasional breakdown—such as in power transmission use where a lightning bolt may cause a capacitor to break down.

Except for the smallest possible capacitors, using two flat plates separated by an insulator is impractical; many common capacitors have effective areas of many square meters and yet are physically quite small. There are two ways of dealing with this problem. Multiple layers of conductors separated by insulators can be stacked up in one pile. If alternate conductors are connected together, then each face of the conductors acts as a capacitor with the next layer (except for the outermost layer), and the entire stack acts as a number of small capacitors in parallel (Figure 48-1).

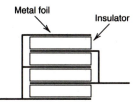

FIGURE 48-1 A simple multilayer capacitor.

A thin, flexible insulator, such as a sheet of mylar, can be coated on one side with a conductor—aluminum is quite common. Two layers of this material can then be stacked up, rolled, and perhaps folded up into a small package. Microscopically, the geometry has not changed; this is still a parallel-plate capacitor. But macroscopically, it has been crammed into a much smaller space.

With either of these techniques, the **leads**, the wires coming out of the capacitor, must be attached in some way, and then the whole package is sealed inside some protective case. The technique used to do this and the geometrical details of the assembly of the capacitor will contribute to the **parasitic inductance** of the capacitor.

A second geometrical fact is that one of the two conductors of any rolled or multilayer capacitor is the **outer** conductor while the other conductor is shielded by the outer one. The outer conductor, or outer foil, can pick up or transmit signals as an antenna; in noise-sensitive circuits, it is usual to connect the outer foil to ground or, at least, to the point in the circuit that is closest to ground. Most capacitors have some method of indicating which is the outer foil. The most common marking on cylindrical capacitors is a bar at the end corresponding to the outer foil. However, this marking depends on both the type of capacitor and the manufacturer—so be careful.

No insulator is perfect. Some **leakage** current will flow through the dielectric in any capacitor. Although few dielectrics are really ohmic, it is common to express the leaking current in a capacitor in terms of an equivalent resistance. Typical resistances quoted for capacitors range from megohms to 10^{12} ohms or even more. Obviously, other than any possible heating effects, a little leakage current is not important in the filter capacitors of a high-current dc power supply. Just as obviously, if a capacitor is being used in an integrator that is accumulating charge over a period of a week, the leakage current is very important. Typically, a great deal of information about the size and properties of the leakage current is available for low-leakage capacitors whereas very little is available for normal capacitors.

The reason why a dielectric inserted between two plates increases the capacitance of the capacitor is that the dielectric **polarizes** under the influence of the electric field between the two plates. Because of this polarization, the electric field and, hence, the voltage between the two plates is decreased for a given charge placed on the two plates—this is just another way of saying that the capacitance has increased.

On a microscopic scale, saying that the dielectric polarizes under the influence of the electric field means that the atoms making up the dielectric change shape slightly. As a capacitor is charged, energy is stored in the capacitors. Some of this energy is stored in the electric field between the two plates, and some of the energy is stored in the distorted shapes of the atoms.

There are two important results of the polarization. First, as the electric field changes, the shape of the atoms changes. Because of atomic friction, some energy is lost every time the atoms change shape. Thus some of the energy put into charging a capacitor is lost; that is, it appears as heat in the dielectric. This is the **dielectric loss** of the capacitor. The amount of dielectric loss increases with frequency and depends on the type of insulator. Some dielectrics are very good for high-frequency uses while others are not.

The second result of the polarization is the occurrence of two time constants in a capacitive circuit. The energy stored in the electric field between the two plates and most of the energy stored in the polarization of the dielectric can be removed instantaneously from the capacitor, that is, as fast as the *RC* time constant of the circuit will allow the capacitor to discharge. But some of the polarization energy is released on a time scale determined by atomic processes in the material, this can have time constants measured in milliseconds to days. As a result, a large capacitor that has a highly polarized dielectric can be discharged and then left with its leads open; some time later, it is found that there is some voltage between the two terminals. This is because some of the energy stored in the polarization has been slowly returned to the electric field. This phenomenon can be easily measured in the laboratory with a power supply, a voltmeter, and a medium-sized electrolytic capacitor. It can be very serious when working with high-powered, high-voltage circuits—people have been killed by "discharged" capacitors.

A modern discussion of the phenomena taking place in dielectrics must be conducted in terms of quantum mechanics and wave functions and has no room for concepts such as the shape of an atom or atomic friction. The phenomena of polarization and dielectric loss described above exist and can be analyzed in terms of quantum mechanics, thermodynamics, and statistical mechanics. However, the classical model used above gives us a way of thinking about these phenomena and, if not taken too literally, does not lead to any incorrect conclusions.

It is inevitable that any real capacitor will have some parasitic inductance. Because of the difference in the frequency response of inductive and capacitive impedance, any capacitor will look inductive at a high enough frequency. In general, it is hard to make a large capacitor that looks like a capacitor at high frequencies. Thus, if it is necessary to have a capacitor that will operate from low to extremely high frequencies, it is common to put several capacitors in parallel: one large capacitor to take care of the low frequencies, one intermediate-size capacitor to dominate the impedance at intermediate frequencies, and another small one to operate at the highest frequencies. Because of the dielectric loss and dielectric-related time constant, it is very difficult to draw a general equivalent circuit for a real capacitor that shows the parasitic elements in any realistic way. Hence, no circuit analogous to Figure 47-1 will be given here. (Anyone who has need of such a drawing is far from being a beginner and should consult appropriate advanced references.)

CAPACITOR SPECIFICATION

Most of the important capacitor specifications have been mentioned at least implicitly in the foregoing discussions. However, at the cost of some repetition, a brief summary of most of the important capacitor specification is given below.

Value

This is both the most important and the most obvious specification of a capacitor. It simply tells the capacitance.

Tolerance

This gives the possible error in the nominal value of the capacitor. In general, the tolerances on capacitors are large and frequently asymmetrical. Typical values might be -50% and $+100\%$. This tolerance would mean that a nominal 100-μF capacitor could have a value in the range of 50 μF to 200 μF.

Voltage Rating

The maximum voltage that can be applied to a capacitor without breakdown is the voltage rating. It depends on the dielectric thickness and material. Generally lower voltage ratings make it much easier to make very-high-capacitance capacitors. Some capacitors will have two voltage ratings, one for short-time periods (maybe as short as a millisecond) and another working rating (the voltage that the capacitor can withstand continuously).

Insulation Resistance

This is a measure of the (assumed) ohmic resistance of the dielectric layer of the capacitor.

Dissipation Factor or Power Factor

This is a measure of the energy loss in the capacitor due to resistive leakage and dielectric loss. Any specification of this parameter must include a detailed description of how the parameter is measured to be useful.

Quality Factor

The quality factor, Q, is defined as

$$Q = \frac{1}{\delta}$$

where δ is the loss factor. A near-perfect capacitor, with no losses, has a quality factor approaching infinity. The quality factor will depend on the frequency at which it is measured, since the dissipation factor depends on the frequency.

Temperature Coefficient

The properties of the dielectric will change as a function of temperature. As a result, the capacitance of a capacitor changes as a function of temperature.

Voltage Coefficient

Very high polarized dielectrics—very high dielectric constants—generally have a dielectric constant that decreases as a function of the total voltage applied to the capacitor. As a result, the capacitance of the capacitor decreases slowly as the voltage increases. This decrease is not fast enough to cause the capacitor to quit acting like a capacitor, but it is important enough to be taken into account in some designs.

Aging Specification

Some of the more exotic dielectrics change as a function of time. This means that the capacitor may change or even fail as time passes. Obviously, this is a very important specification if long-term stability is important.

CAPACITOR MARKINGS

Over a period of years, there have been hundreds of different schemes for marking the value of capacitance on the capacitor. Most modern capacitors that are large enough have the capacitance and other relevant information printed directly on the outside of the capacitor. Some of the newer, smaller capacitors are so small that there is not room enough on the capacitor for the printing. Sometimes the decimal marking scheme mentioned for resistors is used; sometimes even more cryptic schemes are used. Given a capacitor, it is often true that the only safe way of determining its capacitance is to use some sort of capacitance meter.

FIXED CAPACITORS—TYPES AND PROPERTIES

There are two main classes of fixed capacitors: electrolytic and nonelectrolytic. The simplest version of the **electrolytic capacitor** consists of a metal plate inserted into a conducting fluid called an electrolyte. The metal plate (most often tantalum or aluminum) and the electrolyte initially act as if they were connected together and current can flow through the structure. However, as current flows through the device in one direction, an electrochemical reaction takes place on the surface of the metal plate, and a thin layer of insulator, generally an oxide, forms on the plate. As a result, a structure is formed that consists of a metal plate, an oxide insulating layer, and a conducting liquid. This structure is a capacitor with the liquid acting as one of the conducting plates. Since the oxide layer can be extremely thin—maybe even as thin as one atomic layer—very large capacitances can be obtained in very small packages.

The thickness of the dielectric layer that is **formed** is dependent on the voltage applied during the forming process. Once the layer is formed, the capacitor will not conduct if lower voltages are applied. Thus higher-voltage electrolytic capacitors can be produced by forming thicker insulating layers and applying higher voltages during the forming process.

A simple electrolytic capacitor made in the way describe above would be **polarized**, that is, one plate of the capacitor would always have to be positive with respect to the other. If the polarity of the applied voltage were different from that used during the forming process, the reverse electrochemical reaction would take place, and eventually the oxide layer would be gone and there would no longer be a capacitor. More complex techniques can be used to make nonpolarized electrolytics, but these are larger and cost more. Most electrolytic capacitors are polarized and will have some markings to indicate which lead must be the positive one.

Unused electrolytic capacitors have a tendency to fail when they are put into service because the oxide layer has slowly disappeared. Sometimes these failures are reversible; that is, after an initial surge of current, the oxide will reform and the capacitor will function again. In any case, failed electrolytic capacitors are one of the main causes of failure of modern electronic equipment. Unfortunately, there is no other way to get so much capacitance in so small a space and, as a result, electrolytics must be used in most equipment. There are two main types of electrolytic capacitors.

Electrolytic Capacitors

Aluminum

Aluminum electrolytic capacitors are the most common electrolytic capacitors in use. They are available with capacitances ranging from 1 to 1,000,000 μF and in voltage ranges from 100 to 700 V.

Tantalum

Tantalum electrolytic capacitors are smaller than aluminum for equivalent ratings. They have somewhat better characteristics in all respects, have a longer life expectancy, and they cost more. As a result, tantalum capacitors are used in high-quality equipment and military equipment but are not common in lower-cost equipment. Tantalum capacitors are available in the capacitance range of 0.1 to 1000 μF (1 to 10 μF is the most common range) and in voltage ranges of 3 to 150 V (somewhat higher for special models).

Nonelectrolytic Capacitors

There have been dramatic changes in nonelectrolytic capacitors in recent years. The availability of thin sheets of plastic materials has introduced a whole new range of wonder materials into the manufacturing of capacitors. Many of these new materials have given rise to much better capacitors than the traditional materials. In the following listing of materials used to make capacitors, the new materials are the "polys" and the ceramic materials.

Polypropylene

Polypropylene is often used to make rope, but as an insulating material in capacitors it has very low dielectric absorption, high insulation resistance and a low dissipation factor. It has a high dielectric strength, which permits economical, compact capacitor design.

Polyimide

Polyimide is most often known as Kapton, or H film. It has excellent properties in capacitors, including the ability to operate at very high temperatures. As a result, it is often used in environmentally difficult situations.

Polystyrene

Polystyrene is the basis of fiberglass, of styrofoam, and also of excellent capacitors. It has very low dielectric absorption and good temperature stability and is available in very thin sheets, which make for large capacitance in small spaces. However, it cannot be used in warm situations; its upper temperature limit is 85°C.

Polycarbonate

This material has most of the good properties of polystyrene, but in addition it can be used at temperatures running up to 125°C. It also is available in sheets as thin as 0.00008 inch, allowing for very compact capacitors.

Polyester

Besides being the basis for the clothes that the engineers wear, polyester has also become the most common dielectric material for capacitors. It has good, but not outstanding, characteristics, is readily available, and is reasonably cheap.

Paper

For may years, capacitors were made from aluminum foil separated by layers of paper. To keep the paper from absorbing moisture from the air, the paper was impregnated with some material such as paraffin, mineral oil, or PCB.

Only the cheapest equipment uses paper capacitors any more, although much antique equipment is filled with them.

Mica

Mica is a natural insulating mineral that has excellent insulating properties. Mica may still be the best material for high-frequency applications. Mica capacitors can have very high Qs, but they are only available in small values, with 0.1 μF being a very large mica capacitor.

Glass

Glass is a substitute for mica. Glass is only used in capacitors that have to have the highest order of reliability and stability. Most people will never see a glass capacitor other than in a laboratory demonstration model.

Ceramic

Many of the modern ceramic materials have very large dielectric constants. As a result, they are suitable for making compact, reasonably large capacitors. The general-application ceramic capacitors are available in the range of 100 pF to 0.15 μF, with reasonable tolerances and voltage ratings up to 1000 V or so. The small disk ceramic capacitors are the most common capacitors in most equipment.

VARIABLE CAPACITORS

There are two main uses for variable capacitors: (1) tuning circuits such as radio receivers over a wide range of frequencies, and (2) trimming the frequency response of a fixed tuned circuit. There are two main types of variable capacitors designed for these uses; as might be expected, these are called tuning and trimming capacitors.

The most common **tuning capacitor** is an air-insulated variable capacitor. One set of plates is fixed in place; the other set is attached to a shaft that can be rotated so that more or less of the rotating plates are inserted between the fixed plates. In this way, a capacitor is made that can be varied by a factor of 5 or more. A traditional model runs from 50 to 350 pF. Tuning capacitors have many problems: they fill up with dust and their capacitance depends on the temperature and the humidity. Because the temperature of the capacitor often changes as the equipment warms up, special circuits are required to stabilize the operation of the circuits.

Whereas tuning capacitors are adjusted with a shaft that generally is turned by hand, **trimming capacitors** usually require a screwdriver to adjust. For a trimmer, stability is more important than easy adjustment. Some trimming capacitors are built like small tuning capacitors; turning the adjusting screw changes the amount of meshing of two sets of plates. For some of these, insulating material other than air is used. A second type of trimming capacitor simply depends on squeezing a stack of conductors and insulators more tightly together and thus changing the spacing slightly. Typical trimmers vary by a factor of 2 or less and have values of a few hundred picofarads.

Chapter 49

INDUCTORS AND TRANSFORMERS

OBJECTIVES

1. Become familiar with the major types and uses of inductors and transformers.

INTRODUCTION

This chapter presents an introduction to practical inductors and transformers and their various uses, specifications, and related terminology. This chapter is similar to, but somewhat shorter and more superficial than, the previous two chapters for two reasons. First, except for common power transformers, the use of inductors and transformers is rapidly disappearing from electronics. Second, the remaining uses are mainly in specialty circuits, not something a beginner is likely to design and build.

INDUCTORS

A search of the catalog of a major electronic parts distributor, such as Newark or Allied Electronics (see Appendix A for full references) reveals only a few pages of inductors and transformers as opposed to 70 pages of capacitors and 40 pages of resistors. These inductors are called **coils**, **choke coils**, **chokes**, or just **inductors**. There are two main types of coils: the large, massive chokes to be used in power supplies, and the small inductors intended to be used in low-powered frequency discriminating circuitry. A few comments about each type follow.

The large **chokes** intended for power supply use consist of heavy copper windings on a metal core, just as do power transformers. Their primary use is as replacement parts in older equipment, since most modern equipment uses active filter/regulator circuits. For these chokes, the main parameters are the **inductance** (usually in the range of a few mH to 50 H), the nominal **resistance** of the winding, the **maximum current** (usually in the range of a few mA to several A, which is set by heating limits), the **size** (often surprisingly large), and finally the **cost**. Little more needs to be said.

The small inductors intended for frequency discriminating circuits come in several sizes and shapes. Some look like low-powered resistors, have inductances in the range of 0.1 μH to 100 mH, and have color code bands similar to those of resistors, which indicate the value of the inductance. The lower-valued inductors are air core coils, whereas the larger-valued inductors are made with ferrite cores. **Ferrites** are feromagnetic ceramics having many very nice properties that render them very attractive for making coils and permanent magnets. It takes a certain amount of experience to recognize one of these inductors as being an inductor and not a resistor in a circuit. Many interesting errors have occurred in repairing equipment when inductors have not been recognized.

Other small inductors in the range of 10 to 1000 μH look like small tire-shaped windings typically of the order of an inch or less in diameter. These windings are usually mounted on some sort of a cylindrical core. This core can be made from some ceramic material or Bakelite. Again, the larger values of inductance are obtained by inserting a piece of ferrite material inside the core. For situations where noise radiation or noise pickup may be a problem, these coils are available in small metal cases that provide some shielding when the case is grounded.

Finally, **variable inductors** can be made by providing a mechanism whereby the ferrite core can be moved partially into or out of the coil. By controlling how much of the ferrite material is actually inserted into the winding of the coil, it is possible to control the inductance of the coil. Usually this is done by threading the ferrite core, taping the central hole of the cylinder, and simply screwing the ferrite plug in and out of the core. When adjusting these coils, the adjustment must be done with a nonmagnetic screwdriver, since a metallic screwdriver will change the inductance of the coil while it is being used.

For all these small inductors, the interesting facts are the **inductance** (usually in μH), the **resistance** of the coil, the **current carrying** limits (usually in the mA range), the **maximum Q** that can be obtained with the coil at some specified frequency, perhaps some indication of the maximum (or minimum) frequency at which this coil is intended to be used, the size, and a one or two-word description of the type of coil.

TRANSFORMERS

There are three main types of transformers in use today: power transformers, radio frequency coupling transformers, and audio coupling transformers. Of these, the second two types are rapidly vanishing from the market and require very little comment here. These two will be discussed first.

Speaker systems typically have impendances of 4, 8, or 16 Ω. To get the maximum power transferred from an amplifier to the speakers, it is necessary for the output impedance of the amplifier to be matched to the speaker impedance (see Chapter 11). Although it is possible to build amplifier circuits with op-amps and transistors that have output impedances which are comparable to 4 Ω, older or very high-quality audio equipment is built with vacuum tubes. The typical vacuum tube circuit has an output impedance in the range of a few thousand ohms. **Audio transformers** were and are used to provide the correct impedance match between the speakers and the vacuum tube amplifiers—see Chapter 11 for details about impedance transformation. Obviously, these transformers are of interest only for the repair of existing equipment and for specialists in extreme-high-fidelity design.

RF coupling transformers are used in making very selective, tuned high-frequency circuits. Typically, these transformers are built by either making one winding on top of the other, or by mounting one winding next to the other on a ferrite core. These transformers are of interest to builders of radio amateur gear and to people who are repairing existing equipment. Manufacturers of original equipment typically have specially designed transformers made for each specific purpose. Consequently not too many of these are available from the standard parts suppliers.

Power transformers are used in making power supplies. Their main function is to transform the existing ac voltage (115/220 Vac) to the voltages required to make the power supplies needed. A fairly wide variety of these exist, as any distributor's catalog will show. The variation is primarily to service the replacement market for supplies for vacuum tubes, for specialty circuits requiring high voltages, and for modern low-voltage semiconductor equipment.

The primary parameters to be specified for a transformer include the voltage of the primary winding, the number of secondary winding, the voltage of each secondary winding, whether or not there are center taps on the secondary windings, the current-carrying limits on each winding, and some information about the maximum voltage that can be applied between any two windings. The wide variety of transformers available is mainly because several different secondaries can be combined in one transformer, thus providing many different options as well as variations in size, mounting style, and connector types.

Chapter 50

TRANSDUCERS

OBJECTIVES

1. Become familiar with a few of the many types of transducers that are currently available.

INTRODUCTION

Tranducers provide the input and output of electronic systems—the ways by which information is transferred between the outside world and the electronics world. Some transducers provide the means by which electronic systems can perform measurements of quantities such as temperature, light intensity, and pressure. Other transducers provide the means by which electronic systems can perform control functions and can transmit information to human beings. The purpose of this chapter is to provide an awareness of the existence of the major classes of transducers. The many different types, the variety within each type, and the vast amount of detailed information available about each type precludes doing anything more in an introduction.

Most surveys of transducers focus entirely on **input transducers**, those devices that convert some characteristic of the external world into an electrical signal that can then be processed and used in an electronic circuit. However, there are a whole class of **output transducers**, which allow information to be taken from the electronic circuit and delivered to the outside world. This chapter will look briefly at both of these classes of transducers.

OUTPUT TRANSDUCERS

Most of the common output transducers are so often encountered that it is rarely necessary to discuss them. However, as a vital link in most electronic systems, they need to be considered to present a logical whole. Some of the most important output transducers are described below.

419

Speakers

Speakers in radios, televisions, computers, PA systems, and so on are one of the most common types of output transducers. They provide the means by which electric currents can be converted into sound energy that people can hear. However, the use of speakerlike devices is spreading far beyond just communications; sonar, sonic biological probes, sonic surgical techniques, and the like are all applications that use speakerlike transducers as part of an overall system. The old-fashioned voice coil speaker operated by passing a current through a cylindrical coil mounted inside a magnetic field. The coil was restrained by a spring and, as a result, the amount of current passing through the coil determined how far the coil pulled into the magnetic field. This type of device has been used in many other situations. Examples are the voice coil-positioned heads in a computer disk system. New speaker systems make use of pizoelectric materials and other phenomena.

Solenoids

The **solenoid** is a very similar structure to the voice coil speaker. In the basic solenoid design, a metallic rod restrained by a spring can be pulled into a cylindrical coil by the magnetic field created by a current flowing in the coil. With the current on, the rod is pulled to one end by the magnetic field; with the current off, the spring pulls the rod to the other end. This device forms the basis of relays that are electrically controlled switches, or electrically controlled valves, or for countless other applications in electrical control systems (now rapidly becoming computer control systems), such as valves and shutter controls.

Stepping Motors

Another electromechanical transducer that has recently become very important in many different areas is the **stepping motor**. A stepping motor is a motor that turns a fixed fraction of a revolution (say, 2 degrees) for each electrical pulse it receives at its input. Thus a control system can cause a stepping motor to turn a precisely determined number (or fraction) of revolutions by simply sending a fixed number of pulses to the motor. Likewise, the motor may run continuously at a variable speed by simply sending it a continuous pulse train of varying frequency. Obviously, stepping motors are almost perfectly suited to computer control systems.

Lights

Ordinary lights and especially LEDs can be used as transducers. The intensity may be modulated or the lights can be blinked at varying rates to transmit information.

Computer Displays and Printouts

Computer display systems and **computer printers** are also forms of transducers. They convert electrical signals into forms of information that can be used by human beings, either immediately from the displays or later from the printouts.

INPUT TRANSDUCERS

A number of different input transducers will be discussed in slightly more detail than were the output transducers. This discussion will be organized by the nature of the input signal that is being converted into electrical form.

Electrochemical Transducers

There are many **electrochemical probes** that can be put into chemical systems and will give out an electrical signal, generally a voltage, which is proportional to some feature in the chemical system. The most common of these is the **pH probe**, a probe that generates a voltage that is related to the hydrogen ion concentration (the pH) in a solution. Typically, these probes generate signals of a few millivolts at most and have very high internal impedances, that is, they cannot supply much current at all. These are devices that are used with high-grade instrumentation amplifier systems and with much care devoted to noise elimination.

Humidity

Many hygroscopic materials can be made into detectors of atmospheric humidity. The resistance of the material depends on the amount of moisture it has absorbed, which in turn depends on the relative humidity of the surrounding atmosphere. Thus a simple circuit that measures the resistance of the probe can be used to measure the humidity of the air. The main problems are calibration and long-term drift.

Light

There are a number of different devices that can be used to convert light intensity into electrical signals. The most sensitive of these are **photomultiplier tubes**, fairly complex vacuum tube systems that can be made sufficiently sensitive so that they can respond to a single photon of light. Much less sensitive are **photodiodes**, which were discussed previously. These are diodes that are used with a reverse bias; the current flowing in this condition is proportional to the light intensity striking the diode. Photodiodes can be combined with another *pn* junction to form **phototransistors**, which have greater sensitivity than do photodiodes because the built-in transistor gives them some additional gain. However, they typically have a slower response than do photodiodes. Even simpler devices are the **photoresistors**, which are simply resistors whose resistance changes as the light intensity striking them changes.

Position

There are many different ways of converting position into electrical signals. A **linear potentiometer** can be used as a measuring device for measuring how far a shaft has been pulled out. Likewise, a normal **rotary potentiometer** can be used to measure rotations. Slightly more complex systems measure the spacing between two objects by measuring the capacitance between them. Since the capacitance depends on the spacing between the two objects, this gives rise to a whole class of **capacitance transducers**.

Strain gauges have been a traditional way of measuring changes of shape and, hence, the strain or load on some object. The traditional strain gauge is a set of four resistors arranged in the form of a Wheatstone bridge. Some of the resistors are made from an elastic material that can be bent and stretched. One of these resistors is fastened onto the object (say, a rod) to be measured; then, when the rod changes length under its load, the resistor is stretched or compressed. Since the resistance of the resistor depends on its length and cross-sectional area, changing its shape changes its resistance. If the bridge circuit was originally balanced, as the load on the rod changes, the bridge becomes unbalanced and the resultant output voltage can be related to the load on the rod. By using a bridge arrangement, errors such as temperature effects in the resistors are canceled out.

A **linear variable differential transformer**, or **LVDT**, is another more elegant way to measure the position of a shaft. In this case, a transformer is wound on a cylinder and a rod of a suitable type of magnetic material is slid in and out of the cylinder. The coupling, the mutual inductance, between the two windings of the transformer depends on how much of the rod is inserted into the transformer. Thus if a constant-amplitude signal is fed to the primary, the amplitude of the secondary signal is a measure of how far the rod is inserted into the transformer.

Pressure

The position-measuring transducers discussed above can be converted into pressure-measuring transducers in many different ways. It is possible to buy small packages with fittings for liquids and gases and a connector for electrical signals that are very accurate **pressure transducers**. Such devices are in widespread use in computer-controlled industrial applications as well as in continuous blood pressure monitors in hospitals.

Sound

Microphones are the input transducers equivalent to speakers. Many different systems are used to make microphones. Speakers can be used as microphones—many intercom systems use this fact. **Pizoelectric** crystals can be used as parts of microphones; capacitive position transducers can be used as microphones. All of these techniques are used in high-grade microphone systems.

Temperature

A very wide range of transducers is available for measuring temperature. One of the reasons for this is that there is a very wide range of temperatures to be measured. No single transducer can be used over the entire range of temperatures being measured at present. The most common **temperature transducer** is the **thermocouple**, which is simply a junction between two different types of wire. Because of differences in the electrochemical potential of the two different metals, there is a voltage developed across such a junction. Typically, the voltages are millivolts and change at the rate of about 50 μV/°C. The junctions have relatively low impedance, so simple systems can be used to measure the voltage. A little thought about the circuit shows that there must always be a pair of these junctions, so one of the junctions is kept at a reference temperature—typically in an ice water bath—and the other is the temperature probe. Thus the entire device measures the temperature difference between the two junctions.

Thermistors are resistors that have negative coefficients of resistance, typically about $-4\%/°C$. They are available in a wide variety of sizes, shapes, mountings, and resistance values and can be used as simple temperature probes in a wide variety of systems. Because of their simplicity, they are very common in noncritical control situations.

There are several integrated circuit temperature transducers. A typical **voltage reference** IC can be operated in a mode in which the voltage varies as a function of the temperature; a typical variation is 2.1 mV/°C. (Of course, the voltage reference circuit can also be operated in a mode such that the temperature variations are corrected, and the output voltage is essentially constant. This is their normal purpose.)

The Analog Devices AD 590 is a very interesting and useful temperature transducer. It is a two-lead IC. When 5 to 20 V is applied, it draws 1 μA/°K current. Thus at room temperature, 20°C, it draws 293 μA. Because it is relatively insensitive to the applied voltage, it is relatively insensitive to the resistance in the wires running to the device. Thus it is ideally suited for use in situations in which the temperature probe is located a long distance from the point at which the temperature is to be read out.

PART EIGHT

MATHEMATICS

This portion of this book provides an introduction to a number of mathematical techniques that are useful in electronics and that may not be familiar to all readers. These introductions are short and to the point; the techniques and procedures are described and illustrated and some sample problems are provided. No proofs or extended mathematical discussions are included.

PREREQUISITES

CHAPTER 51 COMPLEX NUMBERS
  1. A knowledge of simple algebra.

CHAPTER 52 SOLVING LINEAR EQUATIONS
  1. A knowledge of simple algebra.
  2. Some of the problems use complex numbers—Chapter 51—as a preparation for ac circuit analysis.

CHAPTER 53 DECIBELS
  1. A knowledge of logarithms.
  2. A calculator that can calculate logarithms and antilogarithms, all to the base 10.

CHAPTER 54 BINARY NUMBERS
  1. A slight familiarity with computers will provide the motivation for some of the topics discussed in this chapter.

CHAPTER 55 FOURIER SERIES
  1. A knowledge of ac circuit analysis—Chapter 8.

Chapter 51

COMPLEX NUMBERS

OBJECTIVES

1. Be able to do simple complex arithmetic and algebra.

2. Be able to convert complex numbers from one form to another.

3. Know Euler's formula.

INTRODUCTION

This chapter is a very brief and intuitive introduction to complex numbers; no proofs are included. The goal is to present (or review) only those topics necessary to be able to use complex numbers in elementary circuit analysis. The extension of the material covered in this chapter to the study of functions of complex variables is one of the most interesting and useful topics in mathematics. Many good textbooks exist on functions of complex variables for those who wish to learn more.

REAL NUMBERS

The set of all real numbers, consisting of the rational and irrational numbers, may be placed in one-to-one correspondence with the set of points on a line called the **real number line**. This correspondence is such that each real number is represented by a unique point on the line, and each point on the line represents a unique number. In this correspondence, terms such as "greater than" are changed into geometrical terms such as "to the right of." A schematic representation of the real number line is presented in Figure 51-1.

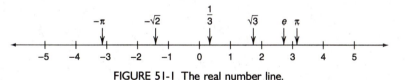

FIGURE 51-1 The real number line.

Any element of the set of real numbers may be combined with any other element of the set via the operations of addition, subtraction, multiplication, and division, and the result is another member of the set of real numbers. The square root of any positive real number is also a real number, but no real number is the square root of a negative number. In geometrical terms, the square root of any number lying to the left of zero is not on the real number line.

IMAGINARY NUMBERS

The square root of a negative real number is called an **imaginary number**, or sometimes a pure imaginary number. Thus $\sqrt{-2}$, $\sqrt{-15}$, and $\sqrt{-\pi}$ are all pure imaginary numbers. It is customary to designate

$$j = \sqrt{-1}$$
$$\sqrt{-1} = j\sqrt{2}$$
$$\sqrt{-4} = 2j, \text{ etc.}$$

Furthermore, it follows immediately from the definition that

$$j^2 = -1$$
$$j^3 = -j$$
$$j^4 = 1$$
$$j^5 = j, \text{ etc.}$$

The set of all pure imaginary numbers can also be represented by the set of points on a line called the **imaginary number line**. A representation of this line is given in Figure 51-2.

Note: Mathematicians and physicists usually use the notation

$$i = \sqrt{-1}$$

but electrical engineers use j to avoid confusion with currents. j will be used throughout this book.

Finally, the name imaginary is most unfortunate, since it leaves the impression that these numbers do not exist. These numbers do exist. The easiest way to understand their meaning is geometrical. In the next section, it will be shown that complex numbers are a logical way of specifying the location of a point in a plane.

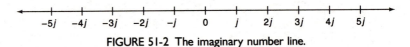

FIGURE 51-2 The imaginary number line.

COMPLEX NUMBERS

A complex number Z is any number of the form

$$Z = x + jy$$

where both x and y are real numbers. The expressions jy and yj are used interchangeably to aid legibility; mathematically, they are the same. In a complex number, x is called the real part and y the imaginary part. If $y = 0$, the number is a real number; if $x = 0$, the number is pure imaginary. Thus the sets of real numbers and pure imaginary numbers are subsets of the set of complex numbers.

If the real number line is set at right angles to the imaginary number line (j axis), as shown in Figure 51-3, then any complex number represents a unique point in the plane and any point in the plane represents a unique complex number. Thus a one-to-one correspondence exists between the complex numbers and the points of the complex plane. This correspondence includes, as subsets, the correspondence between the real numbers and the real number line and between the imaginary numbers and the imaginary number line. In the figure, six complex numbers are plotted in the complex plane. These are

$$Z_1 = 7 \qquad Z_2 = 3 + 2j \qquad Z_3 = 5j$$
$$Z_4 = -4 + 5j \qquad Z_5 = -5 - 3j \qquad Z_6 = -5j$$

Although the concepts of greater than or less than do not apply to complex numbers, it is meaningful to compare their magnitudes. If

$$Z = x + jy \tag{51-1}$$

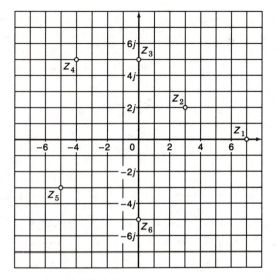

FIGURE 51-3 The complex plane.

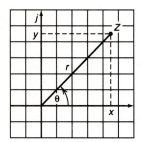

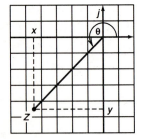

FIGURE 51-4 Two different complex numbers plotted in the complex plane.

the the **absolute value**, **modulus**, or **magnitude**, of Z is defined as

$$r = |Z| = (x^2 + y^2)^{1/2} \tag{51-2}$$

It is clear that this represents the distance of the point Z from the origin of the complex plane.

The line from the origin of the complex plane to the complex number makes the angle θ with respect to the positive real axes

$$\theta = \tan^{-1}\left(\frac{y}{x}\right) \tag{51-3}$$

This quantity is called the **amplitude**, the **argument**, or the **phase**, of the complex number Z. Both the magnitude and the phase are illustrated in Figure 51-4.

NOTATION

There are many different ways to write a complex number. Some of these different notations are discussed briefly here before complex arithmetic is discussed. The form $Z = x + jy$ is called the **rectangular** or **cartesian** form for a complex number.

From Figure 51-4, it is easy to verify that

$$x = r \cos \theta \tag{51-4}$$

$$y = r \sin \theta \tag{51-5}$$

and, hence,

$$Z = x + jy = r(\cos \theta + j \sin \theta) \tag{51-6}$$

This is called the **trigonometric** form for the complex number. With a little care in the use of the signs of the trigonometric functions (sin, cos, and arctan), these same expressions work for both complex numbers shown in Figure 51-4. Thus this form is in no way limited to the first quadrant.

Euler's formula,

$$e^{j\theta} = \cos \theta + j \sin \theta \tag{51-7}$$

has been called one of the most remarkable expressions in all of mathematics. When one encounters this expression, it seems almost magical. It expresses the fundamental relationship between the trigonometric functions and the exponential functions. This relationship is one of the many important results from the study of functions of complex variables. For a proof (really a heuristic argument) of Euler's formula, see the last section of this chapter. Using Euler's formula, the trigonometric form can be written as

$$Z = r(\cos \theta + j \sin \theta) = re^{j\theta} \tag{51-8}$$

This is the **exponential** form.

These last two forms show that the magnitude and argument of the complex number are the important factors; the differences in these forms are just notation. The simplest possible form is the **polar**, or **Steinmetz** form (the form favored by electrical engineers):

$$Z = r\angle\theta \qquad\qquad (51\text{-}9)$$

Complex numbers written in this form are sometimes called **phasors**.

Thus there are four ways to write a complex number. They are all equivalent. To summarize:

Rectangular form:

$$Z = x + jy$$

Polar or Steinmetz form:

$$Z = r\angle\theta$$

Exponential form:

$$Z = re^{j\theta}$$

Trigonometric form:

$$Z = r(\cos\theta + j\sin\theta)$$

It is usual to express θ in degrees in the polar form but in radians in the exponential form—a detail that depends on your calculator more than anything else.

MATHEMATICAL OPERATIONS

The usual mathematical operations—addition, subtraction, multiplication, and division—can all be defined for complex numbers. These operations are discussed below. Since the details of carrying out some of these operations depend on the form in which the complex number is written, this section has a sort of rambling nature. The important point to remember is that although the calculation may look different depending on the form of the numbers, the result is the same. Thus the product of two complex numbers is the same whether they are written in the rectangular form or in the exponential form.

Two complex numbers are **equal** if and only if their real and imaginary parts are equal. Thus if

$$Z_1 = a + jb$$
$$Z_2 = c + jd$$

then

$$Z_1 = Z_2$$

if and only if $a = c$ and $b = d$. $\qquad\qquad (51\text{-}10)$

The **conjugate** Z^*, sometimes called the **complex conjugate** of a complex number $Z = x + jy$, is defined as

$$Z^* = x - jy \tag{51-11}$$

That is, the conjugate of a complex number has the same real part but the opposite sign on the imaginary part. Thus the conjugate of $Z = 3 + 2j$ is $Z^* = 3 - 2j$. In terms of the complex plane, the conjugate of Z is its reflection with respect to the real axis. This is shown in Figure 51-5. Since $\cos(-\theta) = \cos\theta$ and $\sin(-\theta) = -\sin\theta$, it is easy to verify that if Z has an argument of θ, then Z^* has an argument of $-\theta$. The four different notations give four different ways of writing the conjugate. These are as follows:

Rectangular:

$$Z = x + jy \qquad Z^* = x - jy$$

Polar:

$$Z = r\angle\theta \qquad Z^* = r\angle-\theta$$

Exponential:

$$Z = re^{j\theta} \qquad Z^* = re^{-j\theta}$$

Trigonometric:

$$Z = r(\cos\theta + j\sin\theta) \qquad Z^* = r(\cos\theta - j\sin\theta)$$

Addition of complex numbers is defined similarly to addition of vectors; in fact, complex numbers and two-dimensional vectors are essentially identical. To add two complex numbers, add the real parts and then add the imag-

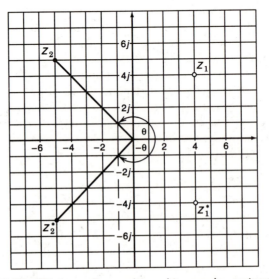

FIGURE 51-5 A complex number and its complex conjugate.

inary parts. Similarly for **subtraction**, subtract the real and imaginary parts separately. Thus for

$$Z_1 = x_1 + jy_1 \quad \text{and} \quad Z_2 = x_2 + jy_2$$

$$Z_1 + Z_2 = (x_1 + x_2) + j(y_1 + y_2) \tag{51-12}$$

$$Z_1 - Z_2 = (x_1 - x_2) + j(y_1 - y_2) \tag{51-13}$$

For example, if

$$Z_1 = 3 + 5j \quad \text{and} \quad Z_2 = 6 - 8j$$

then

$$Z_1 + Z_2 = (3 + 6) = j(5 - 8) = 9 - 3j$$

$$Z_1 - Z_2 = (3 - 6) + j(5 + 8) = -3 + 13j$$

This rule is easily extended to the sum of three or more complex numbers. Addition and subtraction are abelian, that is, they are commutative and associative, as are addition and subtraction of real numbers.

From a purely practical standpoint, addition and subtraction of complex numbers can be easily performed only when the numbers are in rectangular form. If they are in other forms, they have to be converted before adding or subtracting. Some calculators have the ability to do complex arithmetic on complex numbers written in polar form. In that case, conversion is not needed—the calculator does it.

The **multiplications** of two complex numbers written in rectangular or trigonometric form is calculated by treating the numbers as binomials. Thus

$$Z_1 Z_2 = (x_1 + jy_1)(x_2 + jy_2)$$

$$= x_1 x_2 + jx_1 y_2 + jy_1 x_2 - y_1 y_2$$

$$= (x_1 x_2 - y_1 y_2) + j(x_1 y_2 + y_1 x_2) \tag{51-14}$$

For example, if

$$Z_1 = 3 + 6j \quad \text{and} \quad Z_2 = 5 - 2j$$

then

$$Z_1 Z_2 = (3 + 6j)(5 - 2j)$$

$$= (15 + 12) + j(-6 + 30) = 27 + 24j$$

When the complex numbers are written in the exponential form, the product is defined in terms of the multiplication of exponentials:

$$Z_1 Z_2 = (r_1 e^{i\theta})(r_2 e^{j\phi})$$

$$= r_1 r_2 e^{j(\theta + \phi)} \tag{51-15}$$

For example, if

$$Z_1 = 3e^{j\pi/3} \quad \text{and} \quad Z_2 = 5e^{-j\pi/6}$$

then

$$Z_1 Z_2 = 15e^{j(\pi/3 - \pi/6)} = 15e^{j\pi/6}$$

The multiplication of complex numbers in polar form follows immediately from this:

$$Z_1 = r_1 \angle \theta_1$$
$$Z_2 = r_2 \angle \theta_2$$
$$Z_1 Z_2 = (r_1 r_2) \angle (\theta_1 + \theta_2) \tag{51-16}$$

For example, if

$$Z_1 = 6 \angle 35° \quad \text{and} \quad Z_2 = 3 \angle -50°$$

then

$$Z_1 Z_2 = 18 \angle -15°$$

It is an easy task to show that these various definitions are all equivalent; all it takes is a little skill with trigonometric identities. It is also clear that multiplication of complex number is commutative and associative.

Finally, note that

$$|Z|^2 = ZZ^* = x^2 + y^2 \tag{51-17}$$

Division of complex numbers is defined in terms of multiplication. For complex numbers in exponential or polar form, division is relatively easy:

$$\frac{Z_1}{Z_2} = \frac{r_1 e^{j\theta_1}}{r_2 e^{j\theta_2}} = \frac{r_1}{r_2} e^{j(\theta_1 - \theta_2)} \tag{51-18}$$

and

$$\frac{Z_1}{Z_2} = \frac{r_1 \angle \theta_1}{r_2 \angle \theta_2} = \frac{r_1}{r_2} \angle (\theta_1 - \theta_2) \tag{51-19}$$

Examples:

$$Z_1 = 6e^{j\pi/2} \quad \text{and} \quad Z_2 = 3e^{j\pi/4}$$

$$\frac{Z_1}{Z_2} = \frac{6e^{j\pi/2}}{3e^{j\pi/4}} = 2e^{j\pi/4}$$

$$Z_1 = 5 \angle 35° \quad \text{and} \quad Z_2 = 5 \angle -45°$$

$$\frac{Z_1}{Z_2} = \frac{5 \angle 35°}{5 \angle -45°} = 1 \angle 80°$$

For complex numbers expressed in rectangular form, it is necessary to **rationalize** the denominator by multiplying both the numerator and the denominator by the conjugate of the denominator. This leaves the denominator a real number and the result in rectangular form. Thus

$$\frac{Z_1}{Z_2} = \frac{x_1 + jy_1}{x_2 + jy_2} = \frac{x_1 + jy_1}{x_2 + jy_2} \frac{x_2 - jy_2}{x_2 - jy_2}$$

$$= \frac{(x_1x_2 + y_1y_2) + j(y_1x_2 - x_1y_2)}{x_2^2 + y_2^2} \tag{51-20}$$

An example:

$$\frac{Z_1}{Z_2} = \frac{3 - 5j}{4 + 6j} = \frac{3 - 5j}{4 + 6j} \frac{4 - 6j}{4 - 6j}$$

$$= \frac{(12 - 30) + j(-20 - 18)}{16 + 36} = \frac{-18 - 38j}{52}$$

$$= -0.346 - 0.731j$$

One very important result of this is

$$\frac{1}{j} = -j \tag{51-21}$$

It is easy to extend the concepts of roots, exponentials, logarithms, trigonometric functions, and so on, to the case of complex numbers. This is far beyond the scope of this book and will not be discussed here. For details, see any of the standard texts that treat functions of complex variables.

A PROOF OF EULER'S FORMULA

Assume that a function $f(x)$ can be represented by a power series in x and that this series must be of the form of the Maclaurin series:

$$f(x) = f(0) + xf'(0) + \frac{x^2}{2!}f''(0) + \cdots + \frac{x^{(n-1)}}{(n-1)!}f^{(n-1)}(0) + \cdots$$

The function and all its derivatives must exist at $x = 0$. The Maclaurin series for $\cos\theta$, $\sin\theta$, and $e^{j\theta}$ are

$$\cos\theta = 1 - \frac{\theta^2}{2!} + \frac{\theta^4}{4!} - \frac{\theta^6}{6!} + \cdots$$

$$\sin\theta = \theta - \frac{\theta^3}{3!} + \frac{\theta^5}{5!} - \frac{\theta^7}{7!}$$

$$e^{j\theta} = 1 + j\theta - \frac{\theta^2}{2!} - \frac{j\theta^3}{3!} + \frac{\theta^4}{4!} + \frac{j\theta^5}{5!} - \frac{\theta^6}{6!} - \frac{j\theta^7}{7!} + \cdots$$

Rearranging the terms of the series for $e^{j\theta}$ yields

$$e^{j\theta} = \left(1 - \frac{\theta^2}{2!} + \frac{\theta^4}{4!} - \frac{\theta^6}{6!} + \cdots\right) = j\left(\theta - \frac{\theta^3}{3!} + \frac{\theta^5}{5!} - \frac{\theta^7}{6!} + \cdots\right)$$

$$= \cos\theta = j\sin\theta$$

The last step follows from comparing the various expansions. To complete this proof, it would be necessary to prove the Maclaurin expansion theorem and to prove that it is permissible to rearrange the series without changing its value. These steps would take us far beyond the scope of this book, but they can be done. For further details, see a textbook on the functions of complex variables.

PROBLEMS

Convert the following to polar form:

51-1 $3 + 6j$

51-2 $5 - 8j$

51-3 $-4 + 18j$

51-4 $-6 - 5j$

Convert to rectangular form:

51-5 $3e^{j\pi/35}$

51-6 $18\angle 16°$

51-7 $4e^{-j\pi/4}$

51-8 $6\angle -50°$

Calculate $Z_1 + Z_2$, $Z_1 - Z_2$, and $Z_2 - Z_1$:

51-9 $Z_1 = 3 + 2j \quad Z_2 = 5 - 8j$

51-10 $Z_1 = -4 + 18j \quad Z_2 = -6 - 5j$

51-11 $Z_1 = 15 + 21j \quad Z_2 = 5\angle 30°$

51-12 $Z_1 = 3 - 6j \quad Z_2 = 15 + 18j$

Calculate $Z_1 Z_2$, Z_1/Z_2, and Z_2/Z_1. Where necessary, rationalize the denominator:

51-13 $Z_1 = 8\angle 30° \quad Z_2 = 4\angle 51°$

51-14 $Z_1 = 15\angle -21° \quad Z_2 = 5\angle -42°$

51-15 $Z_1 = 10e^{j\pi} \quad Z_2 = 8e^{-j\pi/4}$

51-16 $Z_1 = 3e^{-j\pi/6} \quad Z_2 = 3e^{j\pi/6}$

51-17 $Z_1 = 3 + 6j \quad Z_2 = 5 - 8j$

51-18 $Z_1 = -4 + 18j \quad Z_2 = -6 - 6j$

ANSWERS TO ODD-NUMBERED PROBLEMS

51-1 $6.708\angle63.43°$
51-3 $18.439\angle102.53°$
51-5 $1.5 + 2.598j$
51-7 $2.828 - 2.828j$
51-9

$Z_1 + Z_2$	$Z_1 - Z_2$	$Z_2 - Z_1$
$8 - 6j$	$-2 + 10j$	$2 - 10j$

51-11 $19.33 + 23.5j$ $10.67 + 18.5j$ $-10.67 - 18.5j$
51-13

$Z_1 Z_2$	Z_1/Z_2	Z_2/Z_1
$32\angle81°$	$2\angle-21°$	$0.5\angle21°$

51-15 $80e^{j3\pi/4}$ $1.25e^{j5\pi/4}$ $0.80e^{-j5\pi/4}$
51-17 $63 + 6j$ $-0.371 + 0.607j$ $-0.733 - 1.2j$

Chapter 52

SOLVING LINEAR EQUATIONS

OBJECTIVES

1. Be able to evaluate determinants.
2. Be able to solve systems of linear equations using Cramer's rule.

INTRODUCTION

This chapter presents a cookbook approach to solving the simultaneous linear equations encountered in circuit analysis. No proofs will be given, and many interesting topics will be omitted. For further details, see any standard textbook on linear algebra. The procedure described in this chapter, called Cramer's rule, is a brute force procedure. It will often be true that a little insight and perhaps a trick or two will solve a set of equations more easily than an application of Cramer's rule. But Cramer's rule will always work and does not require any insight. So, when no easy way presents itself, use this method. Because Cramer's rule uses determinants, this chapter begins with a discussion of determinants.

DEFINITION OF A MATRIX

A **matrix** is a rectangular array of real or complex numbers or functions. When explicitly written out, it is enclosed in brackets as shown below:

$$
A = \begin{bmatrix}
a_{11} & a_{12} & a_{13} & \cdots & a_{1n} \\
a_{21} & a_{22} & a_{23} & \cdots & a_{2n} \\
\cdot & \cdot & \cdot & & \cdot \\
\cdot & \cdot & \cdot & & \cdot \\
\cdot & \cdot & \cdot & & \cdot \\
a_{m1} & a_{m2} & a_{m3} & & a_{mn}
\end{bmatrix}
$$

3rd column $\downarrow$

$\leftarrow$ 2nd row

438

Each individual element in the matrix is identified by its position in the matrix, that is, by its row and column. Thus a_{ij} is in the ith row and jth column. The particular matrix above is said to be of **order m by n**, or **$m \times n$**, where m is the number of rows and n is the number of columns in the matrix. A **square matrix**, that is, one of order n by n, is often simply called a square matrix of order n, or a matrix of order n.

A complete matrix algebra can be defined. This algebra is very useful for proving theorems in vector analysis and circuit analysis. For this reason, a working knowledge of matrix algebra is essential in a formal study of circuit theory. However, the circuit analysis done in this book is sufficiently simple that only a knowledge of determinants is needed.

DEFINITION OF A DETERMINANT

The **determinant** is only defined for square matrices. It is a single real or complex number that is calculated by a carefully defined sum of products of all the elements of the matrix. This definition is quite complex and will be given in several steps. The determinant of a matrix A is written as $|A|$. For a square matrix of order 1, the determinant is defined as

$$|A| = |a_{11}| = a_{11} \tag{52-1}$$

Beware of the possible confusion here. The notation might seem to indicate that the determinant is the absolute value of a_{11}, but this is not correct. The determinant of a square matrix of order 1 is a signed quantity, that is, it may be negative.

For a matrix of order 2,

$$|A| = \begin{vmatrix} a_{11} & a_{12} \\ a_{21} & a_{22} \end{vmatrix} = a_{11}a_{22} - a_{12}a_{21} \tag{52-2}$$

The sketch below illustrates an easy way to remember the calculation of the determinant of a matrix of order 2. As shown, it is the product along the downward diagonal less the product along the upward diagonal:

For orders greater than 2, an inductive procedure is used to define the determinant. Two additional terms are used in this definition. The **minor of an element a_{ij}** of a determinant of order n is the determinant of order $n - 1$ obtained by deleting the ith row and the jth column from the original determinant— these are the row and column containing a_{ij}. The minor of a_{ij} is denoted as $|M_{ij}|$. The signed minor

$$(-1)^{i+j}|M_{ij}| \tag{52-3}$$

is called the **cofactor of** a_{ij} and is sometimes denoted as Δ_{ij}. A simple example will illustrate these definitions. For the determinant

$$|A| = \begin{vmatrix} a_{11} & a_{12} & a_{13} \\ a_{21} & a_{22} & a_{23} \\ a_{31} & a_{32} & a_{33} \end{vmatrix}$$

$$|M_{32}| = \begin{vmatrix} a_{11} & a_{13} \\ a_{21} & a_{23} \end{vmatrix}$$

and

$$\Delta_{32} = (-1)^{3+2}|M_{32}| = -\begin{vmatrix} a_{11} & a_{13} \\ a_{21} & a_{23} \end{vmatrix} = -(a_{11}a_{23} - a_{21}a_{13})$$

The value of the determinant $|A|$ of order n is the sum of the n products formed by multiplying each element in one row or column by its cofactor. Thus for the example discussed above, expanding along the bottom row gives

$$|A| = \begin{vmatrix} 5 & 1 & 3 \\ 6 & 10 & 8 \\ 8 & 2 & 1 \end{vmatrix} = (-1)^{3+1}8\begin{vmatrix} 1 & 3 \\ 10 & 8 \end{vmatrix} + (-1)^{3+2}2\begin{vmatrix} 5 & 3 \\ 6 & 8 \end{vmatrix} + (-1)^{3+3}\begin{vmatrix} 5 & 1 \\ 6 & 10 \end{vmatrix}$$

$$= 8(8 - 30) - 2(40 - 18) + (50 - 6)$$

$$= -176$$

It can be shown that all possible ways of expanding the determinant yield the same result. For the particular example presented here, it is easy to verify that other possible expansions give the same value by simply doing them. It is also possible to show that the general definition is consistent with the definition of a determinant of a matrix of order 2 that was given earlier (see problem 52-1).

By use of this definition, a determinant of any order can be reduced to a sum of terms involving determinants of one lower order. By repeated applications of this process, a determinant can be reduced to a sum of determinants of order 1 or 2; then, the final value can be calculated. Just as a number of theorems can be proved about matrices, so also can a number of theorems be proved about determinants. But here, again, these will not be needed for this book.

This section ends with several more examples. The first one again illustrates the possible notational problem:

$$|A| = |-3| = -3$$

$$|A| = \begin{vmatrix} 3 & 4 \\ 2 & -6 \end{vmatrix} = -18 - 8 = -26$$

$$|A| = \begin{vmatrix} 5 & 1 & 3 \\ 6 & 10 & 8 \\ 8 & 2 & 1 \end{vmatrix} = 5\begin{vmatrix} 10 & 8 \\ 2 & 1 \end{vmatrix} - 1\begin{vmatrix} 6 & 8 \\ 8 & 1 \end{vmatrix} + 3\begin{vmatrix} 6 & 10 \\ 8 & 2 \end{vmatrix}$$

$$= 5(10 - 16) - 1(6 - 64) + 3(12 - 80)$$

$$= -176$$

which is just the result obtained earlier by a different expansion. One last example, which is slightly more complex, is

$$|A| = \begin{vmatrix} 1 & 6 & 8 & 3 \\ 2 & 16 & 6 & 2 \\ 3 & 5 & 8 & 1 \\ 4 & 3 & 5 & 6 \end{vmatrix}$$

$$= \begin{vmatrix} 16 & 6 & 2 \\ 5 & 8 & 1 \\ 3 & 5 & 6 \end{vmatrix} - 6\begin{vmatrix} 2 & 6 & 2 \\ 3 & 8 & 1 \\ 4 & 5 & 6 \end{vmatrix} + 8\begin{vmatrix} 2 & 16 & 2 \\ 3 & 5 & 1 \\ 4 & 3 & 6 \end{vmatrix} - 3\begin{vmatrix} 2 & 16 & 6 \\ 3 & 5 & 8 \\ 4 & 3 & 5 \end{vmatrix}$$

$$= 528 + 192 - 1536 - 624$$

$$= -1440$$

SOLVING LINEAR EQUATIONS

Consider the two simultaneous linear equations below:

$$a_{11}x + a_{12}y = c_1$$
$$a_{21}x + a_{22}y = c_2 \tag{52-4}$$

One way to solve these equations is by elimination of variables. To do this, multiply the first equation by a_{22} and the second by a_{12}, obtaining

$$a_{11}a_{22}x + a_{12}a_{22}y = c_1a_{22}$$
$$a_{12}a_{21}x + a_{12}a_{22}y = c_2a_{12}$$

Next, subtract the second equation from the first, obtaining

$$(a_{11}a_{22} - a_{12}a_{21})x = a_{22}c_1 - a_{12}c_2$$

Finally, solving this for x yields

$$x = \frac{a_{22}c_1 - a_{12}c_2}{a_{11}a_{22} - a_{12}a_{21}} \tag{52-5}$$

Note that this is the same as

$$x = \frac{\begin{vmatrix} c_1 & a_{12} \\ c_2 & a_{22} \end{vmatrix}}{\begin{vmatrix} a_{11} & a_{12} \\ a_{21} & a_{22} \end{vmatrix}} \tag{52-6}$$

This second way of obtaining the solution is an example of the general procedure known as Cramer's rule.

Cramer's rule provides a mechanical procedure for solving any set of n simultaneous independent linear equations in n unknowns. The general rule will be stated here by a more complex example. For any set of linear equations (in this example, four equations in four unknowns) written in the form below,

$$a_{11}w + a_{12}x + a_{13}y + a_{14}z = c_1$$
$$a_{21}w + a_{22}x + a_{23}y + a_{24}z = c_2$$
$$a_{31}w + a_{32}x + a_{33}y + a_{34}z = c_3 \qquad (52\text{-}7)$$
$$a_{41}w + a_{42}x + a_{43}y + a_{44}z = c_4$$

the solution is given by

$$w = \frac{\begin{vmatrix} c_1 & a_{12} & a_{13} & a_{14} \\ c_2 & a_{22} & a_{23} & a_{24} \\ c_3 & a_{32} & a_{33} & a_{34} \\ c_4 & a_{42} & a_{43} & a_{44} \end{vmatrix}}{\text{denom}} \qquad (52\text{-}8)$$

$$x = \frac{\begin{vmatrix} a_{11} & c_1 & a_{13} & a_{14} \\ a_{21} & c_2 & a_{23} & a_{24} \\ a_{31} & c_3 & a_{33} & a_{34} \\ a_{41} & c_4 & a_{43} & a_{44} \end{vmatrix}}{\text{denom}} \qquad (52\text{-}9)$$

$$y = \frac{\begin{vmatrix} a_{11} & a_{12} & c_1 & a_{14} \\ a_{21} & a_{22} & c_2 & a_{24} \\ a_{31} & a_{32} & c_3 & a_{34} \\ a_{41} & a_{42} & c_4 & a_{44} \end{vmatrix}}{\text{denom}} \qquad (52\text{-}10)$$

$$z = \frac{\begin{vmatrix} a_{11} & a_{12} & a_{13} & c_1 \\ a_{21} & a_{22} & a_{23} & c_2 \\ a_{31} & a_{32} & a_{33} & c_3 \\ a_{41} & a_{42} & a_{43} & c_4 \end{vmatrix}}{\text{denom}} \qquad (52\text{-}11)$$

where

$$\text{denom} = \begin{vmatrix} a_{11} & a_{12} & a_{13} & a_{14} \\ a_{21} & a_{22} & a_{23} & a_{24} \\ a_{31} & a_{32} & a_{33} & a_{34} \\ a_{41} & a_{42} & a_{43} & a_{44} \end{vmatrix} \qquad (52\text{-}12)$$

The generalization to any number of equations and unknowns should be obvious. If the determinant in the denominator is 0, the equations are not independent and there is no unique solution to the set of equations.

A FINAL COMMENT

Nothing that has been done here precludes either the coefficients in the linear equations or the elements of the matrix from being complex. The evaluation of the determinant or the solution of the equations proceeds just the same; only, it is more work if the elements are complex.

PROBLEMS

52-1 Show that expanding a determinant of order 2 into two determinants of order 1 by the iterative scheme gives the same result as Equation 52-2.

Evaluate the following determinants:

52-2 $\begin{vmatrix} 5 & 6 \\ 3 & 8 \end{vmatrix}$

52-3 $\begin{vmatrix} 5 & 8 & 1 \\ 3 & 9 & 2 \\ 1 & 10 & 3 \end{vmatrix}$

52-4 $\begin{vmatrix} 10 & 6 & 5 \\ 5 & -3 & 1 \\ -2 & 8 & 2 \end{vmatrix}$

52-5 $\begin{vmatrix} 2 + 3j & 5 \\ 6j & -2j \end{vmatrix}$

Solve the following equations:

52-6 $3x + 6y = 5$

$-8x - 2y = 6$

52-7 $x + y + z = 4$

$-5x + y - z = 2$

$2x - 6y - 6z = 0$

52-8 $2i_1 - i_2 + 4i_3 = 6$

$i_1 + i_2 + i_3 = 0$

$-8i_1 + 4i_2 + 4i_3 = 4$

52-9 $2jx - jy = 0$

$x + y = 3 + 4j$

ANSWERS TO ODD-NUMBERED PROBLEMS

52-3 0

52-5 $6 - 34j$

52-7 $D = -16, x = 3, y = 9, z = -8$

52-9 $D = 3j, x = 1 + 1.333j, y = 2 + 2.667j$

Chapter 53

DECIBELS

OBJECTIVES

1. Be able to calculate power, voltage, and current gains in decibels.

2. Be able to convert multiplicative gains into gains in decibels, and vice versa.

3. Be able to use decibels to calculate the gains for several circuits connected in series.

INTRODUCTION

The gains of electronic circuits are usually measured in the unfamiliar units of decibels, or db. These units are defined and discussed in this chapter. Although the language is that of electronics, this chapter is mainly mathematics.

DEFINITION OF A DECIBEL

The power gain of any circuit is defined as

$$\text{Gain} = \frac{\text{power output}}{\text{power input}} \tag{53-1}$$

Gain is always positive, but it can be greater than or less than 1; a gain less than 1 simply means that there is less power available at the output than at the input. For instance, a voltage divider has a gain less than 1.

As defined above, gain is a dimensionless multiplicative factor. More often, however, gain is measured in dimensionless logarithmic units called **bells**. The gain, in bells, is given by

$$\text{Gain (in bells)} = \log_{10}\left(\frac{\text{power output}}{\text{power input}}\right) \tag{53-2}$$

Bells turn out to be too large for practical use; in fact, they are never used except in textbooks as a first step in the introduction of decibels. The units of **0.1 bells**, **decibels**, or **db** are used:

$$\text{Gain (in db)} = 10 \times \log_{10}\left(\frac{\text{power output}}{\text{power input}}\right) \tag{53-3}$$

Since power is proportional to V^2 or I^2, this can also be written as

$$\text{Gain (in db)} = 20 \times \log_{10}\left(\frac{\text{voltage output}}{\text{voltage input}}\right) \tag{53-4}$$

or

$$\text{Gain (in db)} = 20 \times \log_{10}\left(\frac{\text{current output}}{\text{current input}}\right) \tag{53-5}$$

Because of the nature of the logarithm, gains in db that are greater than zero correspond to multiplicative gains greater than 1, whereas gains in db that are less than zero correspond to multiplicative gains less than 1. Naturally, a gain of 0 db corresponds to a multiplicative gain of 1. The db values shown in Table 53-1 should be useful.

COMMENTS ON THE USEFULNESS OF DECIBELS

There are several reasons why it is customary to measure gains in units of db rather than as multiplicative values. The main reason for using decibels is the ease with which the total gain of several successive circuits can be calcu-

TABLE 53-1 Multiplicative Gains Equivalent to Various Decibel Gains

| Gain in db | Multiplicative Gain | |
	For Power	For Voltage or Current
0	1	1
1	1.26	1.12
2	1.58	1.26
3	2.0	1.41 ($\sqrt{2}$)
6	4.0	2.0
10	10.0	3.16
12	16.0	4.00
20	100.	10.0
40	10,000.	100.
60	1,000,000.	1000.
−3	0.5	0.707 ($1/\sqrt{2}$)
−6	0.25	0.5
−10	0.1	0.316
−20	0.01	0.1
−40	0.0001	0.01

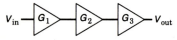

FIGURE 53-1 Three amplifiers in series.

lated. If several circuits are connected together so that the output of the first is the input of the second, the output of the second is the input of the third, and so on, then the overall multiplicative gain of the combination is found by multiplying together the gains of the individual circuits. Thus for the case shown in Figure 53-1, the overall gain is given by

$$G_{total} = G_1 \times G_2 \times G_3 \quad \text{(multiplicative)} \quad (53\text{-}6)$$

Thus if the circuits have gains of 38,000, 0.0000435 and 47000, the overall gain of the circuit is 77,700. If the gains are expressed as decibels, then the overall gain is found by adding the individual gains. That is,

$$G_{total} = G_1 + G_2 + G_3 \quad \text{(db)} \quad (53\text{-}7)$$

For the example above, the individual gains are 45.8, −43.6, and 46.7 db, respectively, and the overall gain is the sum of these, or 48.9 db. This is clearly easier to calculate. Of course, it may not mean as much to someone to say the gain is 48.9 db as to say it is 77,700, but experience makes these units much easier to understand. The advantage of adding decibels becomes much clearer when dealing with sets of calibrated attenuators. Thus it is a trivial exercise to figure out what the effect is of 10-db, 6-db, and 1-db attenuators placed in series; it is a 17-db attenuation.

A second reason for using decibels is that the human senses are roughly logarithmic. That is, the minimum increment in sound or light intensity that is just detectable is more or less a constant fraction of the intensity and corresponds to about a 1-db increase in intensity. Thus a sound that is 10 db louder than another sound roughly corresponds to 10 times the smallest increment in intensity that can be discerned (at the intensity of interest). Thus gains expressed in decibels correspond fairly directly to physiological units.

A final reason often given in the literature is that the numbers required to express gains as multiplicative factors cover a very large range, say, from 1 to several million. Since the same range covered in decibels is only 0 to 60, the numbers are not so large and awkward. Of course, the same people who write this, happily deal with picofarads and megohms with no further problems.

Hence, for whatever combination of the reasons given above, gains are traditionally measured in decibels. There is no hope of avoiding decibels when working in electronics. A calculator that can be used to convert numbers into and out of decibels, that is, take logarithms and antilogarithms to the base 10, is necessary. (When checking out a calculator, remember that taking an antilogarithm to the base 10 is the same as raising 10 to the appropriate power.)

PROBLEMS

53-1 Convert the following multiplicative voltage gains into decibels: 56,000; 128,000; 0.05.

53-2 Do the same for these multiplicative voltage gains: 1,600,000; 0.0065.

53-3 Convert the following gains in decibels to multiplicative gains; do the calculation twice, once if they are power gains and the second time if they are voltage gains: 45 db; 2.3 db; −5.4 db.

53-4 Do the same for these gains: 18. db; −17.3 db.

53-5 Given amplifiers having voltage gains of 125,000, 4600, 0.045, and 13 db, calculate the gain of the four amplifiers in series. Express the answer both as a multiplicative gain and in decibels.

53-6 Do the same as for Problem 53-5, only the amplifiers have gains of 23 db, −8 db, −14 db, and 183,000.

ANSWERS TO ODD-NUMBERED PROBLEMS

53-1 95 db; 102 db; −26 db

53-3

Power Gain	Voltage Gain
32,000	178
1.7	1.3
0.29	0.54

53-5 161 db, or 1.12×10^8

Chapter 54

BINARY NUMBERS

OBJECTIVES

1. Become familiar with binary numbers and binary arithmetic.

2. Become familiar with the schemes used to indicate signed binary numbers.

3. Become familiar with octal and hexadecimal numbers and understand their use.

INTRODUCTION

This chapter is a quick introduction to binary numbers and to some of the conventions used in computers. This includes a brief discussion of number system notation and the methods of dealing with signed numbers, with very large and very small numbers, and with various notational features. Some of this material is very elementary. It is included here for those who may have never seen it. To provide a common background, the chapter begins with a review of decimal numbers.

DECIMAL NUMBERS

Decimal numbers are a base 10 number system. They are expressed with the 10 symbols 0, 1, 2, 3, 4, 5, 6, 7, 8, and 9, together with the decimal point and the plus and minus signs. The meaning of the notation is so common that it is rarely mentioned, but nevertheless let us state here that the number

879.257

means

$$8 \times 10^2 + 7 \times 10^1 + 9 \times 10^0 + 2 \times 10^{-1} + 5 \times 10^{-2} + 7 \times 10^{-3}$$

This meaning is expressed by the way in which this number is read: eight hundred seventy-nine and two hundred fifty-seven thousandths.

Negative numbers are indicated by putting a minus sign in front of the number. This is called a **sign and magnitude** notation. Although subtraction is defined in terms of addition in formal number theory, in schools two separate operations—addition and subtraction—are defined and learned by young students.

In many calculations, very large and/or very small numbers often occur. For instance, the speed of light is 30,000,000,000 cm/sec. Accurately writing and comprehending such numbers is very difficult. As a result, so-called **scientific notation** is often used. Thus the speed of light is written as 3.0×10^{10} cm/sec. The basic rule in scientific notation is that the number is written as **a mantissa** whose absolute value is less than 10 multiplied by a power of 10, called the **exponent**. There are two signs in scientific notation, one for the number and one for the exponent. Thus numbers such as -4.3×10^{-8} can occur. Techniques for adding, subtracting, multiplying, and dividing numbers expressed in scientific notation are sometimes taught in school systems, much to the bewilderment of the students.

BINARY NUMBERS

Computers deal with digital signals. Thus there are only two values available: H and L, or 0 and 1. The binary number system, which is a base 2 number system having symbols 0 and 1, is ideally suited for computers. A typical binary number is a string of 0's and 1's such as 100101. The meaning of this number is

$$100101 = 1 \times 2^5 + 1 \times 2^2 + 1 \times 2^0$$

where the terms involving 0×2^4, 0×2^3, and 0×2^1 have been omitted. This number can be converted into a decimal number by working out this sum:

$$100101 = 1 \times 2^5 + 1 \times 2^2 + 1 \times 2^0$$
$$= 32 + 4 + 1 = 37$$

The digits in binary numbers are often called **bits**. The leftmost bit of a binary number is called the **most significant bit**, or **MSB**, because it represents the largest part of the binary number. Likewise, the rightmost bit is called the **least significant bit**, or **LSB**. Since computers usually deal with a fixed-length word, that is, with data in fixed-length clumps such as 8, 16, or 32 bits, numbers expressed in computers often have a considerable number of preceding zeros. Thus the decimal number 37 (above) when expressed in binary on a 16-bit computer would look like 0000 0000 0010 0101, where some spaces have been included to make it easier to count the number of places.

Addition and multiplication of binary numbers is defined just as for decimal numbers. The basic addition and multiplication tables are particularly short and simple because there are only two input values:

	Addition				Multiplication	
	0	1			0	1
0	0	1		0	0	0
1	1	10		1	0	1

The addition and multiplication of multidigit binary numbers proceeds just as for decimal numbers. An example of each is shown below:

$$
\begin{array}{r}
101\,101 \\
+\ 001\,011 \\
\hline
111\,000
\end{array}
\qquad\qquad
\begin{array}{r}
101\,101 \\
\times\quad 001\,011 \\
\hline
101\,101 \\
1011\,01 \\
101101 \\
\hline
111101\,111
\end{array}
$$

Obviously, binary arithmetic can be very tedious. The multiplication example shows why binary multiplication is often said to proceed by a shift-and-add process—of course, so does decimal multiplication.

In ordinary decimal notation, symbols outside the number system are used to indicate signs and decimal points. In the computer, only 0's and 1's are available, so some conventions are needed. There are at least three ways of dealing with signed numbers in binary.

The easiest way is to use the first bit (MSB) of a binary number as the sign bit; a 0 means the number is positive, and a 1 means the number is negative. Thus for an eight-bit word size, the following two representations would occur:

$$+25 = 0001\,1001$$

$$-25 = 1001\,1001$$

This is called the **sign and magnitude** method, just as for decimal numbers. The disadvantage of this scheme is that the hardware required to do signed addition and subtraction is relatively complex, since different algorithms will have to be used depending on the sign of the numbers.

A second scheme used to express negative numbers is called **one's complement notation**. This notational scheme is best described by the prescription through which a number is negated that is, has its sign changed. To negate a number, complement all the bits in the number; that is, replace each 1 with a 0 and each 0 with a 1. Thus for an eight-bit computer word and the number +25 from above,

$$+25 = 0001\,1001$$

To form the negative of this number,

$$-25 = 1110\,0110$$

Thus a negative number in this scheme will always start with one or more 1 bits. An interesting feature of this notational scheme is illustrated by adding these two numbers:

$$+25 = 0001\,1001$$

$$-25 = 1110\,0110$$

$$\text{Sum} = 1111\,1111 = -0$$

Thus there are two ways of representing zero—all 0's and all 1's. The computer has to recognize both as zero to execute instructions that involve testing for zero.

Signed addition is relatively simple in 1's complement notation; the same addition process gives correct answers if so-called **end around carry** is used; that is, any overflow from the leading bit is added to the least significant bit. The two examples below illustrate this process:

$$
\begin{array}{ll}
+25 = 0001\,1001 & +16 = 0001\,0000 \\
-16 = \underline{1110\,1111} & -25 = \underline{1110\,0110} \\
\quad 1 \quad 0000\,1000 & \qquad 1111\,0110 = -9 \\
\end{array}
$$

$$\underline{\qquad\qquad 1}\text{ end around carry}$$

$$0000\,1001 = 9$$

Subtracting a number simply requires complementing the number and then adding it with end around carry.

Two's complement notation avoids the problem of having two zeros at the cost of a more complex scheme of negating a number. The negation scheme in 2's complement notation is this; first form the 1's complement of the number, then add 1. Thus

$$
\begin{array}{ll}
& 25 = 0001\,1001 \\
\text{1's complement} & 1110\,0110 \\
\text{add 1} & 1110\,0111 = -25 \text{ in 2's complement notation}
\end{array}
$$

With this scheme, signed addition is even simpler; normal addition gives the correct result. No end around carry is required. The following two examples illustrate this:

$$
\begin{array}{ll}
+25 = 0001\,1001 & +16 = 0001\,0000 \\
-16 = \underline{1111\,0000} & -25 = \underline{1110\,0111} \\
\quad\ 0000\,1001 = 9 & \qquad 1111\,0111 = -9
\end{array}
$$

In summary, two's complement notation permits faster and easier adding but requires a more complex process for negation. One's complement notation takes longer for an addition (it is necessary to wait long enough for a bit to ripple through the adder twice), but it permits much simpler negation. Which system to use is the kind of choice computer designers make.

Finally, for anyone who wants to have fun, it is possible to create signed decimal numbers in a similar fashion. The 9's complement of a number can easily be created by subtracting each digit from 9. This gives an interesting way to represent negative numbers.

The decimal point, or maybe it should be called the binary point, is generally located in a binary number by convention. For instance, the decimal point could be at the LSB end of the number, meaning that all the numbers are integers. Or the decimal point could be to the left of the MSB, resulting in all the numbers being less than 1. Obviously, the decimal point could also be located somewhere in the middle of the number. When binary numbers having implicit decimal points are divided or multiplied together, the decimal point in the answer has moved, and the number will have to be shifted around to return it to the correct form. This is a problem when doing decimal arithmetic on a computer.

Floating point notation is the binary analog to scientific notation. In this notation, all real numbers are represented as an exponent and a mantissa just as above. The **mantissa** is a signed number having an absolute value less than 1.0, and the **exponent** is the signed power of 2, that is needed to express the number. Thus, 8 would have a mantissa of 1 and an exponent of 4, since $8 = \frac{1}{2} \times 2^4$.

Many different conventions exist for representing floating point numbers in a computer word; each computer system has its own variation for the representation of floating point numbers. The details of these schemes are only of interest to computer designers or advanced programmers. Given the representation scheme, specialized hardware or software can be designed to do the various mathematical operations. This hardware tends to be very complex and relatively expensive and is usually available as an option for a computer system. In general, macho computer types will insist on having floating point hardware on their computers.

Finally, one last piece of terminology: In the current era, almost all computers work with units of data that are some multiple of eight bits long—8, 16, 32, 64, and so on. The eight-bit unit is given the name **byte**. Thus a byte of data is eight bits of data. Sometimes, the byte is considered to be divided into two four-bit **nibbles**. All sorts of jargon using these words are common in the computer literature.

OTHER NUMBER SYSTEMS

Human beings have a great deal of difficulty dealing with 32- or 64-bit-long strings of 0's and 1's. However, these strings can be represented as much shorter hexadecimal or octal strings. Three binary bits can represent numbers ranging from 0 to 7; thus, three binary bits can represent one octal digit. Octal numbers are a base 8 number system. The eight symbols used are 0, 1, 2, 3, 4, 5, 6, and 7. Four binary bits can represent numbers ranging from 0 to 15; thus, four binary bits can represent one hexadecimal digit. Hexadecimal numbers are a base 16 number system. The 16 symbols used are 0, 1, 2, 3, 4, 5, 6, 7, 8, 9, A, B, C, D, E, and F. There are obscure IBM documents that explain how to pronounce hexadecimal numbers so that you can distinguish between 18 and 1A. Hexadecimal numbers are almost universally called **hex**.

The conversion of numbers from binary to either octal or hex, and vice versa, is quite easy. Table 54-1 shows the equivalence for the first 19 numbers. Conversion of longer numbers proceeds easily. To convert from binary to

TABLE 54-1 Binary, Octal, Hexadecimal, and Decimal Equivalents

Binary	Octal	Hex	Decimal
00000	00	00	0
00001	01	01	1
00010	02	02	2
00011	03	03	3
00100	04	04	4
00101	05	05	5
00110	06	06	6
00111	07	07	7
01000	10	08	8
01001	11	09	9
01010	12	0A	10
01011	13	0B	11
01100	14	0C	12
01101	15	0D	13
01110	16	0E	14
01111	17	0F	15
10000	20	10	16
10001	21	11	17
10010	22	12	18

either octal or hex, simply mark off the binary number in groups of three (four bits) starting at the right (LSB); then, convert these group by group. Two examples follow:

Binary to octal:

$$100100101001 = 100\,100\,101\,001 = 4451$$

Binary to hexadecimal:

$$100100101001 = 1001\,0010\,1001 = 929$$

Conversion the other way simply requires replacing each octal (hex) digit by its binary equivalent. One example of hex to binary follows:

$$AF1C = 1010\,1111\,0001\,1100$$

Addition and multiplication tables for octal and hex numbers are easy to set up and not impossible to learn. Octal arithmetic has been taught in some school systems in recent years. Many programmers have actually learned how to do some octal and/or hex arithmetic. The purpose is to be able to check the results of the computer programs. **But it must be kept in mind that although hex or octal representations for the numbers are used, the computer is using binary arithmetic.** The hex and/or octal representations are simply to make life easier for the human users.

BINARY-CODED DECIMAL

The final topic in this chapter is binary-coded decimal numbers. As shown above, one hex digit or four binary bits can represent the decimal numbers between 0 and 15. If these binary digits are constrained to be 1001 or less, then they represent the decimal numbers between 0 and 9. This is called **binary-coded decimal**, or **BCD**. As was shown in Chapter 24, circuits can be built that count in binary; these circuits can also be made into modulo 10 counters, counters that count in BCD. For such circuits, the counting sequence becomes

0000 0000	0	0000 0111	7
0000 0001	1	0000 1000	8
0000 0010	2	0000 1001	9
0000 0011	3	0001 0000	10
0000 0100	4	0001 0001	11
0000 0101	5	0001 0010	12
0000 0110	6	⋮	⋮

Chapter 55

FOURIER SERIES

OBJECTIVES

1. Become familiar with the power and utility of Fourier series as a tool for advanced circuit analysis.

2. Understand the utility of square waves for measuring frequency response.

INTRODUCTION

Because the use of Fourier analysis is very important in formal circuit analysis, a few comments about it are included here, even though a detailed discussion of its use is beyond the scope of this book. This is a "gee-whiz" chapter. Its purpose is to expose some of the power and utility of Fourier analysis.

FOURIER SERIES

The Fourier theorems are special cases of more general mathematical theorems that concern the expansion of any function in terms of a family of more familiar functions. These expansions are of great importance in many physical, chemical, and engineering calculations. The familiar functions include the sine and cosine functions, hermite polynominals, Bessel functions, and many others that are even less well known. Although these functions may seem quite exotic, they are well known to mathematicians and have many nice properties. Fourier series use only trigonometric functions.

The **Fourier theorem** for periodic functions says that any periodic function, $g(t)$ having a period T can be replaced by a constant term and a sum of sine and cosine terms having frequencies $f, 2f, 3f, \ldots$, where $f = 1/T$. That is,

$$g(t) = u_0 + a_1 \sin(2\pi f t) + a_2 \sin(4\pi f t) + a_3 \sin(6\pi f t) + \cdots$$
$$+ b_1 \cos(2\pi f t) + b_2 \cos(4\pi f t) + b_3 \cos(6 f t) + \cdots$$

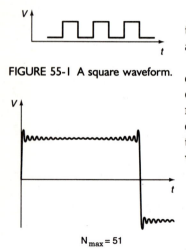

FIGURE 55-1 A square waveform.

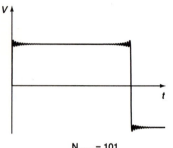

$N_{max} = 51$

The Fourier theorem for nonperiodic functions says that any smooth function can be replaced by a constant term and a sum of sine and cosine terms of all frequencies. In this case, the sum is done by means of an integral.

The proof of these theorems and the methods of calculating the coefficients in the expansions are covered in detail in textbooks on Fourier series or transform calculus. For simple periodic waveforms, the calculations are not very difficult, although they are often tedious. The square wave is a very common waveform encountered in circuit analysis (Figure 55-1). A problem that is assigned in every course on Fourier analysis is to show that a square wave of frequency f and amplitude V may be expanded in the form

$$\frac{4V}{\pi} \sin \omega t + \frac{4V}{3\pi} \sin 3\omega t + \frac{4V}{5\pi} \sin 5\omega t + \cdots \qquad (55\text{-}1)$$

where

$$\omega = 2\pi f$$

The series of graphs in Figure 55-2 shows the waveforms generated by evaluating various numbers of terms in this expansion. As can be seen from these plots, as the number of terms gets larger, the expansion gets closer to the original waveform. There will always be an overshoot at the leading and trailing edges of the square wave. This is because the square wave has a step-wise discontinuity at each edge.

$N_{max} = 101$

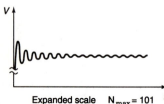

Expanded scale $N_{max} = 101$

FIGURE 55-2 Various approximations to the square wave.

THE USE OF FOURIER SERIES

With the use of Fourier analysis, the response of a circuit to any complex waveform can be calculated if the responses of the circuit to dc and to pure sine waves are known. The procedure is as follows:

1. Decompose the input waveform into its Fourier series (or integral).
2. Multiply each term in the expansion by the gain of the circuit for that particular frequency. Remember, the gain includes phase shift information.
3. Synthesize the output of the circuit from this new Fourier series.

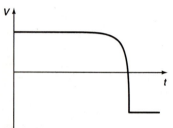

FIGURE 55-3 Response of a low-pass filter to a square wave. (The highest term kept in this expansion is 51 f.)

Although this may be very hard to do in practice it, at least, gives a procedure for obtaining an exact solution for any input. Two examples of this process are given here. In the first example, a square wave having a frequency f is passed through a low-pass filter that has its 3-db point at $3f$. The output of the circuit is shown in Figure 55-3.

In the second example, a square wave is passed through a high-pass filter; again, the 3-db point of the filter is at $3f$. The output of this circuit is shown in Figure 55-4.

THE USE OF SQUARE WAVES

The two examples above together with Equation 55-1 explain why square
waves are often used for testing high-fidelity audio amplifiers. Because a
square wave is made up of sine waves having many different frequencies, a
great deal can be learned about the entire frequency response of an amplifier
by a single observation of its response to a square wave. Thus, if the output
has the trailing edges rounded off similar to those in Figure 55-3, the amplifier has problems at high frequencies. Likewise, if the output looks similar to
the waveform in Figure 55-4, the amplifier has low-frequency problems.

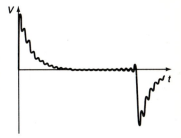

FIGURE 55-4 Response of a high-
pass filter to a square wave.

REFERENCES

TEXTBOOKS

This appendix is a collection of annotated references to sources of information and equipment. The first class of reference materials are standard textbooks. Many introductory electronics textbooks have been published, as well as many texts in related subjects. Some of these are listed below.

Electronics

The Art of Electronics by Paul Horowitz and Winfield Hill, Cambridge University Press, New York, 1980 and 1989.

This is by far the best general-purpose textbook currently available. When it was published, it was almost up to date, and it continues to be an excellent reference; the new edition makes it even better. Although it is often a difficult book for a true beginner, anyone who has mastered the material in this book is ready to read Horowitz and Hill.

Basic Electronics for Scientists by James J. Brophy, McGraw-Hill Book Company, New York, 1990.

Principles of Electronic Instrumentation by A. James Diefenderfer, W. B. Saunders Company, Philadelphia, 1972.

Electronics with Digital and Analog Integrated Circuits by Richard J. Higgins, Prentice-Hall, Inc., Englewood Cliffs, NJ, 1983.

Essential Electronics by Warren Fenton Stubbins, John Wiley & Sons, Inc., New York, 1986.

These are all standard introductory texts that have been widely used in recent years. each has a different emphasis and each treats a few topics that none of the other books consider. Any of these might be a productive reference for an introduction to something not covered in this book.

Circuit Analysis

As previously mentioned, there have been many formal textbooks written for the electrical engineering that deal only with the techniques of circuit analysis. Two of these are listed below:

Circuit Theory Fundamentals and Applications by Aram Budak, Prentice-Hall, Inc., Englewood Cliffs, NJ, 1978.

Electric Circuits by James W. Nilsson, Addison-Wesley Publishing Co., Reading, MA, 1975.

The *Schaum's Outline Series* published by McGraw-Hill Book Company also has several books on circuit analysis. *Basic Circuit Analysis* by John O'Malley has proved to be very useful for students who want more practice at simple circuit analysis.

Digital Design

Digital design or, as it is now called, computer design, is another area in which a number of books have been published that might be of interest to a person just starting to work in electronics. Two books in this area are the following:

Computer Design by Glen G. Langdon, Jr. Computeach Press Inc., P.O. Box 20851, San Jose, CA 95160, 1982.

Computer Engineering Hardware Design by M. Morris Mano, Prentice-Hall, Inc., Englewood Cliffs, NJ, 1988.

Logic

There are many books on logic, symbolic logic, and related topics. Two of the standard books are

Logic, Techniques of Formal Reasoning, 2nd edition, by Donald Kalish and Richard Montague, Harcourt, Brace, Javonovich, Inc., New York, 1980.

Methods of Logic by W. V. Quine, Harvard University Press, Cambridge, MA, 1982.

OTHER TRADE BOOKS

There are thousands of speciality books and monographs on electronics. These range from the useless to the very useful, from the completely obscure to the very clear. Only a few of them can be listed here, but they will give the flavor of the variety available.

Cookbooks

The Howard W. Sams & Company, a Division of Macmillan, Inc., 4300 West 62nd Street, Indianapolis, IN 46268, has published a whole series of small paperback "cookbooks" on electronics. These include:

TTL Cookbook by Don Lancaster.

CMOS Cookbook by Don Lancaster and Howard M. Berlin.

IC Timer Cookbook by Walter G. Jung.

IC Op-Amp Cookbook by Walter G. Jung.

Active Filter Cookbook by Don Lancaster.

Also by the same publisher, there are a number of similar books with slightly different titles. These include:

Active Filter Design by Carson Chen.

The 555 Timer Applications Source Book with Experiments by Howard M. Berlin.

Design of Op-Amp Circuits with Experiments by Howard M. Berlin.

Despite the titles and the relatively low-key presentations of these books, they are often very useful. They offer a source of ideas about how to approach certain problems and also present enough practical information to prevent many common errors. These books will be found on the desks of a surprisingly large number of electronics designers.

Manufacturers' Handbooks

Besides the data books discussed below, many manufacturers publish handbooks about speciality areas in electronics. These are generally excellent sources of information, containing discussions, examples, and tables of data that are available nowhere else. One example is Analog Devices, Two Technology Way, P.O. Box 280, Norwood, MA 02062, which has published a number of these handbooks. These include:

Analog-Digital Conversion Handbook

Nonlinear Circuits Handbook

Both edited by Daniel H. Sheingold, these are excellent books containing a wealth of information about normal op-amp circuits as well as nonlinear circuits and analog-to-digital conversion.

Computer Books

With the production of each new microprocessor and associated auxiliary chip, books have appeared on the market explaining how these computers work and how to program them. These books exist for all popular computer systems. A number of examples follow:

The 80386/387 Architecture by Stephen P. Morse, Eric J. Isaacson, and Douglas J. Albert.

The MCS-80/85 Family Users Manual published by Intel, the manufacturer.

68000, 68010, and 68020 Primer by Stan Kelly-Bootle.

8087/80287/80387 for the IBM PC & Compatible by Richard Startz.

65816/65802 Assembly Language Programming by Michael Fischer.

DATA BOOKS AND CATALOGS

Electronics is such a rapidly changing field that one of the main problems faced by anyone working in the field is keeping up with the flood of new devices and techniques. To facilitate the use of their products, the manufacturers of modern electronic gear produce large data books or catalogs or both that contain a wealth of information about their products. These publications often include tutorials on the use of the circuits and the definitions of the terms used to describe the products, as well as applications notes or addresses where applications notes for the products can be obtained.

As an example of the extensive nature of these publications, Motorola, Literature Distribution Center, P.O. Box 20912, Phoenix, Az. 85036, publishes data books for at least the following product lines (these are all on the shelf in my electronics laboratory):

Fast and LS TTL
High-Speed CMOS
Power MOSFET Transistors
MECL Devices
MECL System Design
Linear and Interface ICs
Optoelectronics Devices
Linear/Switchmode Voltage Regulators
CMOS Logic
Small-Signal Semiconductors
TTL

The other manufacturers all publish similar sets of data books and catalogs. What follows is simply a list of some of the major manufacturers and their addresses. These companies are continually producing new updated books.

Analog Devices
(See above)

Harris Semiconductor
Information Center
P.O. Box 2021
Cathedral Station
Boston, MA 02118

Intersil, Inc.
10710 N. Tantan Ave.
Cupertino, CA 95014
408-986-5000

Motorola
(See above)

National Semiconductors
2900 Semiconductor Drive
Santa Clara, CA 95051
408-737-5000

Texas Instruments
Literature Response Center
P.O. Box 401560
Dallas, TX 75240

MAGAZINES

There are a number of general-purpose electronic magazines that are intended for people working in the field. The product announcements and advertisements are often as useful or more useful than the actual articles. Some of these magazines are:

EDN, Cahners Publishing Co., a Division of Reed Publishing, 275 Washington Street, Newton, MA 02158.

Electronic Design, Hayden Publishing Co. Inc., 50 Essex Street, Rochelle Park, NJ 07662.

Electronic Products, Hearst Business Communications, Inc., UTP Division, 645 Stewart Ave., Garden City, NY 11530.

Also, several popular electronics magazines for the home hobbyist often have useful articles. These are:

Popular Electronics, Gemsback Publications, Inc., 500-B Bi-County Blvd., Farmingdale, NY 11735.

Radio-Electronics, Gemsback Publications, Inc., New York.

Byte, One Phoenix Mill Lane, Peterborough, NH 03458, a magazine for small-computer users, sometimes has interesting articles on hardware topics.

EQUIPMENT SUPPLIERS

There are four main classes of electronics equipment and supplies. These are (1) the large local or national general-purpose suppliers, (2) the local or national distributors for individual companies, (3) the electronics "department stores," which have been established in a number of major cities, and (4) the mail order stores that deal with hobbyists. Although several references are given here, your local yellow pages are your best source of information—look under Electronic Equipment & Supplies.

General Suppliers

In an area like the Boston area, there are many different full-line electronics distributors. In other areas, they are not so plentiful. However, two of the major national distributors are:

Allied Radio
401 E. 8th Street
Fort Worth, TX 76102
800-433-5700

Newark Electronics
Administration Offices
4801 N. Ravenswood Street
Chicago, IL 60640
313-784-5100

Electronics "Department" Stores

In many major cities and their environs there are now stores that specialize in electronics equipment and components. In the Boston area, the store has the name *You-Do-It Electronics Center* and is located at 40 Franklin Street, Nedham, MA, 02194; (617)449-1005. Others have less suggestive names. In any case, look in the yellow pages.

    Radio Shack, which has a national chain of outlets, carries a fairly complete line of the most commonly needed electronic components. Even if the local outlet does not have what you need, they may be able to get it for you.

Mail Order

Several companies that carry extensive lines of electronic and computer equipment also engage in mail order sales with the general public. These companies generally have prominent advertisements in the electronics and computer magazines. Two of these companies are:

Jameco Electronics
1355 Shoreway Road
Belmont, CA 94002
415-592-8097

Digi-Key Corporation
701 Brooks Ave South
P.O. Box 677
Thief river Falls, MN 56701
1-800-344-4539

SAMPLE DATA SHEETS

A few data sheets from several manufacturers' data books are presented in this appendix. These are included only as examples of types of data sheets, or for ready reference while reading this book. This appendix is not intended to substitute for the actual data books. Anyone studying or working with electronics must have several of the manufacturers' data books available to refer to.

A SHORT SELECTION TABLE

To show the richness of choice in integrated circuits, the following selection table has been prepared. It is a list of some of the simple gates available in TTL by functional category. This chart is not complete; only simple gates have been listed in the chart, and not all of those available in this family have been included. Furthermore, there are many more complex units—flip-flops, registers, counters, decoders, multiplexers, ALUs, and the like—available in this family that are not listed here. Finally, most of the ICs listed in this chart are available in several different TTL families as well as in several different pin-for-pin compatible CMOS families. The chart lists only the generic "7400" numbers, but the more specific "74LS00" or "74C00" type numbers could equally well be used for most of the entries.

AND

7408	Quad two-input AND gate
7409	Quad two-input AND gate, open-collector
7411	Triple three-input AND gate
7415	Triple three-input AND gate, open-collector
7421	Dual four-input AND gate

Inverters(NOT)

7404	Hex inverter
7405	Hex inverter, open-collector

NAND

7400	Quad two-input NAND gate
7437	Quad two-input NAND buffer
7403	Quad two-input NAND gate, open-collector

7410	Triple three-input NAND gate
7412	Triple three-input NAND gate, open-collector
7420	Dual four-input NAND gate
7440	Dual four-input NAND buffer
7422	Dual four-input NAND gate, open-collector
7430	Eight-input NAND gate

NOR

7402	Quad two-input NOR gate
7428	Quad two-input NOR buffer
7433	Quad two-input NOR buffer, open-collector
7427	Triple three-input NOR gate

OR

7432	Qaud two-input OR gate

XOR

7488	Quad exclusive OR gate

DATA SHEETS

Data sheets for eight different integrated circuits are included here. First, a few comments are given to indicate why each circuit is included and to indicate some specific points that should be noticed in the data sheet. The actual data sheets follow these comments. The circuits included here are:

74LS00

A Qaud two-input NAND gate in the Low Power Schottky series. This sheet illustrates the basic parameters of this, currently the most popular, subfamily of the TTL line; it is also representative of the data sheets for simple gates. This sheet was taken from the Motorola *Fast and LS TTL Data Book,* copyright © by Motorola, Inc., used by permission.

74C00

The CMOS pin-compatible equivalent of the 7400. This sheet allows a direct comparison of the parameters of this particular series of CMOS gates with the LS TTL series. This sheet was taken from the Motorola *High-Speed CMOS Logic Data Book,* copyright © by Motorola, Inc., used by permission.

74LS363

This is a fairly typical data sheet for a flip-flop. Special attention should be paid to the functional table and the description to fully understand the operation of this flip-flop and how it differs from the other flip-flops described on the same pages. This technique of presenting information for several related but different circuits in one set of data sheets is both very common and useful. Several more examples of this will be seen below. This sheet was taken from the Texas Instruments *TTL Data Book for Design Engineers,* used by permission.

74150

This is a one-of-sixteen multiplexer, which is used in Chapter 25 as a way to construct arbitrary truth tables with 16 lines. This data sheet is included here both as a reference for use and as a sample of the amount and type of information that is given about a typical MSI/LSI chip. Again, note how information about several different related circuits is presented in one set of data sheets. Also note the schematics for the internal operations as well as for the typical inputs and outputs. This sheet was taken from the Texas Instruments *TTL Data Book for Design Engineers,* used by permission.

74LS682

This is an eight-bit magnitude comparator, which is used in the computer interface design example in Chapter 39. This sheet is included as reference for that design and again to show the amount and type of information available about this type of circuit. Again, a number of different, related circuits are presented in one data sheet to make the selection of the correct circuit easier. These sheets were taken from the Motorola *Fast and LS TTL Data Book,* copyright © Motorola, Inc., used by permission.

741

The data sheets for the MC1741, which is the Motorola number for the 741. This is probably the most common operational amplifier in use today. This data sheet shows the types of information available for an operational amplifier; it also indicates the ranges the various parameters may take. For analog circuits, much information is commonly presented in the forms of graphs; this case is no exception. This sheet was taken from the Motorola *Linear and Interface Integrated Circuit* data book, copyright © by Motorola, Inc., used by permission.

555

The data sheet for the Motorola MC1455, which is the Motorola part number for the 555, the basic timer circuit. This unit is described in Chapter 23. This data sheet shows the basic nature of the circuit as well as many of the uses of the circuit. Again, notice how much information is presented in the form of graphs. For this, a very versatile circuit, a fairly large number of possible applications is shown on the data sheet. This sheet was also taken from the Motorola *Linear and Interface Integrated Circuit* data book, copyright © Motorola, Inc., used by permission.

AD570

An eight-bit analog-to-digital converter. This converter is used in the computer interface design example in Chapter 39. This data sheet is included here both as a reference for that design and as an example of the types of information that are available for more complex analog circuits. Besides the tables of specifications, note how much useful information about the use and operation of this circuit is presented in the data sheet. These sheets were taken from the Analog Devices catalog, courtesy of Analog Devices, Inc., Norwood, Massachusetts.

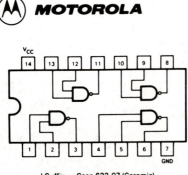

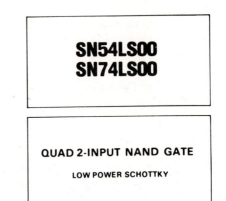

J Suffix — Case 632-07 (Ceramic)
N Suffix — Case 646-05 (Plastic)

SN54LS00
SN74LS00

QUAD 2-INPUT NAND GATE

LOW POWER SCHOTTKY

GUARANTEED OPERATING RANGES

SYMBOL	PARAMETER		MIN	TYP	MAX	UNIT
V_{CC}	Supply Voltage	54	4.5	5.0	5.5	V
		74	4.75	5.0	5.25	
T_A	Operating Ambient Temperature Range	54	−55	25	125	°C
		74	0	25	70	
I_{OH}	Output Current — High	54, 74			−0.4	mA
I_{OL}	Output Current — Low	54			4.0	mA
		74			8.0	

DC CHARACTERISTICS OVER OPERATING TEMPERATURE RANGE (unless otherwise specified)

SYMBOL	PARAMETER		LIMITS MIN	TYP	MAX	UNITS	TEST CONDITIONS
V_{IH}	Input HIGH Voltage		2.0			V	Guaranteed Input HIGH Voltage for All Inputs
V_{IL}	Input LOW Voltage	54			0.7	V	Guaranteed Input LOW Voltage for All Inputs
		74			0.8		
V_{IK}	Input Clamp Diode Voltage			−0.65	−1.5	V	V_{CC} = MIN, I_{IN} = −18 mA
V_{OH}	Output HIGH Voltage	54	2.5	3.5		V	V_{CC} = MIN, I_{OH} = MAX, V_{IN} = V_{IH} or V_{IL} per Truth Table
		74	2.7	3.5		V	
V_{OL}	Output LOW Voltage	54,74		0.25	0.4	V	I_{OL} = 4.0 mA $\quad V_{CC}$ = V_{CC} MIN,
		74		0.35	0.5	V	I_{OL} = 8.0 mA $\quad V_{IN}$ = V_{IL} or V_{IH} per Truth Table
I_{IH}	Input HIGH Current				20	µA	V_{CC} = MAX, V_{IN} = 2.7 V
					0.1	mA	V_{CC} = MAX, V_{IN} = 7.0 V
I_{IL}	Input LOW Current				−0.4	mA	V_{CC} = MAX, V_{IN} = 0.4 V
I_{OS}	Short Circuit Current		−20		−100	mA	V_{CC} = MAX
I_{CC}	Power Supply Current Total, Output HIGH				1.6	mA	V_{CC} = MAX
	Total, Output LOW				4.4		

AC CHARACTERISTICS: T_A = 25°C

SYMBOL	PARAMETER	LIMITS MIN	TYP	MAX	UNITS	TEST CONDITIONS
t_{PLH}	Turn Off Delay, Input to Output		9.0	15	ns	V_{CC} = 5.0 V
t_{PHL}	Turn On Delay, Input to Output		10	15	ns	C_L = 15 pF

A Quad two-input NAND gate in the Low Power Schottky series.

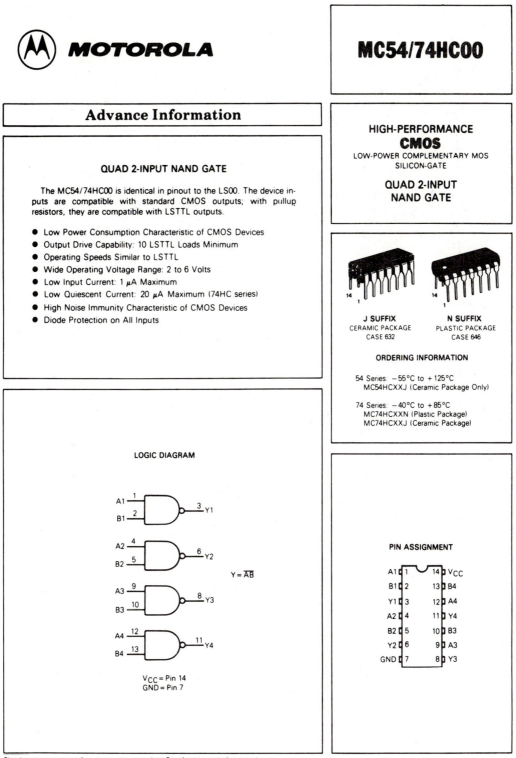

MOTOROLA

MC54/74HC00

Advance Information

QUAD 2-INPUT NAND GATE

The MC54/74HC00 is identical in pinout to the LS00. The device inputs are compatible with standard CMOS outputs; with pullup resistors, they are compatible with LSTTL outputs.

- Low Power Consumption Characteristic of CMOS Devices
- Output Drive Capability: 10 LSTTL Loads Minimum
- Operating Speeds Similar to LSTTL
- Wide Operating Voltage Range: 2 to 6 Volts
- Low Input Current: 1 μA Maximum
- Low Quiescent Current: 20 μA Maximum (74HC series)
- High Noise Immunity Characteristic of CMOS Devices
- Diode Protection on All Inputs

HIGH-PERFORMANCE
CMOS
LOW-POWER COMPLEMENTARY MOS
SILICON-GATE

QUAD 2-INPUT
NAND GATE

J SUFFIX
CERAMIC PACKAGE
CASE 632

N SUFFIX
PLASTIC PACKAGE
CASE 646

ORDERING INFORMATION

54 Series: −55°C to +125°C
MC54HCXXJ (Ceramic Package Only)

74 Series: −40°C to +85°C
MC74HCXXN (Plastic Package)
MC74HCXXJ (Ceramic Package)

LOGIC DIAGRAM

A1 1
B1 2
3 Y1

A2 4
B2 5
6 Y2

$Y = \overline{AB}$

A3 9
B3 10
8 Y3

A4 12
B4 13
11 Y4

V_{CC} = Pin 14
GND = Pin 7

PIN ASSIGNMENT

A1	1	14	V_{CC}
B1	2	13	B4
Y1	3	12	A4
A2	4	11	Y4
B2	5	10	B3
Y2	6	9	A3
GND	7	8	Y3

This document contains information on a new product. Specifications and information herein are subject to change without notice

The CMOS pin-compatible equivalent of the 7400.

MAXIMUM RATINGS*

Symbol	Parameter	Value	Unit
V_{CC}	DC Supply Voltage (Referenced to GND)	−0.5 to +7.0	V
V_{in}	DC Input Voltage (Referenced to GND)	−1.5 to V_{CC} +1.5	V
V_{out}	DC Output Voltage (Referenced to GND)	−0.5 to V_{CC} +0.5	V
I_{in}	DC Input Current, per Pin	±20	mA
I_{out}	DC Output Current, per Pin	±25	mA
I_{CC}	DC Supply Current, V_{CC} and GND Pins	±50	mA
P_D	Power Dissipation, per Package†	500	mW
T_{stg}	Storage Temperature	−65 to +150	°C
T_L	Lead Temperature (10-Second Soldering)	300	°C

*Maximum Ratings are those values beyond which damage to the device may occur
†Power Dissipation Temperature Derating:
    Plastic "N" Package: −12mW/°C from 65°C to 85°C
    Ceramic "J" Package: −12mW/°C from 100°C to 125°C

This device contains circuitry to protect the inputs against damage due to high static voltages or electric fields; however, it is advised that normal precautions be taken to avoid applications of any voltage higher than maximum rated voltages to this high impedance circuit. For proper operation it is recommended that V_{in} and V_{out} be constrained to the range GND ≤ (V_{in} or V_{out}) ≤ V_{CC}.

Unused inputs must always be tied to an appropriate logic voltage level (e.g., either GND or V_{CC}).

RECOMMENDED OPERATING CONDITIONS

Symbol	Parameter	Min	Max	Unit
V_{CC}	DC Supply Voltage (Referenced to GND)	2.0	6.0	V
V_{in}, V_{out}	DC Input Voltage, Output Voltage (Referenced to GND)	0	V_{CC}	V
T_A	Operating Temperature − 74HC Series	−40	+85	°C
	54HC Series	−55	+125	
t_r, t_f	Input Rise and Fall Time (Figure 1)	−	500	ns

ELECTRICAL CHARACTERISTICS (Voltages Referenced to GND)

Symbol	Parameter	Test Conditions	V_{CC}	25°C 54HC and 74HC Typical	25°C 54HC and 74HC Guaranteed	85°C 74HC Guaranteed	125°C 54HC Guaranteed	Unit
V_{IH}	Minimum High-Level Input Voltage	V_{out} = 0.1 V or V_{CC} − 0.1 V $|I_{out}|$ = 20 µA	2.0	1.2	1.5	1.5	1.5	V
			4.5	2.4	3.15	3.15	3.15	
			6.0	3.2	4.2	4.2	4.2	
V_{IL}	Maximum Low-Level Input Voltage	V_{out} = 0.1 V or V_{CC} − 0.1 V $|I_{out}|$ = 20 µA	2.0	0.6	0.3	0.3	0.3	V
			4.5	1.8	0.9	0.9	0.9	
			6.0	2.4	1.2	1.2	1.2	
V_{OH}	Minimum High-Level Output Voltage	V_{in} = V_{IH} or V_{IL} I_{out} = −20 µA	2.0	1.998	1.9	1.9	1.9	V
			4.5	4.499	4.4	4.4	4.4	
			6.0	5.999	5.9	5.9	5.9	
		V_{in} = V_{IH} or V_{IL} I_{out} = −4.0 mA	4.5	4.20	3.98	3.84	3.70	V
		I_{out} = −5.2 mA	6.0	5.80	5.48	5.34	5.20	
V_{OL}	Maximum Low-Level Output Voltage	V_{in} = V_{IH} or V_{IL} I_{out} = 20 µA	2.0	0.002	0.1	0.1	0.1	V
			4.5	0.001	0.1	0.1	0.1	
			6.0	0.001	0.1	0.1	0.1	
		V_{in} = V_{IH} or V_{IL} I_{out} = 4.0 mA	4.5	0.22	0.26	0.33	0.40	V
		I_{out} = 5.2 mA	6.0	0.18	0.26	0.33	0.40	
I_{in}	Maximum Input Leakage Current	V_{in} = V_{CC} or GND	6.0	0.00001	±0.1	±1.0	±1.0	µA
I_{CC}	Maximum Quiescent Supply Current (Per Package)	V_{in} = V_{CC} or GND I_{out} = 0 µA	6.0	−	2	20	40	µA

The CMOS pin-compatible equivalent of the 7400 (continued).

SWITCHING CHARACTERISTICS ($V_{CC} = 5$ V, $T_A = 25°C$, $C_L = 15$ pF, Input $t_r = t_f = 6$ ns)

Symbol	Parameter	54HC and 74HC Typical	54HC and 74HC Guaranteed Limit	Unit
t_{PLH}	Maximum Propagation Delay, Input A or B to Output Y	8	15	ns
t_{PHL}	(Figures 1 and 2)	8	15	
t_{TLH}, t_{THL}	Maximum Output Transition Time, Any Output (Figures 1 and 2)	5	10	ns

SWITCHING CHARACTERISTICS ($C_L = 50$ pF, Input $t_r = t_f = 6$ ns)

Symbol	Parameter	V_{CC}	25°C 54HC and 74HC Typical	Guaranteed Limit	86°C 74HC	125°C 54HC	Unit
t_{PLH}	Maximum Propagation Delay, Input A or B to Output Y (Figures 1 and 2)	2.0	45	90	113	134	ns
		4.5	9	18	23	27	
		6.0	8	15	19	23	
t_{PHL}		2.0	45	90	113	134	ns
		4.5	9	18	23	27	
		6.0	8	15	19	23	
t_{TLH}, t_{THL}	Maximum Output Transition Time, Any Output (Figures 1 and 2)	2.0	38	75	95	110	ns
		4.5	8	15	19	22	
		6.0	6	13	16	19	
C_{in}	Maximum Input Capacitance	—	5	10	10	10	pF
C_{PD}	Power Dissipation Capacitance*	—	20	—	—	—	pF

* C_{PD} is used to determine the no-load dynamic power consumption per gate:

$$P_D = C_{PD} V_{CC}^2 f + I_{CC} V_{CC}$$

FIGURE 1 — SWITCHING WAVEFORMS

FIGURE 2 — TEST CIRCUIT

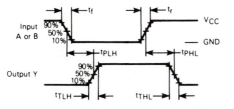

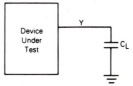

The CMOS pin-compatible equivalent of the 7400 (continued).

TTL
MSI

TYPES SN54LS363, SN54LS364, SN74LS363, SN74LS364
OCTAL D-TYPE TRANSPARENT LATCHES AND
EDGE-TRIGGERED FLIP-FLOPS

BULLETIN NO. DL-S 7612466, OCTOBER 1976

- High V_{OH} . . . 3.65 V Min (74LS')
- Choice of 8 Latches or 8 D-Type Flip-Flops In a Single Package
- 3-State Bus-Driving Outputs
- Full Parallel-Access for Loading and Reloading
- Buffered Control Inputs
- Clock/Enable Input Has Hysteresis to Improve Noise Rejection and P-N-P Inputs To Reduce D-C Loading
- SN54LS373/SN74LS373 and SN54LS374/ SN74LS374 Are Similar But Have Standard V_{OH} of 2.4 V Min

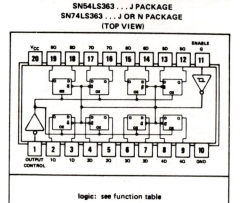

SN54LS363 . . . J PACKAGE
SN74LS363 . . . J OR N PACKAGE
(TOP VIEW)

logic: see function table

'LS363 FUNCTION TABLE

OUTPUT CONTROL	ENABLE G	D	OUTPUT
L	H	H	H
L	H	L	L
L	L	X	Q_0
H	X	X	Z

'LS364 FUNCTION TABLE

OUTPUT CONTROL	CLOCK	D	OUTPUT
L	↑	H	H
L	↑	L	L
L	L	X	Q_0
H	X	X	Z

See explanation of function tables on page 3-8.

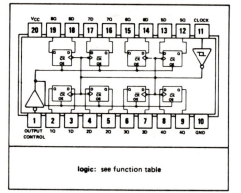

SN54LS364 . . . J PACKAGE
SN74LS364 . . . J OR N PACKAGE
(TOP VIEW)

logic: see function table

description

These 8-bit registers feature totem-pole three-state outputs designed specifically for driving highly-capacitive or relatively low-impedance loads. The high-impedance third state and increased high-logic-level drive provide these registers with the capability of being connected directly to and driving the bus lines in a bus-organized system without need for interface or pull-up components. They are particularly attractive for implementing buffer registers, I/O ports, bidirectional bus drivers, and working registers.

The eight latches of the 'LS363 are transparent D-type latches meaning that while the enable (G) is high the Q outputs will follow the data (D) inputs. When the enable is taken low the outputs will be latched at the level of the data that was setup.

The eight flip-flops of the 'LS364 are edge-triggered D-type flip-flops. On the positive transition of the clock the Q output will be set to the logic state that was setup at the D input. The 'LS363 is particularly useful for interfacing to MOS logic where a higher than normal V_{OH} level is desirable such as that required by the TMS 8080A microprocessor.

Schmitt-trigger buffered inputs at the enable ('LS363) and clock ('LS364) lines simplify system design as ac and dc noise rejection is improved by typically 400 mV due to the input hysteresis. A buffered output control input can be used to place the eight outputs in either a normal logic state (high or low logic levels) or a high-impedance state. In the high-impedance state the outputs neither load nor drive the bus line significantly.

DESIGN GOAL

This page provides tentative information on a product in the developmental stage. Texas Instruments reserves the right to change or discontinue this product without notice.

TEXAS INSTRUMENTS
INCORPORATED
POST OFFICE BOX 5012 • DALLAS, TEXAS 75222

A Typical data sheet for a flip-flop.

TYPES SN54LS363, SN54LS364, SN74LS363, SN74LS364
OCTAL D-TYPE TRANSPARENT LATCHES AND
EDGE-TRIGGERED FLIP-FLOPS

functional block diagram

Same as SN54LS373/SN74LS373 and SN54LS374/SN74LS374

schematics of inputs and outputs

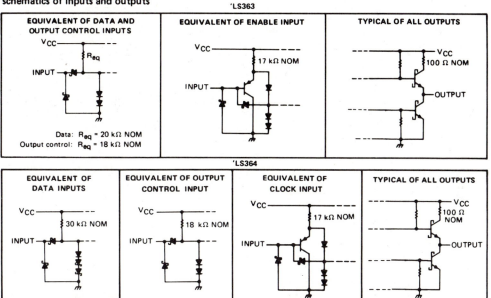

'LS363

| EQUIVALENT OF DATA AND OUTPUT CONTROL INPUTS | EQUIVALENT OF ENABLE INPUT | TYPICAL OF ALL OUTPUTS |

Data: R_{eq} = 20 kΩ NOM
Output control: R_{eq} = 18 kΩ NOM

'LS364

| EQUIVALENT OF DATA INPUTS | EQUIVALENT OF OUTPUT CONTROL INPUT | EQUIVALENT OF CLOCK INPUT | TYPICAL OF ALL OUTPUTS |

absolute maximum ratings over operating free-air temperature range (unless otherwise noted)

Supply voltage, V_{CC} (see Note 1) . . . . 7 V
Input voltage . . . . 7 V
Off-state output voltage . . . . 7 V
Operating free-air temperature range: SN54LS' . . . . −55°C to 125°C
                  SN74LS' . . . . 0°C to 70°C
Storage temperature range . . . . −65°C to 150°C

NOTE 1: Voltage values are with respect to network ground terminal.

recommended operating conditions

		SN54LS'			SN74LS'			UNIT
		MIN	NOM	MAX	MIN	NOM	MAX	
Supply voltage, V_{CC}		4.5	5	5.5	4.75	5	5.25	V
High-level output voltage, V_{OH}				5.5			5.5	V
High-level output current, I_{OH}				−1			−2.6	mA
Width of clock/enable pulse, t_w	High	15			15			ns
	Low	15			15			
Data setup time, t_{su}	'LS363	0↓			0↓			ns
	'LS364	20↑			20↑			
Data hold time, t_h	'LS363	10↓			10↓			ns
	'LS364	0↑			0↑			
Operating free-air temperature, T_A		−55		125	0		70	°C

↑↓ The arrow indicates the transition of the clock/enable input used for reference: ↑ for the low-to-high transition, ↓ for the high-to-low transition.

DESIGN GOAL

This page provides tentative information on a product in the developmental stage. Texas Instruments reserves the right to change or discontinue this product without notice.

Texas Instruments
INCORPORATED
POST OFFICE BOX 5012 • DALLAS, TEXAS 75222

A Typical data sheet for a flip-flop *(continued)*.

TYPES SN54LS363, SN54LS364, SN74LS363, SN74LS364
OCTAL D-TYPE TRANSPARENT LATCHES AND
EDGE-TRIGGERED FLIP-FLOPS

electrical characteristics over recommended operating free-air temperature range (unless otherwise noted)

PARAMETER		TEST CONDITIONS[†]		SN54LS'			SN74LS'			UNIT
				MIN	TYP[‡]	MAX	MIN	TYP[‡]	MAX	
V_{IH}	High-level input voltage			2			2			V
V_{IL}	Low-level input voltage					0.7			0.8	V
V_{IK}	Input clamp voltage	V_{CC} = MIN, I_I = −18 mA				−1.5			−1.5	V
V_{OH}	High-level output voltage	V_{CC} = MIN, V_{IH} = 2 V, V_{IL} = V_{IL}max, I_{OH} = MAX		3.45			3.65			V
V_{OL}	Low-level output voltage	V_{CC} = MIN, V_{IH} = 2 V, V_{IL} = V_{IL}max	I_{OL} = 12 mA		0.25	0.4		0.25	0.4	V
			I_{OL} = 24 mA					0.35	0.5	
I_{OZH}	Off-state output current, high-level voltage applied	V_{CC} = MAX, V_{IH} = 2 V, V_O = 3.65 V				20			20	µA
I_{OZL}	Off-state output current, low-level voltage applied	V_{CC} = MAX, V_{IH} = 2 V, V_O = 0.4 V				−20			−20	µA
I_I	Input current at maximum input voltage	V_{CC} = MAX, V_I = 7 V				0.1			0.1	mA
I_{IH}	High-level input current	V_{CC} = MAX, V_I = 2.7 V				20			20	µA
I_{IL}	Low-level input current	V_{CC} = MAX, V_I = 0.4 V				−400			−400	µA
I_{OS}	Short-circuit output current[§]	V_{CC} = MAX		−30		−130	−30		−130	mA
I_{CC}	Supply current	V_{CC} = MAX, Output control at 4.5 V			42	70		42	70	mA

[†]For conditions shown as MIN or MAX, use the appropriate value specified under recommended operating conditions.

[‡]All typical values are at V_{CC} = 5 V, T_A = 25°C.

[§]Not more than one output should be shorted at a time and duration of the short circuit should not exceed one second.

switching characteristics, V_{CC} = 5 V, T_A = 25°C

PARAMETER	FROM (INPUT)	TO (OUTPUT)	TEST CONDITIONS	'LS363			'LS364			UNIT
				MIN	TYP	MAX	MIN	TYP	MAX	
f_{max}							35	50		MHz
t_{PLH}	Data	Any Q	C_L = 45 pF, R_L = 667 Ω, See Notes 2 and 3		15	23				ns
t_{PHL}					18	27				
t_{PLH}	Clock or enable	Any Q			19	30		21	33	ns
t_{PHL}					24	36		22	34	
t_{PZH}	Output Control	Any Q			16	28		16	28	ns
t_{PZL}					22	36		22	36	
t_{PHZ}	Output Control	Any Q	C_L = 5 pF, R_L = 667 Ω, See Note 3		12	20		10	18	ns
t_{PLZ}					16	25		14	24	

NOTES: 2. Maximum clock frequency is tested with all outputs loaded.

3. See load circuits and waveforms on page 3-11.

f_{max} ≡ maximum clock frequency

t_{PLH} ≡ propagation delay time, low-to-high-level output

t_{PHL} ≡ propagation delay time, high-to-low-level output

t_{PZH} ≡ output enable time to high level

t_{PZL} ≡ output enable time to low level

t_{PHZ} ≡ output disable time from high level

t_{PLZ} ≡ output disable time from low level

TEXAS INSTRUMENTS
INCORPORATED
POST OFFICE BOX 5012 • DALLAS, TEXAS 75222

A Typical data sheet for a flip-flop (*continued*).

TYPES SN54LS363, SN54LS364, SN74LS363, SN74LS364
OCTAL D-TYPE TRANSPARENT LATCHES AND
EDGE-TRIGGERED FLIP-FLOPS

TYPICAL APPLICATION DATA

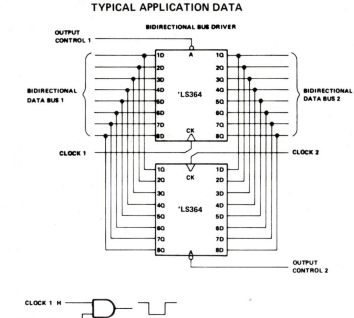

BIDIRECTIONAL BUS DRIVER

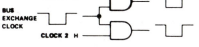

CLOCK CIRCUIT FOR BUS EXCHANGE

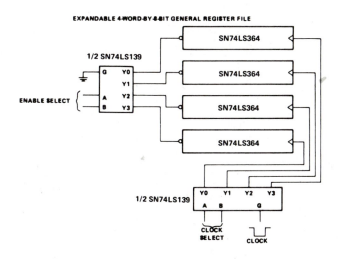

EXPANDABLE 4-WORD-BY-8-BIT GENERAL REGISTER FILE

TEXAS INSTRUMENTS
INCORPORATED
POST OFFICE BOX 5012 • DALLAS, TEXAS 75222

A Typical data sheet for a flip-flop (continued).

TYPES SN54150, SN54151A, SN54152A. SN54LS151, SN54LS152, SN54S151, SN74150, SN74151A, SN74LS151, SN74S151
DATA SELECTORS/MULTIPLEXERS

BULLETIN NO. DL-S 7611819, DECEMBER 1972–REVISED OCTOBER 1976

- **'150 Selects One-of-Sixteen Data Sources**
- **Others Select One-of-Eight Data Sources**
- **Performs Parallel-to-Serial Conversion**
- **Permits Multiplexing from N Lines to One Line**
- **Also For Use as Boolean Function Generator**
- **Input-Clamping Diodes Simplify System Design**
- **Fully Compatible with Most TTL and DTL Circuits**

TYPE	TYPICAL AVERAGE PROPAGATION DELAY TIME DATA INPUT TO W OUTPUT	TYPICAL POWER DISSIPATION
'150	11 ns	200 mW
'151A	8 ns	145 mW
'152A	8 ns	130 mW
'LS151	11 ns†	30 mW
'LS152	11 ns†	28 mW
'S151	4.5 ns	225 mW

†Tentative data

description

These monolithic data selectors/multiplexers contain full on-chip binary decoding to select the desired data source. The '150 selects one-of-sixteen data sources; the '151A, '152A, 'LS151, 'LS152, and 'S151 select one-of-eight data sources. The '150, '151A, 'LS151, and 'S151 have a strobe input which must be at a low logic level to enable these devices. A high level at the strobe forces the W output high, and the Y output (as applicable) low.

The '151A, 'LS151, and 'S151 feature complementary W and Y outputs whereas the '150, '152A, and 'LS152 have an inverted (W) output only.

The '151A and '152A incorporate address buffers which have symmetrical propagation delay times through the complementary paths. This reduces the possibility of transients occurring at the output(s) due to changes made at the select inputs, even when the '151A outputs are enabled (i.e., strobe low).

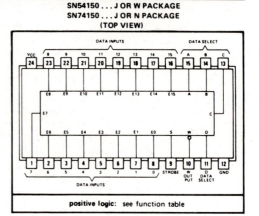

SN54150 . . . J OR W PACKAGE
SN74150 . . . J OR N PACKAGE
(TOP VIEW)

positive logic: see function table

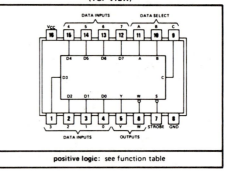

SN54151A, SN54LS151, SN54S151 . . . J OR W PACKAGE
SN74151A, SN74LS151, SN74S151 . . . J OR N PACKAGE
(TOP VIEW)

positive logic: see function table

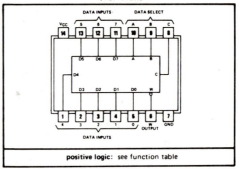

SN54152A, SN54LS152 . . . W PACKAGE
(TOP VIEW)

positive logic: see function table

TEXAS INSTRUMENTS
INCORPORATED

POST OFFICE BOX 5012 • DALLAS, TEXAS 75222

This is a one-of-sixteen multiplexer. It is included both as a reference for use and as a sample of the amount and type of information that is given about a typical MSI/LSI chip.

TYPES SN54150, SN54151A, SN54152A, SN54LS151, SN54LS152, SN54S151, SN74150, SN74151A SN74LS151, SN74S151
DATA SELECTORS/MULTIPLEXERS
REVISED OCTOBER 1976

logic

'150
FUNCTION TABLE

SELECT				STROBE	OUTPUT
D	C	B	A	S	W
X	X	X	X	H	H
L	L	L	L	L	$\overline{E0}$
L	L	L	H	L	$\overline{E1}$
L	L	H	L	L	$\overline{E2}$
L	L	H	H	L	$\overline{E3}$
L	H	L	L	L	$\overline{E4}$
L	H	L	H	L	$\overline{E5}$
L	H	H	L	L	$\overline{E6}$
L	H	H	H	L	$\overline{E7}$
H	L	L	L	L	$\overline{E8}$
H	L	L	H	L	$\overline{E9}$
H	L	H	L	L	$\overline{E10}$
H	L	H	H	L	$\overline{E11}$
H	H	L	L	L	$\overline{E12}$
H	H	L	H	L	$\overline{E13}$
H	H	H	L	L	$\overline{E14}$
H	H	H	H	L	$\overline{E15}$

The column header row above reads: INPUTS (SELECT D C B A) | STROBE S | OUTPUT W

'151A, 'LS151, 'S151
FUNCTION TABLE

SELECT			STROBE	OUTPUTS	
C	B	A	S	Y	W
X	X	X	H	L	H
L	L	L	L	D0	$\overline{D0}$
L	L	H	L	D1	$\overline{D1}$
L	H	L	L	D2	$\overline{D2}$
L	H	H	L	D3	$\overline{D3}$
H	L	L	L	D4	$\overline{D4}$
H	L	H	L	D5	$\overline{D5}$
H	H	L	L	D6	$\overline{D6}$
H	H	H	L	D7	$\overline{D7}$

'152A, 'LS152
FUNCTION TABLE

SELECT INPUTS			OUTPUT
C	B	A	W
L	L	L	$\overline{D0}$
L	L	H	$\overline{D1}$
L	H	L	$\overline{D2}$
L	H	H	$\overline{D3}$
H	L	L	$\overline{D4}$
H	L	H	$\overline{D5}$
H	H	L	$\overline{D6}$
H	H	H	$\overline{D7}$

H = high level, L = low level, X = irrelevant
E0, E1 . . . E15 = the complement of the level of the respective E input
D0, D1 . . . D7 = the level of the D respective input

functional block diagrams

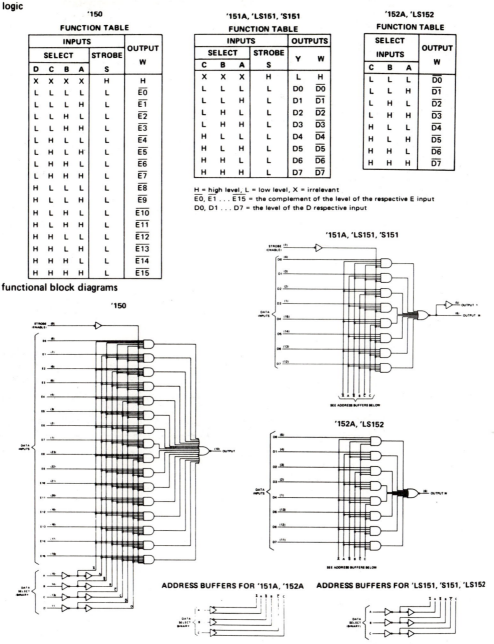

ADDRESS BUFFERS FOR '151A, '152A   ADDRESS BUFFERS FOR 'LS151, 'S151, 'LS152

TEXAS INSTRUMENTS
INCORPORATED
POST OFFICE BOX 5012 • DALLAS, TEXAS 75222

A one-of-sixteen multiplexer (*continued*).

TYPES SN54150, SN54151A, SN54152A, SN74150, SN74151A
DATA SELECTORS/MULTIPLEXERS

REVISED OCTOBER 1976

absolute maximum ratings over operating free-air temperature range (unless otherwise noted)

Supply voltage, V$_{CC}$ (see Note 1) . . . . 7 V
Input voltage (see Note 2) . . . . 5.5 V
Operating free-air temperature range: SN54' Circuits . . . . −55°C to 125°C
             SN74' Circuits . . . . 0°C to 70°C
Storage temperature range: . . . . −65°C to 150°C

NOTES: 1. Voltage values are with respect to network ground terminal.
    2. For the '150, input voltages must be zero or positive with respect to network ground terminal.

recommended operating conditions

	SN54'			SN74'			UNIT
	MIN	NOM	MAX	MIN	NOM	MAX	
Supply voltage, V$_{CC}$	4.5	5	5.5	4.75	5	5.25	V
High-level output current, I$_{OH}$			−800			−800	μA
Low-level output current, I$_{OL}$			16			16	mA
Operating free-air temperature, T$_A$	−55		125	0		70	°C

electrical characteristics over recommended operating free-air temperature range (unless otherwise noted)

	PARAMETER	TEST CONDITIONS[†]	'150			'151A, '152A			UNIT
			MIN	TYP[‡]	MAX	MIN	TYP[‡]	MAX	
V$_{IH}$	High-level input voltage		2			2			V
V$_{IL}$	Low-level input voltage				0.8			0.8	V
V$_{IK}$	Input clamp voltage	V$_{CC}$ = MIN, I$_I$ = −8 mA						−1.5	V
V$_{OH}$	High-level output voltage	V$_{CC}$ = MIN, V$_{IH}$ = 2 V, V$_{IL}$ = 0.8 V, I$_{OH}$ = −800 μA	2.4	3.4		2.4	3.4		V
V$_{OL}$	Low-level output voltage	V$_{CC}$ = MIN, V$_{IH}$ = 2 V, V$_{IL}$ = 0.8 V, I$_{OL}$ = 16 mA		0.2	0.4		0.2	0.4	V
I$_I$	Input current at maximum input voltage	V$_{CC}$ = MAX, V$_I$ = 5.5 V			1			1	mA
I$_{IH}$	High-level input current	V$_{CC}$ = MAX, V$_I$ = 2.4 V			40			40	μA
I$_{IL}$	Low-level input current	V$_{CC}$ = MAX, V$_I$ = 0.4 V			−1.6			−1.6	mA
I$_{OS}$	Short-circuit output current[§]	V$_{CC}$ = MAX SN54'	−20		−55	−20		−55	mA
		SN74'	−18		−55	−18		−55	
I$_{CC}$	Supply current	V$_{CC}$ = MAX, '150		40	68				mA
		See Note 3 '151A					29	48	
		'152A					26	43	

[†]For conditions shown as MIN or MAX, use the appropriate value specified under recommended operating conditions for the applicable device type.
[‡]All typical values at V$_{CC}$ = 5 V, T$_A$ = 25°C.
[§]Not more than one output of the '151A should be shorted at a time.
NOTE 3: I$_{CC}$ is measured with the strobe and data select inputs at 4.5 V, all other inputs and outputs open.

TEXAS INSTRUMENTS
INCORPORATED
POST OFFICE BOX 5012 • DALLAS, TEXAS 75222

A one-of-sixteen multiplexer (continued).

TYPES SN54150, SN54151A, SN54152A, SN74150, SN74151A
DATA SELECTORS/MULTIPLEXERS

switching characteristics, $V_{CC} = 5$ V, $T_A = 25°C$

PARAMETER¶	FROM (INPUT)	TO (OUTPUT)	TEST CONDITIONS	'150 MIN	'150 TYP	'150 MAX	'151A, '152A MIN	'151A, '152A TYP	'151A, '152A MAX	UNIT
t_{PLH}	A, B, or C	Y						25	38	ns
t_{PHL}	(4 levels)							25	38	
t_{PLH}	A, B, C, or D	W			23	35		17	26	ns
t_{PHL}	(3 levels)				22	33		19	30	
t_{PLH}	Strobe	Y	$C_L = 15$ pF,					21	33	ns
t_{PHL}			$R_L = 400$ Ω,					22	33	
t_{PLH}	Strobe	W	See Note 4		15.5	24		14	21	ns
t_{PHL}					21	30		15	23	
t_{PLH}	D0 thru D7	Y						13	20	ns
t_{PHL}								18	27	
t_{PLH}	E0 thru E15, or	W			13	20		8	14	ns
t_{PHL}	D0 thru D7				8.5	14		8	14	

¶$t_{PLH} \equiv$ propagation delay time, low-to-high-level output
$t_{PHL} \equiv$ propagation delay time, high-to-low-level output
NOTE 4: Load circuit and voltage waveforms are shown on page 3-10.

schematics of inputs and outputs

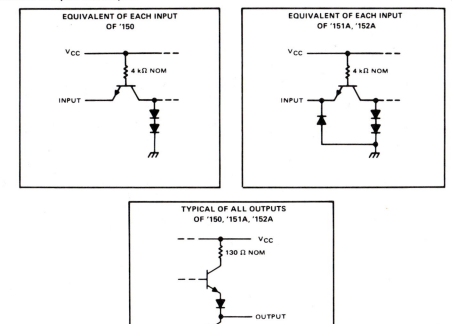

TEXAS INSTRUMENTS
INCORPORATED
POST OFFICE BOX 5012 • DALLAS, TEXAS 75222

A one-of-sixteen multiplexer (*continued*).

TYPES SN54LS151, SN54LS152, SN74LS151
DATA SELECTORS/MULTIPLEXERS

absolute maximum ratings over operating free-air temperature range (unless otherwise noted)

Supply voltage, V_{CC} (see Note 1) . . . . 7 V
Input voltage . . . . 7 V
Operating free-air temperature range: SN54LS' Circuits . . . . −55°C to 125°C
                                SN74LS' Circuits . . . . 0°C to 70°C
Storage temperature range . . . . −65°C to 150°C

NOTE 1: Voltage values are with respect to network ground terminal.

recommended operating conditions

	SN54LS'			SN74LS'			UNIT
	MIN	NOM	MAX	MIN	NOM	MAX	
Supply voltage, V_{CC}	4.5	5	5.5	4.75	5	5.25	V
High-level output current, I_{OH}			−400			−400	µA
Low-level output current, I_{OL}			4			8	mA
Operating free-air temperature, T_A	−55		125	0		70	°C

electrical characteristics over recommended operating free-air temperature range (unless otherwise noted)

PARAMETER	TEST CONDITIONS[†]		SN54LS'			SN74LS'			UNIT
			MIN	TYP[‡]	MAX	MIN	TYP[‡]	MAX	
V_{IH} High-level input voltage			2			2			V
V_{IL} Low-level input voltage					0.7			0.8	V
V_{IK} Input clamp voltage	V_{CC} = MIN, I_I = −18 mA				−1.5			−1.5	V
V_{OH} High-level output voltage	V_{CC} = MIN, V_{IH} = 2 V, V_{IL} = V_{IL} max, I_{OH} = −400 µA		2.5	3.4		2.7	3.4		V
V_{OL} Low-level output voltage	V_{CC} = MIN, V_{IH} = 2 V, V_{IL} = V_{IL} max	I_{OL} = 4 mA		0.25	0.4		0.25	0.4	V
		I_{OL} = 8 mA					0.35	0.5	
I_I Input current at maximum input voltage	V_{CC} = MAX, V_I = 7 V				0.1			0.1	mA
I_{IH} High-level input current	V_{CC} = MAX, V_I = 2.7 V				20			20	µA
I_{IL} Low-level input current	V_{CC} = MAX, V_I = 0.4 V				−0.4			−0.4	mA
I_{OS} Short-circuit output current[§]	V_{CC} = MAX		−20		−100	−20		100	mA
I_{CC} Supply current	V_{CC} = MAX, Outputs open, All inputs at 4.5 V	'LS151		6.0	10		6.0	10	mA
		'LS152		5.6	9				

[†]For conditions shown as MIN or MAX, use the appropriate value specified under recommended operating conditions for the applicable device type.

[‡]All typical values are at V_{CC} = 5 V, T_A = 25°C.

[§]Not more than one output should be shorted at a time and duration of short-circuit should not exceed one second.

TEXAS INSTRUMENTS
INCORPORATED
POST OFFICE BOX 5012 • DALLAS TEXAS 75222

A one-of-sixteen multiplexer (*continued*).

TYPES SN54LS151, SN54LS152, SN74LS151
DATA SELECTORS/MULTIPLEXERS

switching characteristics, $V_{CC} = 5$ V, $T_A = 25°C$

PARAMETER¶	FROM (INPUT)	TO (OUTPUT)	TEST CONDITIONS	SN54LS', SN74LS'			UNIT
				MIN	TYP	MAX	
t_{PLH}	A, B, or C	Y			27	43	ns
t_{PHL}	(4 levels)				18	30	
t_{PLH}	A, B, or C	W			14	23	ns
t_{PHL}	(3 levels)				20	32	
t_{PLH}	Strobe	Y	$C_L = 15$ pF,		26	42	ns
t_{PHL}			$R_L = 2$ kΩ,		20	32	
t_{PLH}	Strobe	W	See Note 5		15	24	ns
t_{PHL}					18	30	
t_{PLH}	Any D	Y			20	32	ns
t_{PHL}					16	26	
t_{PLH}	Any D	W			13	21	ns
t_{PHL}					12	20	

¶ t_{PLH} ≡ Propagation delay time, low-to-high-level output
  t_{PHL} ≡ Propagation delay time, high-to-low-level output
NOTE 5: See load circuits and waveforms on page 3-11.

schematics of inputs and outputs

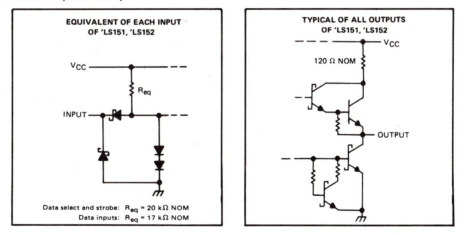

EQUIVALENT OF EACH INPUT
OF 'LS151, 'LS152

Data select and strobe: $R_{eq} = 20$ kΩ NOM
Data inputs: $R_{eq} = 17$ kΩ NOM

TYPICAL OF ALL OUTPUTS
OF 'LS151, 'LS152

120 Ω NOM

TEXAS INSTRUMENTS
INCORPORATED
POST OFFICE BOX 5012 • DALLAS, TEXAS 75222

A one-of-sixteen multiplexer (*continued*).

TYPES SN54S151, SN74S151
DATA SELECTORS/MULTIPLEXERS

absolute maximum ratings over operating free-air temperature range (unless otherwise noted)

Supply voltage, V_{CC} (see Note 1)	7 V
Input voltage	5.5 V
Operating free-air temperature range: SN54S151 Circuits	−55°C to 125°C
SN74S151 Circuits	0°C to 70°C
Storage temperature range	−65°C to 150°C

NOTE 1: Voltage values are with respect to network ground terminal.

recommended operating conditions

	SN54S151			SN74S151			UNIT
	MIN	NOM	MAX	MIN	NOM	MAX	
Supply voltage, V_{CC}	4.5	5	5.5	4.75	5	5.25	V
High-level output current, I_{OH}			−1			−1	mA
Low-level output current, I_{OL}			20			20	mA
Operating free-air temperature, T_A	−55		125	0		70	°C

electrical characteristics over recommended operating free-air temperature range (unless otherwise noted)

	PARAMETER	TEST CONDITIONS[†]		MIN	TYP[‡]	MAX	UNIT
V_{IH}	High-level input voltage			2			V
V_{IL}	Low-level input voltage					0.8	V
V_{IK}	Input clamp voltage	V_{CC} = MIN, I_I = −18 mA				−1.2	V
V_{OH}	High-level output voltage	V_{CC} = MIN, V_{IH} = 2 V,	SN54S'	2.5	3.4		V
		V_{IL} = 0.8 V, I_{OH} = −1 mA	SN74S'	2.7	3.4		
V_{OL}	Low-level output voltage	V_{CC} = MIN, V_{IH} = 2 V, V_{IL} = 0.8 V, I_{OL} = 20 mA				0.5	V
I_I	Input current at maximum input voltage	V_{CC} = MAX, V_I = 5.5 V				1	mA
I_{IH}	High-level input current	V_{CC} = MAX, V_I = 2.7 V				50	μA
I_{IL}	Low-level input current	V_{CC} = MAX, V_I = 0.5 V				−2	mA
I_{OS}	Short-circuit output current§	V_{CC} = MAX		−40		−100	mA
I_{CC}	Supply current	V_{CC} = MAX, All inputs at 4.5 V, All outputs open			45	70	mA

[†]For conditions shown as MIN or MAX, use the appropriate value specified under recommended operating conditions for the applicable device type.

[‡]All typical values are at V_{CC} = 5 V, T_A = 25°C.

§Not more than one output should be shorted at a time, and duration of the short-circuit should not exceed one second.

TEXAS INSTRUMENTS
INCORPORATED
POST OFFICE BOX 5012 · DALLAS, TEXAS 75222

A one-of-sixteen multiplexer (*continued*).

TYPES SN54S151, SN74S151
DATA SELECTORS/MULTIPLEXERS

switching characteristics, $V_{CC} = 5$ V, $T_A = 25°C$

PARAMETER¶	FROM (INPUT)	TO (OUTPUT)	TEST CONDITIONS	SN54S151, SN74S151 MIN	TYP	MAX	UNIT
t_{PLH}	A, B, or C	Y			12	18	ns
t_{PHL}	(4 levels)				12	18	
t_{PLH}	A, B, or C	W			10	15	ns
t_{PHL}	(3 levels)				9	13.5	
t_{PLH}	Any D	Y	$C_L = 15$ pF,		8	12	ns
t_{PHL}			$R_L = 280$ Ω,		8	12	
t_{PLH}	Any D	W	See Note 4		4.5	7	ns
t_{PHL}					4.5	7	
t_{PLH}	Strobe	Y			11	16.5	ns
t_{PHL}					12	18	
t_{PLH}	Strobe	W			9	13	ns
t_{PHL}					8.5	12	

¶$t_{PLH} \equiv$ Propagation delay time, low-to-high-level output
$t_{PHL} \equiv$ Propagation delay time, high-to-low-level output
NOTE 4: See load circuits and waveforms on page 3-10.

schematics of inputs and outputs

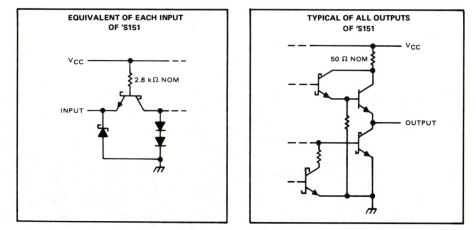

TEXAS INSTRUMENTS
INCORPORATED
POST OFFICE BOX 5012 • DALLAS, TEXAS 75222

A one-of-sixteen multiplexer (*continued*).

 **MOTOROLA**

SN54LS/74LS682
thru
SN54LS/74LS689

8-BIT MAGNITUDE COMPARATORS

LOW POWER SCHOTTKY

DESCRIPTION — The SN54LS/74LS682 thru SN54LS/74LS689 are 8-bit magnitude comparators. These device types are designed to perform comparisons between two eight-bit binary or BCD words. All device types provide $\overline{P=Q}$ outputs and the LS682 thru LS687 have $\overline{P>Q}$ outputs also.

The LS682, LS684, LS686 and LS688 are totem pole devices. The LS683, LS685, LS687 and LS689 are open-collector devices.

The LS682 and LS683 have a 20 kΩ pullup resistor on the Q inputs for analog or switch data.

TYPE	$\overline{P=Q}$	$\overline{P>Q}$	OUTPUT ENABLE	OUTPUT CONFIGURATION	PULLUP
LS682	yes	yes	no	totem-pole	yes
LS683	yes	yes	no	open-collector	yes
LS684	yes	yes	no	totem-pole	no
LS685	yes	yes	no	open-collector	no
LS686	yes	yes	yes	totem-pole	no
LS687	yes	yes	yes	open-collector	no
LS688	yes	no	yes	totem-pole	no
LS689	yes	no	yes	open-collector	no

FUNCTION TABLE

INPUTS				OUTPUTS	
DATA	ENABLES				
P, Q	$\overline{G}, \overline{G1}$	$\overline{G2}$		$\overline{P=Q}$	$\overline{P>Q}$
P = Q	L	L		L	H
P > Q	L	L		H	L
P < Q	L	L		H	H
X	H	H		H	H

H = high level, L = low level, X = irrelevant

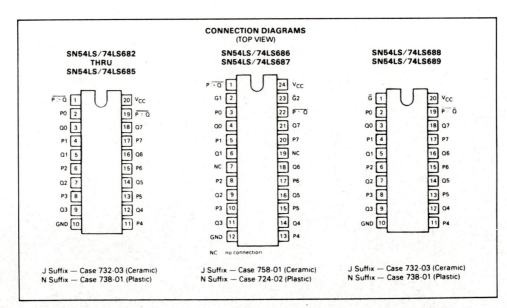

CONNECTION DIAGRAMS
(TOP VIEW)

SN54LS/74LS682 THRU SN54LS/74LS685

$\overline{P>Q}$	1	20	V_{CC}
P0	2	19	$\overline{P=Q}$
Q0	3	18	Q7
P1	4	17	P7
Q1	5	16	Q6
P2	6	15	P6
Q2	7	14	Q5
P3	8	13	P5
Q3	9	12	Q4
GND	10	11	P4

J Suffix — Case 732-03 (Ceramic)
N Suffix — Case 738-01 (Plastic)

SN54LS/74LS686 SN54LS/74LS687

$P>Q$	1	24	V_{CC}
G1	2	23	$\overline{G2}$
P0	3	22	$\overline{P=Q}$
Q0	4	21	Q7
P1	5	20	P7
Q1	6	19	NC
NC	7	18	Q6
P2	8	17	P6
Q2	9	16	Q5
P3	10	15	P5
Q3	11	14	Q4
GND	12	13	P4

NC no connection

J Suffix — Case 758-01 (Ceramic)
N Suffix — Case 724-02 (Plastic)

SN54LS/74LS688 SN54LS/74LS689

$\overline{G}$	1	20	V_{CC}
P0	2	19	$\overline{P=Q}$
Q0	3	18	Q7
P1	4	17	P7
Q1	5	16	Q6
P2	6	15	P6
Q2	7	14	Q5
P3	8	13	P5
Q3	9	12	Q4
GND	10	11	P4

J Suffix — Case 732-03 (Ceramic)
N Suffix — Case 738-01 (Plastic)

This is an eight-bit magnitude comparator, which is used in the computer interface design example in Chapter 39.

SN54LS/74LS682 • SN54LS/74LS684
SN54LS/74LS686 • SN54LS/74LS688

GUARANTEED OPERATING RANGES

SYMBOL	PARAMETER			MIN	TYP	MAX	UNIT
V_{CC}	Supply Voltage	54	4.5	5.0	5.5	V
		74	4.75	5.0	5.25	
T_A	Operating Ambient Temperature Range	54	−55	25	125	°C
		74	0	25	70	
I_{OH}	Output Current — High	54,74			−0.4	mA
I_{OL}	Output Current — Low	54			12	mA
		74			24	

DC CHARACTERISTICS OVER OPERATING TEMPERATURE RANGE (unless otherwise specified)

SYMBOL	PARAMETER		LIMITS			UNITS	TEST CONDITIONS	
			MIN	TYP	MAX			
V_{IH}	Input HIGH Voltage		2.0			V	Guaranteed Input HIGH Voltage for All Inputs	
V_{IL}	Input LOW Voltage	54			0.7	V	Guaranteed Input LOW Voltage for All Inputs	
		74			0.8			
V_{IK}	Input Clamp Diode Voltage			−0.65	−1.5	V	V_{CC} = MIN, I_{IN} = −18 mA	
V_{OH}	Output HIGH Voltage	54	2.5	3.5		V	V_{CC} = MIN, I_{OH} = MAX, V_{IN} = V_{IH} or V_{IL} per Truth Table	
		74	2.7	3.5		V		
V_{OL}	Output LOW Voltage	54,74		0.25	0.4	V	I_{OL} = 12 mA	V_{CC} = V_{CC} MIN, V_{IN} = V_{IL} or V_{IH} per Truth Table
		74		0.35	0.5	V	I_{OL} = 24 mA	
I_{IH}	Input HIGH Current				20	µA	V_{CC} = MAX, V_{IN} = 2.7 V	
		LS682-Q Inputs			0.1	mA	V_{CC} = MAX, V_{IN} = 5.5 V	
		Others			0.1	mA	V_{CC} = MAX, V_{IN} = 7.0 V	
I_{IL}	Input LOW Current	LS682-Q Inputs			−0.4	mA	V_{CC} = MAX, V_{IN} = 0.4 V	
		Others			−0.2	mA		
I_{OS}	Short Circuit Current		−30		−130	mA	V_{CC} = MAX	
I_{CC}	Power Supply Current	LS682			70	mA	V_{CC} = MAX	
		LS684			65	mA		
		LS686			75	mA		
		LS688			65	mA		

An eight-bit magnitude comparator (*continued*).

SN54LS/74LS683 ● SN54LS/74LS685
SN54LS/74LS687 ● SN54LS/74LS689

BLOCK DIAGRAMS

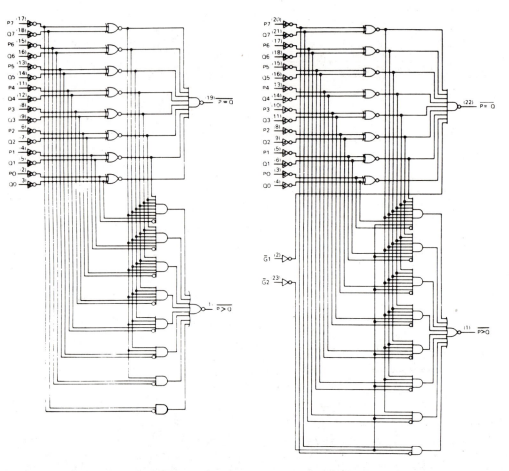

SN54LS/74LS682 thru LS685   SN54LS/74LS686, LS687

An eight-bit magnitude comparator (*continued*).

BLOCK DIAGRAM

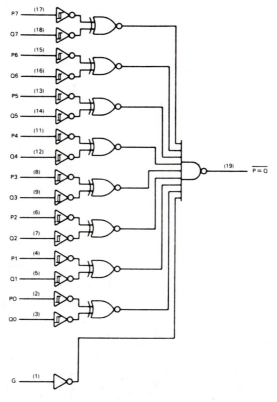

SN54LS/74LS688, LS689

An eight-bit magnitude comparator (*continued*).

SN54LS/74LS683 • SN54LS/74LS685
SN54LS/74LS687 • SN54LS/74LS689

GUARANTEED OPERATING RANGES

SYMBOL	PARAMETER			MIN	TYP	MAX	UNIT
V_{CC}	Supply Voltage	54		4.5	5.0	5.5	V
		74		4.75	5.0	5.25	
T_A	Operating Ambient Temperature Range	54		−55	25	125	°C
		74		0	25	70	
V_{OH}	Output Voltage — High	54,74				5.5	V
I_{OL}	Output Current — Low	54				12	mA
		74				24	

DC CHARACTERISTICS OVER OPERATING TEMPERATURE RANGE (unless otherwise specified)

SYMBOL	PARAMETER		LIMITS			UNITS	TEST CONDITIONS
			MIN	TYP	MAX		
V_{IH}	Input HIGH Voltage		2.0			V	Guaranteed Input HIGH Voltage for All Inputs
V_{IL}	Input LOW Voltage	54			0.7	V	Guaranteed Input LOW Voltage for All Inputs
		74			0.8		
V_{IK}	Input Clamp Diode Voltage			−0.65	−1.5	V	V_{CC} = MIN, I_{IN} = −18 mA
I_{OH}	Output HIGH Current	54			250	μA	V_{CC} = MIN, V_{OH} = MAX
		74			100	μA	
V_{OL}	Output LOW Voltage	54,74		0.25	0.4	V	I_{OL} = 12 mA, V_{CC} = V_{CC} MIN, V_{IN} = V_{IL} or V_{IH} per Truth Table
		74		0.35	0.5	V	I_{OL} = 24 mA
I_{IH}	Input HIGH Current				20	μA	V_{CC} = MAX, V_{IN} = 2.7 V
		LS683-Q Inputs			0.1	mA	V_{CC} = MAX, V_{IN} = 5.5 V
		Others			0.1	mA	V_{CC} = MAX, V_{IN} = 7.0 V
I_{IL}	Input LOW Current	LS683-Q Inputs			−0.4	mA	V_{CC} = MAX, V_{IN} = 0.4 V
		Others			−0.2	mA	
I_{CC}	Power Supply Current	LS683			70	mA	V_{CC} = MAX
		LS685			65	mA	
		LS687			75	mA	
		LS689			65	mA	

An eight-bit magnitude comparator (*continued*).

SN54LS/74LS682 THRU SN54LS/74LS685

AC CHARACTERISTICS: $T_A = 25°C$

SN54LS/74LS682

SYMBOL	PARAMETER	LIMITS			UNITS	TEST CONDITIONS
		MIN	TYP	MAX		
t_{PLH} t_{PHL}	Propagation Delay, P to $\overline{P = Q}$		13 15	25 25	ns	
t_{PLH} t_{PHL}	Propagation Delay, Q to $\overline{P = Q}$		14 15	25 25	ns	$V_{CC} = 5.0$ V $C_L = 45$ pF $R_L = 667$ Ω
t_{PLH} t_{PHL}	Propagation Delay, P to $\overline{P > Q}$		20 15	30 30	ns	
t_{PLH} t_{PHL}	Propagation Delay, Q to $\overline{P > Q}$		21 19	30 30	ns	

SN54LS/74LS683

SYMBOL	PARAMETER	LIMITS			UNITS	TEST CONDITIONS
		MIN	TYP	MAX		
t_{PLH} t_{PHL}	Propagation Delay, P to $\overline{P = Q}$		30 20	45 30	ns	
t_{PLH} t_{PHL}	Propagation Delay, Q to $\overline{P = Q}$		24 23	35 35	ns	$V_{CC} = 5.0$ V $C_L = 45$ pF $R_L = 667$ Ω
t_{PLH} t_{PHL}	Propagation Delay, P to $\overline{P > Q}$		31 17	45 30	ns	
t_{PLH} t_{PHL}	Propagation Delay, Q to $\overline{P > Q}$		30 21	45 30	ns	

SN54LS/74LS684

SYMBOL	PARAMETER	LIMITS			UNITS	TEST CONDITIONS
		MIN	TYP	MAX		
t_{PLH} t_{PHL}	Propagation Delay, P to $\overline{P = Q}$		15 17	25 25	ns	
t_{PLH} t_{PHL}	Propagation Delay, Q to $\overline{P = Q}$		16 15	25 25	ns	$V_{CC} = 5.0$ V $C_L = 45$ pF $R_L = 667$ Ω
t_{PLH} t_{PHL}	Propagation Delay, P to $\overline{P > Q}$		22 17	30 30	ns	
t_{PLH} t_{PHL}	Propagation Delay, Q to $\overline{P > Q}$		24 20	30 30	ns	

SN54LS/74LS685

SYMBOL	PARAMETER	LIMITS			UNITS	TEST CONDITIONS
		MIN	TYP	MAX		
t_{PLH} t_{PHL}	Propagation Delay, P to $\overline{P = Q}$		30 19	45 35	ns	
t_{PLH} t_{PHL}	Propagation Delay, Q to $\overline{P = Q}$		24 23	45 35	ns	$V_{CC} = 5.0$ V $C_L = 45$ pF $R_L = 667$ Ω
t_{PLH} t_{PHL}	Propagation Delay, P to $\overline{P > Q}$		32 16	45 35	ns	
t_{PLH} t_{PHL}	Propagation Delay, Q to $\overline{P > Q}$		30 20	45 35	ns	

An eight-bit magnitude comparator (*continued*).

SN54LS/74LS686 • SN54LS/74LS687
SN54LS/74LS688 • SN54LS/74LS689

AC CHARACTERISTICS: $T_A = 25°C$

SN54LS/74LS686

SYMBOL	PARAMETER	LIMITS			UNITS	TEST CONDITIONS
		MIN	TYP	MAX		
tPLH tPHL	Propagation Delay, P to $\overline{P = Q}$		13 20	25 30	ns	
tPLH tPHL	Propagation Delay, Q to $\overline{P = Q}$		13 21	25 30	ns	
tPLH tPHL	Propagation Delay, $\overline{G}$, $\overline{G1}$ to $\overline{P = Q}$		11 19	20 30	ns	$V_{CC} = 5.0$ V $C_L = 45$ pF $R_L = 667 \, \Omega$
tPLH tPHL	Propagation Delay, P to $\overline{P > Q}$		19 15	30 30	ns	
tPLH tPHL	Propagation Delay, Q to $\overline{P > Q}$		18 19	30 30	ns	
tPLH tPHL	Propagation Delay, $\overline{G2}$ to $\overline{P > Q}$		21 16	30 25	ns	

SN54LS/74LS687

SYMBOL	PARAMETER	LIMITS			UNITS	TEST CONDITIONS
		MIN	TYP	MAX		
tPLH tPHL	Propagation Delay, P to $\overline{P = Q}$		24 20	35 30	ns	
tPLH tPHL	Propagation Delay, Q to $\overline{P = Q}$		24 20	35 30	ns	
tPLH tPHL	Propagation Delay, $\overline{G}$, $\overline{G1}$ to $\overline{P = Q}$		21 18	35 30	ns	$V_{CC} = 5.0$ V $C_L = 45$ pF $R_L = 667 \, \Omega$
tPLH tPHL	Propagation Delay, P to $\overline{P > Q}$		24 16	35 30	ns	
tPLH tPHL	Propagation Delay, Q to $\overline{P > Q}$		24 16	35 30	ns	
tPLH tPHL	Propagation Delay, $\overline{G2}$ to $\overline{P > Q}$		24 15	35 30	ns	

SN54LS/74LS688

SYMBOL	PARAMETER	LIMITS			UNITS	TEST CONDITIONS
		MIN	TYP	MAX		
tPLH tPHL	Propagation Delay, P to $\overline{P = Q}$		12 17	18 23	ns	
tPLH tPHL	Propagation Delay, Q to $\overline{P = Q}$		12 17	18 23	ns	$V_{CC} = 5.0$ V $C_L = 45$ pF $R_L = 667 \, \Omega$
tPLH tPHL	Propagation Delay, $\overline{G}$, $\overline{G1}$ to $\overline{P = Q}$		12 13	18 20	ns	

SN54LS/74LS689

SYMBOL	PARAMETER	LIMITS			UNITS	TEST CONDITIONS
		MIN	TYP	MAX		
tPLH tPHL	Propagation Delay, P to $\overline{P = Q}$		24 22	40 35	ns	
tPLH tPHL	Propagation Delay, Q to $\overline{P = Q}$		24 22	40 35	ns	$V_{CC} = 5.0$ V $C_L = 45$ pF $R_L = 667 \, \Omega$
tPLH tPHL	Propagation Delay, $\overline{G}$, $\overline{G1}$ to $\overline{P = Q}$		22 19	35 30	ns	

An eight-bit magnitude comparator (*continued*).

ORDERING INFORMATION

Device	Alternate	Temperature Range	Package
MC1741CD	—	0°C to +70°C	SO-8
MC1741CG	LM741CH, μA741HC	0°C to +70°C	Metal Can
MC1741CP1	LM741CN, μA741TC	0°C to +70°C	Plastic DIP
MC1741CU	—	0°C to +70°C	Ceramic DIP
MC1741G	—	−55°C to +125°C	Metal Can
MC1741U	—	−55°C to +125°C	Ceramic DIP

MC1741
MC1741C

OPERATIONAL AMPLIFIER
SILICON MONOLITHIC
INTEGRATED CIRCUIT

INTERNALLY COMPENSATED, HIGH PERFORMANCE OPERATIONAL AMPLIFIERS

. . . designed for use as a summing amplifier, integrator, or amplifier with operating characteristics as a function of the external feedback components.

- No Frequency Compensation Required
- Short-Circuit Protection
- Offset Voltage Null Capability
- Wide Common-Mode and Differential Voltage Ranges
- Low-Power Consumption
- No Latch Up

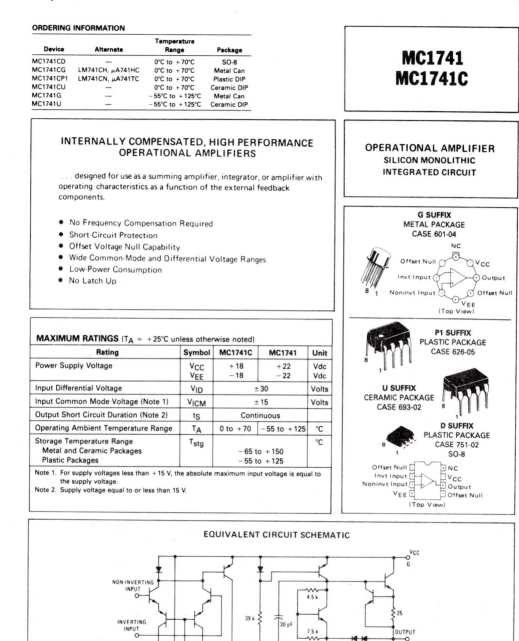

G SUFFIX
METAL PACKAGE
CASE 601-04

P1 SUFFIX
PLASTIC PACKAGE
CASE 626-05

U SUFFIX
CERAMIC PACKAGE
CASE 693-02

D SUFFIX
PLASTIC PACKAGE
CASE 751-02
SO-8

MAXIMUM RATINGS (T_A = +25°C unless otherwise noted)

Rating	Symbol	MC1741C	MC1741	Unit
Power Supply Voltage	V_{CC}	+18	+22	Vdc
	V_{EE}	−18	−22	Vdc
Input Differential Voltage	V_{ID}	±30		Volts
Input Common Mode Voltage (Note 1)	V_{ICM}	±15		Volts
Output Short Circuit Duration (Note 2)	t_S	Continuous		
Operating Ambient Temperature Range	T_A	0 to +70	−55 to +125	°C
Storage Temperature Range	T_{stg}			°C
Metal and Ceramic Packages		−65 to +150		
Plastic Packages		−55 to +125		

Note 1. For supply voltages less than +15 V, the absolute maximum input voltage is equal to the supply voltage.
Note 2. Supply voltage equal to or less than 15 V.

EQUIVALENT CIRCUIT SCHEMATIC

The data sheets for the MC1741, the Motorola number for the 741. This is probably the most common operational amplifier in use today.

MC1741, MC1741C

ELECTRICAL CHARACTERISTICS (V_{CC} = + 15 V, V_{EE} = – 15 V, T_A = 25°C unless otherwise noted).

Characteristic	Symbol	MC1741			MC1741C			Unit
		Min	Typ	Max	Min	Typ	Max	
Input Offset Voltage ($R_S \leqslant$ 10 k)	V_{IO}		1.0	5.0		2.0	6.0	mV
Input Offset Current	I_{IO}		20	200	–	20	200	nA
Input Bias Current	I_{IB}	–	80	500		80	500	nA
Input Resistance	r_i	0.3	2.0		0.3	2.0		MΩ
Input Capacitance	C_i	–	1.4		–	1.4	–	pF
Offset Voltage Adjustment Range	V_{IOR}		+ 15		–	+ 15	–	mV
Common Mode Input Voltage Range	V_{ICR}	+ 12	+ 13		± 12	± 13		V
Large Signal Voltage Gain (V_O = ±10 V, $R_L \geqslant$ 2.0 k)	A_v	50	200		20	200	–	V/mV
Output Resistance	r_o		75			75		Ω
Common Mode Rejection Ratio ($R_S \leqslant$ 10 k)	CMRR	70	90		70	90		dB
Supply Voltage Rejection Ratio ($R_S \leqslant$ 10 k)	PSRR		30	150		30	150	µV/V
Output Voltage Swing	V_O							V
($R_L \geqslant$ 10 k)		+ 12	+ 14		± 12	± 14		
($R_L \geqslant$ 2 k)		+ 10	+ 13		± 10	± 13		
Output Short-Circuit Current	I_{os}		20			20		mA
Supply Current	I_D		1.7	2.8		1.7	2.8	mA
Power Consumption	P_C		50	85		50	85	mW
Transient Response (Unity Gain – Non-Inverting)								
(V_I = 20 mV, $R_L \geqslant$ 2 k, $C_L \leqslant$ 100 pF) Rise Time	t_{TLH}		0.3		–	0.3		µs
(V_I = 20 mV, $R_L \geqslant$ 2 k, $C_L \leqslant$ 100 pF) Overshoot	os		15			15	–	%
(V_I = 10 V, $R_L \geqslant$ 2 k, $C_L \leqslant$ 100 pF) Slew Rate	SR		0.5			0.5		V/µs

ELECTRICAL CHARACTERISTICS (V_{CC} = + 15 V, V_{EE} = – 15 V, T_A = T_{low} to T_{high} unless otherwise noted).

Characteristic	Symbol	MC1741			MC1741C			Unit
		Min	Typ	Max	Min	Typ	Max	
Input Offset Voltage ($R_S \leqslant$ 10 kΩ)	V_{IO}		1.0	6.0			7.5	mV
Input Offset Current	I_{IO}							nA
(T_A = 125°C)			7.0	200				
(T_A = -55°C)			85	500				
(T_A = 0°C to +70°C)							300	
Input Bias Current	I_{IB}							nA
(T_A = 125°C)			30	500				
(T_A = -55°C)			300	1500				
(T_A = 0°C to +70°C)							800	
Common Mode Input Voltage Range	V_{ICR}	+ 12	+ 13					V
Common Mode Rejection Ratio ($R_S \leqslant$ 10 k)	CMRR	70	90					dB
Supply Voltage Rejection Ratio ($R_S \leqslant$ 10 k)	PSRR		30	150				µV/V
Output Voltage Swing	V_O							V
($R_L \geqslant$ 10 k)		+ 12	+ 14					
($R_L \geqslant$ 2 k)		+ 10	+ 13		± 10	± 13		
Large Signal Voltage Gain ($R_L \geqslant$ 2 k, V_{out} = ±10 V)	A_v	25			15	–		V/mV
Supply Currents	I_D							mA
(T_A = 125°C)			1.5	2.5				
(T_A = -55°C)			2.0	3.3				
Power Consumption (T_A = +125°C)	P_C		45	75				mW
(T_A = -55°C)			60	100				

*T_{high} = 125°C for MC1741 and 70°C for MC1741C
T_{low} = -55°C for MC1741 and 0°C for MC1741C

The data sheets for the MC1741 (continued).

MC1741, MC1741C

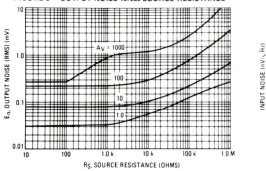

FIGURE 1 – BURST NOISE versus SOURCE RESISTANCE

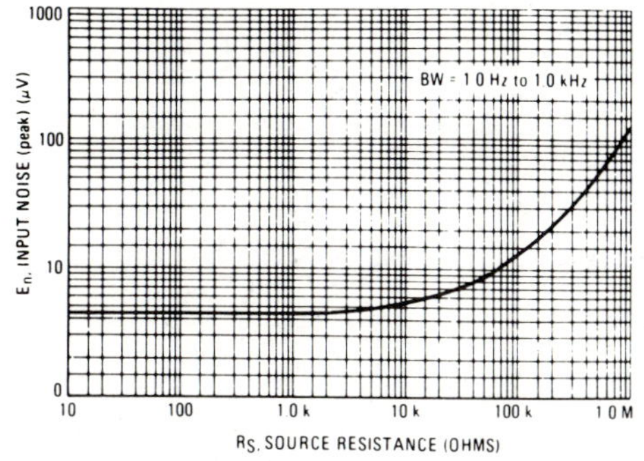

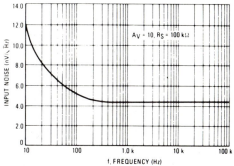

FIGURE 2 – RMS NOISE versus SOURCE RESISTANCE

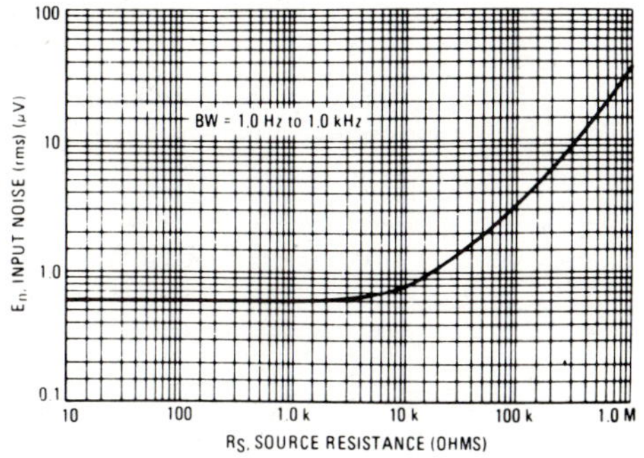

FIGURE 3 – OUTPUT NOISE versus SOURCE RESISTANCE

FIGURE 4 – SPECTRAL NOISE DENSITY

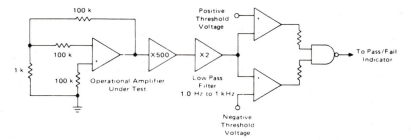

FIGURE 5 – BURST NOISE TEST CIRCUIT

Unlike conventional peak reading or RMS meters, this system was especially designed to provide the quick response time essential to burst (popcorn) noise testing.

The test time employed is 10 seconds and the 20 µV peak limit refers to the operational amplifier input thus eliminating errors in the closed-loop gain factor of the operational amplifier under test

The data sheets for the MC1741 (continued).

MC1741, MC1741C

TYPICAL CHARACTERISTICS

(V_CC = +15 Vdc, V_EE = -15 Vdc, T_A = +25°C unless otherwise noted)

FIGURE 6 – POWER BANDWIDTH
(LARGE SIGNAL SWING versus FREQUENCY)

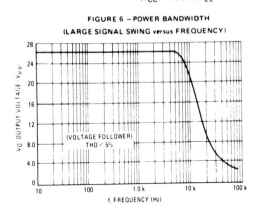

FIGURE 7 – OPEN LOOP FREQUENCY RESPONSE

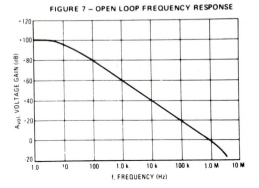

FIGURE 8 – POSITIVE OUTPUT VOLTAGE SWING
versus LOAD RESISTANCE

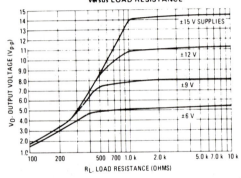

FIGURE 9 – NEGATIVE OUTPUT VOLTAGE SWING
versus LOAD RESISTANCE

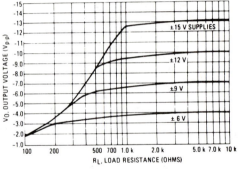

FIGURE 10 – OUTPUT VOLTAGE SWING versus
LOAD RESISTANCE (Single Supply Operation)

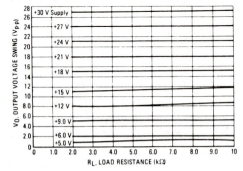

FIGURE 11 – SINGLE SUPPLY INVERTING AMPLIFIER

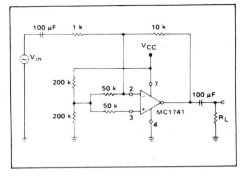

The data sheets for the MC1741 (*continued*).

MC1741, MC1741C

FIGURE 12 — NONINVERTING PULSE RESPONSE

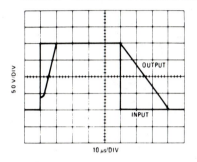

FIGURE 13 — TRANSIENT REPONSE TEST CIRCUIT

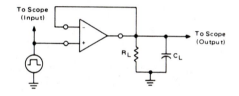

**FIGURE 14 — OPEN LOOP VOLTAGE GAIN
versus SUPPLY VOLTAGE**

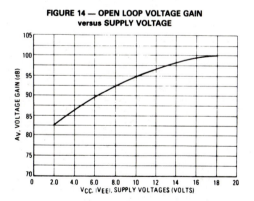

The data sheets for the MC1741 (continued).

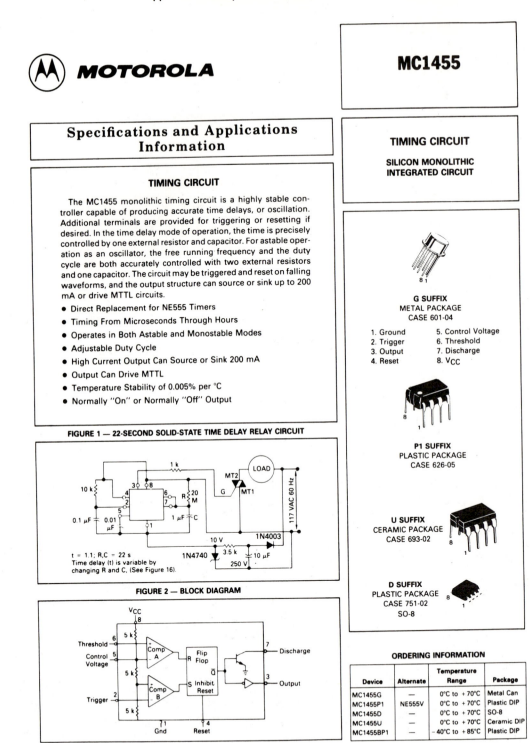

MOTOROLA

MC1455

Specifications and Applications Information

TIMING CIRCUIT

SILICON MONOLITHIC INTEGRATED CIRCUIT

TIMING CIRCUIT

The MC1455 monolithic timing circuit is a highly stable controller capable of producing accurate time delays, or oscillation. Additional terminals are provided for triggering or resetting if desired. In the time delay mode of operation, the time is precisely controlled by one external resistor and capacitor. For astable operation as an oscillator, the free running frequency and the duty cycle are both accurately controlled with two external resistors and one capacitor. The circuit may be triggered and reset on falling waveforms, and the output structure can source or sink up to 200 mA or drive MTTL circuits.

- Direct Replacement for NE555 Timers
- Timing From Microseconds Through Hours
- Operates in Both Astable and Monostable Modes
- Adjustable Duty Cycle
- High Current Output Can Source or Sink 200 mA
- Output Can Drive MTTL
- Temperature Stability of 0.005% per °C
- Normally "On" or Normally "Off" Output

G SUFFIX
METAL PACKAGE
CASE 601-04

1. Ground   5. Control Voltage
2. Trigger   6. Threshold
3. Output   7. Discharge
4. Reset   8. V_{CC}

P1 SUFFIX
PLASTIC PACKAGE
CASE 626-05

U SUFFIX
CERAMIC PACKAGE
CASE 693-02

D SUFFIX
PLASTIC PACKAGE
CASE 751-02
SO-8

FIGURE 1 — 22-SECOND SOLID-STATE TIME DELAY RELAY CIRCUIT

t = 1.1; R,C = 22 s
Time delay (t) is variable by changing R and C. (See Figure 16).

FIGURE 2 — BLOCK DIAGRAM

ORDERING INFORMATION

Device	Alternate	Temperature Range	Package
MC1455G	—	0°C to +70°C	Metal Can
MC1455P1	NE555V	0°C to +70°C	Plastic DIP
MC1455D	—	0°C to +70°C	SO-8
MC1455U	—	0°C to +70°C	Ceramic DIP
MC1455BP1	—	−40°C to +85°C	Plastic DIP

The data sheet for the Motorola MC1455, which is the Motorola part number for the 555, the basic timer circuit. This unit is described in Chapter 23.

MC1455

MAXIMUM RATINGS (T_A = +25°C unless otherwise noted.)

Rating	Symbol	Value	Unit
Power Supply Voltage	V_{CC}	+18	Vdc
Discharge Current (Pin 7)	I_7	200	mA
Power Dissipation (Package Limitation)	P_D		
Metal Can		680	mW
Derate above T_A = +25°C		4.6	mW/°C
Plastic Dual In-Line Package		625	mW
Derate above T_A = +25°C		5.0	mW/°C
Operating Temperature Range (Ambient)	T_A		°C
MC1455B		−40 to +85	
MC1455		0 to +70	
Storage Temperature Range	T_{stg}	−65 to +150	°C

FIGURE 3 — GENERAL TEST CIRCUIT

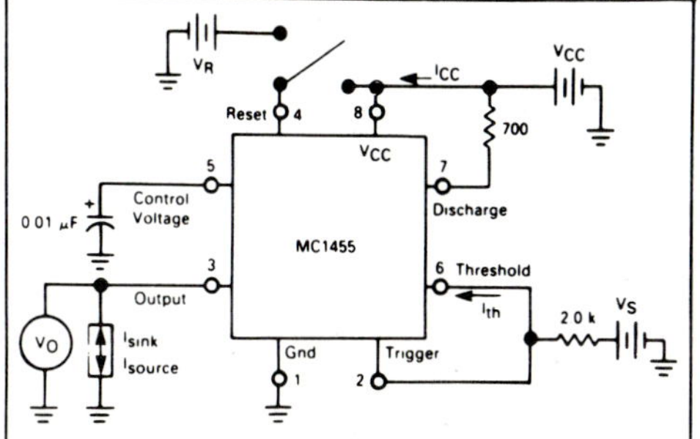

Test Circuit for Measuring dc Parameters (to set output and measure parameters)
a) When V_S · 2 3 V_{CC}, V_O is low
b) When V_S · 1 3 V_{CC}, V_O is high
c) When V_O is low, pin 7 sinks current. To test for Reset, set V_O high, apply Reset voltage, and test for current flowing into pin 7. When Reset is not in use, it should be tied to V_{CC}

ELECTRICAL CHARACTERISTICS (T_A = +25°C, V_{CC} = +5.0 V to +15 V unless otherwise noted.)

Characteristics	Symbol	Min	Typ	Max	Unit
Operating Supply Voltage Range	V_{CC}	4.5	—	16	V
Supply Current	I_{CC}				mA
V_{CC} = 5.0 V, R_L = ∞		—	3.0	6.0	
V_{CC} = 15 V, R_L = ∞		—	10	15	
Low State, (Note 1)					
Timing Error (Note 2)					
R = 1.0 kΩ to 100 kΩ					
Initial Accuracy C = 0.1 μF		—	1.0	—	%
Drift with Temperature		—	50	—	PPM/°C
Drift with Supply Voltage		—	0.1	—	%/Volt
Threshold Voltage	V_{th}	—	2/3	—	xV_{CC}
Trigger Voltage	V_T				V
V_{CC} = 15 V		—	5.0	—	
V_{CC} = 5.0 V		—	1.67	—	
Trigger Current	I_T	—	0.5	—	μA
Reset Voltage	V_R	0.4	0.7	1.0	V
Reset Current	I_R	—	0.1	—	mA
Threshold Current (Note 3)	I_{th}	—	0.1	0.25	μA
Discharge Leakage Current (Pin 7)	I_{dis}	—	—	100	nA
Control Voltage Level	V_{CL}				V
V_{CC} = 15 V		9.0	10	11	
V_{CC} = 5.0 V		2.6	3.33	4.0	
Output Voltage Low	V_{OL}				V
(V_{CC} = 15 V)					
I_{sink} = 10 mA		—	0.1	0.25	
I_{sink} = 50 mA		—	0.4	0.75	
I_{sink} = 100 mA		—	2.0	2.5	
I_{sink} = 200 mA		—	2.5	—	
(V_{CC} = 5.0 V)					
I_{sink} = 8.0 mA		—	—	—	
I_{sink} = 5.0 mA		—	0.25	0.35	
Output Voltage High	V_{OH}				V
(I_{source} = 200 mA)					
V_{CC} = 15 V		—	12.5	—	
(I_{source} = 100 mA)					
V_{CC} = 15 V		12.75	13.3	—	
V_{CC} = 5.0 V		2.75	3.3	—	
Rise Time of Output	t_{OLH}	—	100	—	ns
Fall Time of Output	t_{OHL}	—	100	—	ns

NOTES:
1. Supply current when output is high is typically 1.0 mA less.
2. Tested at V_{CC} = 5.0 V and V_{CC} = 15 V. Monostable mode
3. This will determine the maximum value of R_A + R_B for 15 V operation. The maximum total R = 20 megohms.

The data sheet for the Motorola MC1455 *(continued)*.

MC1455

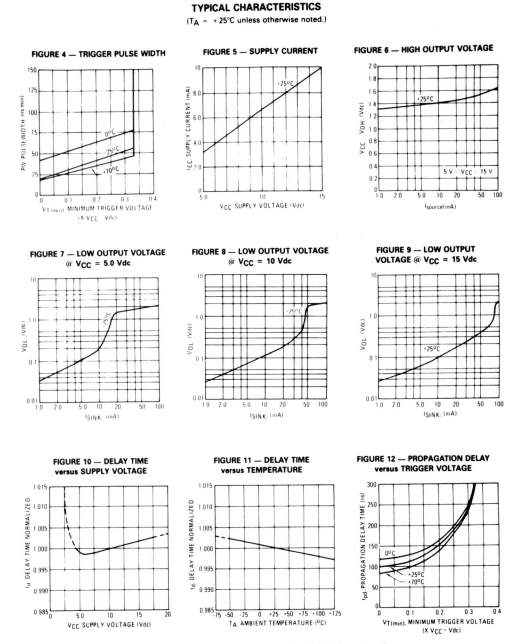

TYPICAL CHARACTERISTICS
(T_A = +25°C unless otherwise noted.)

The data sheet for the Motorola MC1455 (continued).

MC1455

FIGURE 13 — REPRESENTATIVE CIRCUIT SCHEMATIC

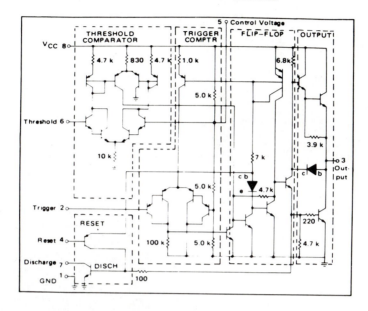

GENERAL OPERATION

The MC1455 is a monolithic timing circuit which uses as its timing elements an external resistor — capacitor network. It can be used in both the monostable (one-shot) and astable modes with frequency and duty cycle controlled by the capacitor and resistor values. While the timing is dependent upon the external passive components, the monolithic circuit provides the starting circuit, voltage comparison and other functions needed for a complete timing circuit. Internal to the integrated circuit are two comparators, one for the input signal and the other for capacitor voltage; also a flip-flop and digital output are included. The comparator reference voltages are always a fixed ratio of the supply voltage thus providing output timing independent of supply voltage.

Monostable Mode

In the monostable mode, a capacitor and a single resistor are used for the timing network. Both the threshold terminal and the discharge transistor terminal are connected together in this mode, refer to circuit Figure 14. When the input voltage to the trigger comparator falls below 1 3 V_{CC} the comparator output triggers the flip-flop so that it's output sets low. This turns the capacitor discharge transistor "off" and drives the digital output to the high state. This condition allows the capacitor to charge at an exponential rate which is set by the RC time constant. When the capacitor voltage reaches 2 3 V_{CC} the threshold comparator resets the flip-flop. This action discharges the timing capacitor and returns the digital output to the low state. Once the flip-flop has been triggered by an input signal, it cannot be retriggered until the present timing period has been completed. The time that the output is high is given by the equation $t = 1.1 R_A C$. Various combinations of R and C and their associated times are shown in Figure 16. The trigger pulse width must be less than the timing period.

A reset pin is provided to discharge the capacitor thus interrupting the timing cycle. As long as the reset pins is low, the capacitor discharge transistor is turned "on" and prevents the capacitor from charging. While the reset voltage is applied the digital output will remain the same. The reset pin should be tied to the supply voltage when not in use.

FIGURE 14 — MONOSTABLE CIRCUIT

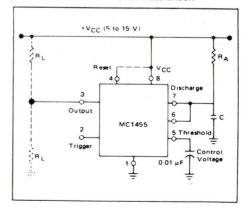

The data sheet for the Motorola MC1455 (*continued*).

GENERAL OPERATION (continued)

FIGURE 15 — MONOSTABLE WAVEFORMS

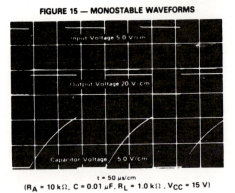

$t = 50 \ \mu s/cm$

$(R_A = 10 \ k\Omega, \ C = 0.01 \ \mu F, \ R_L = 1.0 \ k\Omega, \ V_{CC} = 15 \ V)$

FIGURE 16 — TIME DELAY

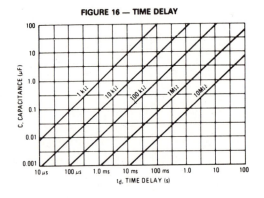

t_d, TIME DELAY (s)

C, CAPACITANCE (μF)

FIGURE 17 — ASTABLE CIRCUIT

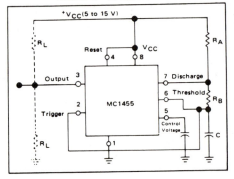

FIGURE 18 — ASTABLE WAVEFORMS

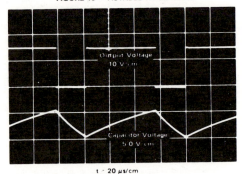

$t = 20 \ \mu s/cm$

$(R_A = 5.1 \ k\Omega, \ C = 0.01 \ \mu F, \ R_L = 1.0 \ k\Omega;$
$R_B = 3.9 \ k\Omega, \ V_{CC} = 15 \ V)$

Astable Mode

In the astable mode the timer is connected so that it will retrigger itself and cause the capacitor voltage to oscillate between $1/3 \ V_{CC}$ and $2/3 \ V_{CC}$. See Figure 17.

The external capacitor charges to $2/3 \ V_{CC}$ through R_A and R_B and discharges to $1/3 \ V_{CC}$ through R_B. By varying the ratio of these resistors the duty cycle can be varied. The charge and discharge times are independent of the supply voltage.

The charge time (output high) is given by: $t_1 = 0.695 \ (R_A + R_B) \ C$

The discharge time (output low) by: $t_2 = 0.695 \ (R_B) \ C$

Thus the total period is given by: $T = t_1 + t_2 = 0.695 \ (R_A + 2R_B) \ C$

The frequency of oscillation is then: $f = \dfrac{1}{T} = \dfrac{1.44}{(R_A + 2R_B) \ C}$

and may be easily found as shown in Figure 19.

The duty cycle is given by: $DC = \dfrac{R_B}{R_A + 2R_B}$

To obtain the maximum duty cycle R_A must be as small as possible; but it must also be large enough to limit the discharge current (pin 7 current) within the maximum rating of the discharge transistor (200 mA).

The minimum value of R_A is given by:

$$R_A \geqslant \frac{V_{CC} \ (Vdc)}{I_7 \ (A)} \geqslant \frac{V_{CC} \ (Vdc)}{0.2}$$

FIGURE 19 — FREE-RUNNING FREQUENCY

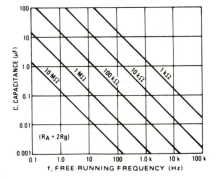

$(R_A + 2R_B)$

f, FREE-RUNNING FREQUENCY (Hz)

C, CAPACITANCE (μF)

The data sheet for the Motorola MC1455 *(continued)*.

MC1455

APPLICATIONS INFORMATION

Linear Voltage Ramp

In the monostable mode, the resistor can be replaced by a constant current source to provide a linear ramp voltage. The capacitor still charges from 0 to 2/3 V_{CC}. The linear ramp time is given by

$$t = \frac{2}{3} \frac{V_{CC}}{I}$$

where $I = \dfrac{V_{CC} - V_B - V_{BE}}{R_E}$. If V_B is much larger than V_{BE}, then t can be made independent of V_{CC}.

Missing Pulse Detector

The timer can be used to produce an output when an input pulse fails to occur within the delay of the timer. To accomplish this, set the time delay to be slightly longer than the time between successive input pulses. The timing cycle is then continuously reset by the input pulse train until a change in frequency or a missing pulse allows completion of the timing cycle, causing a change in the output level.

FIGURE 20 — LINEAR VOLTAGE SWEEP CIRCUIT

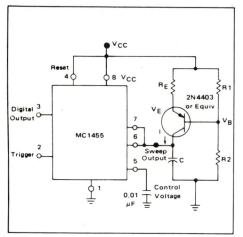

FIGURE 22

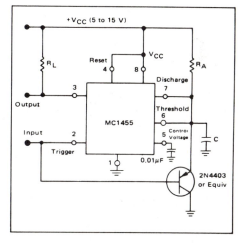

FIGURE 21 — LINEAR VOLTAGE RAMP WAVEFORMS
(R_E = 10 kΩ, R2 = 100 kΩ, R1 = 39 kΩ, C = 0.01 μF, V_{CC} = 15 V)

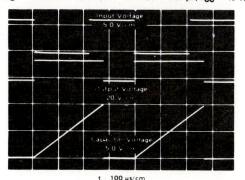

t = 100 μs/cm

FIGURE 23 — MISSING PULSE DETECTOR WAVEFORMS
(R_A = 2.0 kΩ, R_L = 1.0 kΩ, C = 0.1 μF, V_{CC} = 15 V)

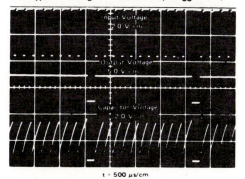

t = 500 μs/cm

The data sheet for the Motorola MC1455 (continued).

MC1455

Pulse Width Modulation

If the timer is triggered with a continuous pulse train in the monostable mode of operation, the charge time of the capacitor can be varied by changing the control voltage at pin 5. In this manner, the output pulse width can be modulated by applying a modulating signal that controls the threshold voltage.

FIGURE 24

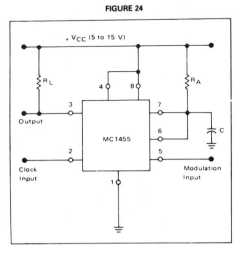

FIGURE 25 — PULSE WIDTH MODULATION WAVEFORMS
(R_A = 10 kΩ, C = 0.02 μF, V_{CC} = 15 V)

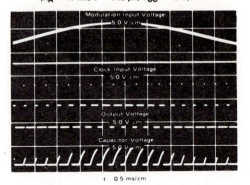

Test Sequences

Several timers can be connected to drive each other for sequential timing. An example is shown in Figure 26 where the sequence is started by triggering the first timer which runs for 10 ms. The output then switches low momentarily and starts the second timer which runs for 50 ms and so forth.

FIGURE 26

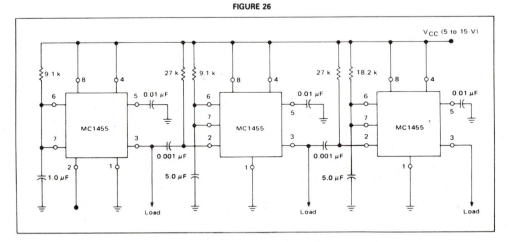

The data sheet for the Motorola MC1455 (*continued*).

ANALOG DEVICES

Low Cost, Complete IC
8-Bit A-to-D Converter

AD570 *

FEATURES
Complete A/D Converter with Reference and Clock
Fast Successive Approximation Conversion — 25μs
No Missing Codes Over Temperature
   0 to +70°C — AD570J
   -55°C to +125°C — AD570S
Digital Multiplexing — 3 State Outputs
18-Pin DIP
Low Cost Monolithic Construction

AD570 FUNCTIONAL BLOCK DIAGRAM

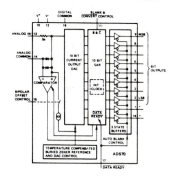

18-PIN DUAL IN LINE PACKAGE

PRODUCT DESCRIPTION
The AD570 is an 8-bit successive approximation A/D converter consisting of a DAC, voltage reference, clock, comparator, successive approximation register and output buffers — all fabricated on a single chip. No external components are required to perform a full accuracy 8-bit conversion in 25μs.

The AD570 incorporates the most advanced integrated circuit design and processing technology available today. I²L (integrated injection logic) processing in the fabrication of the SAR function along with laser trimming of the high stability SiCr thin film resistor ladder network at the wafer stage (LWT) and a temperature compensated, subsurface Zener reference insures full 8-bit accuracy at low cost.

Operating on supplies of +5V and -15V, the AD570 will accept analog inputs of 0 to +10V unipolar or ±5V bipolar, externally selectable. As the BLANK and $\overline{CONVERT}$ input is driven low, the three state outputs will be open and a conversion will commence. Upon completion of the conversion, the DATA READY line will go low and the data will appear at the output. Pulling the BLANK and $\overline{CONVERT}$ input high blanks the outputs and readies the device for the next conversion. The AD570 executes a true 8-bit conversion with no missing codes in approximately 25μs.

The AD570 is available in two versions; the AD570J is specified for the 0 to 70°C temperature range, the AD570S for -55°C to +125°C. Both guarantee full 8-bit accuracy and no missing codes over their respective temperature ranges. The AD570J is also offered in an 18-pin plastic DIP.

*Protected by Patent Nos. 3940760, 4213806 and 4136349.

PRODUCT HIGHLIGHTS
1. The AD570 is a complete 8-bit A/D converter. No external components are required to perform a conversion. Full scale calibration accuracy of ±0.8% (2LSB of 8 bits) is achieved without external trims.

2. The AD570 is a single chip device employing the most advanced IC processing techniques. Thus, the user has at his disposal a truly precision component with the reliability and low cost inherent in monolithic construction.

3. The AD570 accepts either unipolar (0 to +10V) or bipolar (-5V to +5V) analog inputs by simply grounding or opening a single pin.

4. The device offers true 8-bit accuracy and exhibits no missing codes over its entire operating temperature range.

5. Operation is guaranteed with -15V and +5V supplies. The device will also operate with a -12V supply.

An-eight bit analog-to-digital converter that is used in the computer interface design example presented in Chapter 39. This data sheet is included both as a reference for that design and as an example of the types of information available for more complex analog circuits.

SPECIFICATIONS

(T$_A$ = 25°C , V+ = +5V, V− = −12V or −15V,
all voltages measured with respect to digital common, unless otherwise indicated)

Model	AD570J Min	Typ	Max	AD570S Min	Typ	Max	Units
RESOLUTION[1]		8			8		Bits
RELATIVE ACCURACY, T$_A$[1,2,3]			±1/2			±1/2	LSB
T$_{min}$ to T$_{max}$			±1/2			±1/2	LSB
FULL SCALE CALIBRATION[3,4]		±2			±2		LSB
UNIPOLAR OFFSET[3]			±1/2			±1/2	LSB
BIPOLAR OFFSET[3]			±1/2			±1/2	LSB
DIFFERENTIAL NONLINEARITY, T$_A$		8		8			Bits
T$_{min}$ to T$_{max}$		8		8			Bits
TEMPERATURE RANGE	0		+70	−55		+125	°C
TEMPERATURE COEFFICIENTS[3]							
Unipolar Offset			±1			±1	LSB
Bipolar Offset			±1			±1	LSB
Full Scale Calibration[2]			±2			±2	LSB
POWER SUPPLY REJECTION[3]							
Positive Supply							
+4.5≤V+≤+5.5V			±2			±2	LSB
Negative Supply							
−16.00V≤V−≤−13.5V			±2			±2	LSB
ANALOG INPUT IMPEDANCE	3.0	5.0	7.0	3.0	5.0	7.0	kΩ
ANALOG INPUT RANGES							
Unipolar	0		+10	0		+10	V
Bipolar	−5		+5	−5		+5	V
OUTPUT CODING							
Unipolar	Positive True Binary			Positive True Binary			
Bipolar	Positive True Offset Binary			Positive True Offset Binary			
LOGIC OUTPUT							
Output Sink Current							
(V$_{OUT}$ = 0.4V max, T$_{min}$ to T$_{max}$)	3.2			3.2			mA
Output Source Current[6]							
(V$_{OUT}$ = 2.4V max, T$_{min}$ to T$_{max}$)	0.5			0.5			mA
Output Leakage			±40			±40	μA
LOGIC INPUTS							
Input Current			±40			±40	μA
Logic "1"	2.0			2.0			V
Logic "0"			0.8			0.8	V
CONVERSION TIME, T$_A$ and							
T$_{min}$ to T$_{max}$	15	25	40	15	25	40	μs
POWER SUPPLY							
V +	+4.5	+5.0	+7.0	+4.5	+5.0	+7.0	V
V −	−12.0	−15	−16.5	−12.0	−15	−16.5	V
OPERATING CURRENT							
V +		2	10		2	10	mA
V −		9	15		9	15	mA
PACKAGE[7]							
Ceramic DIP		N18A			N18A		mA
Plastic DIP		D18A			D18A		mA

NOTES
[1]The AD570 is selected version of the AD571 10-bit A to D converter. As such, some devices may
exhibit 9 or 10 bits of relative accuracy or resolution, but that is neither tested nor guaranteed.
Only TTL logic inputs should be connected to pins 1 and 18 (or no connection made) or damage
may result.
[2]Relative accuracy is defined as the deviation of the code transition points from the ideal transer
point on straight line from the zero to the full scale of the device.
[3]Specifications given in LSB's refer to the weight of a least significant bit at the 8-bit level, which is
0.39% of full scale.
[4]Full scale calibration is guaranteed trimmable to zero with an external 200Ω potentiometer in place
of the 15Ω fixed resistor. Full scale is defined as 10 volts minus 1LSB or 9.961 volts.
[5]Full Scale Calibration Temperature Coefficient includes effects of unipolar offset drift as well as
gain drift.
[6]The data output lines have active pull-ups to source 0.5mA. The DATA READY line is open collector with
a nominal 6kΩ internal pull-up resistor.
[7]See Section 19 for package outline information.
Specifications subject to change without notice.
Specifications shown in boldface are tested on all production units at final electrical test. Results from those tests are used to calculate outgoing quality levels. All
min and max specifications are guaranteed, although only those shown in
boldface are tested on all production units.

An eight-bit analog-to-digital converter data sheet (*continued*).

ABSOLUTE MAXIMUM RATINGS

V+ to Digital Common .0 to +7V

V− to Digital Common . . . . .0 to −16.5V

Analog Common to Digital Common. . . . .±1V

Analog Input to Analog Common. . . . . ±15V

Control Inputs . . . . 0 to V+

Digital Outputs (Blank Mode). . . . . 0 to V+

Power Dissipation. 800mW

AD570 ORDERING GUIDE

Model	Package Number[1]	Temperature Range
AD570JN	18-Pin Plastic DIP (N18A)	0 to +70°C
AD570JD	18-Pin Ceramic DIP (D18A)	0 to +70°C
AD570SD	18-Pin Ceramic DIP (D18A)	−55°C to +125°C

[1] See Section 19 for package outline information.

CONNECTING THE AD570 FOR STANDARD OPERATION

The AD570 contains all the active components required to perform a complete A/D conversion. Thus, for most situations, all that is necessary is connection of the power supply (+5 and −15), the analog input, and the conversion start pulse. But, there are some features and special connections which should be considered for achieving optimum performance. The functional pin-out is shown in Figure 1.

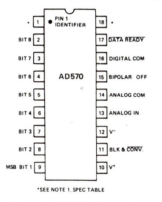

*SEE NOTE 1, SPEC TABLE

Figure 1. AD570 Pin Connections

FULL SCALE CALIBRATION

The 5kΩ thin film input resistor is laser trimmed to produce a current which matches the full scale current of the internal DAC—plus about 0.3%—when a full scale analog input voltage of 9.961 volts (10 volts − 1LSB) is applied at the input. The input resistor is trimmed in this way so that if a fine trimming potentiometer is inserted in series with the input signal, the input current at the full scale input voltage can be trimmed down to match the DAC full scale current as precisely as desired. However, for many applications the nominal 9.961 volt full scale can be achieved to sufficient accuracy by simply inserting a 15Ω resistor in series with the analog input to pin 14. Typical full scale calibration error will then be about ±2LSB or ±0.8%. If a more precise calibration is desired a 200Ω trimmer should be used instead. Set the analog input at 9.961 volts, and set the trimmer so that the output code is just at the transition between 11111110 and 11111111. Each LSB will then have a weight of 39.06mV. If a nominal full scale of 10.24 volts is desired (which makes the LSB have weight of exactly 40.00mV), a 50Ω resistor in series with a 200Ω trimmer (or a 500Ω trimmer with good resolution) should be used. Of course, larger full scale ranges can be ar-

ranged by using a larger input resistor, but linearity and full scale temperature coefficient may be compromised if the external resistor becomes a sizeable percentage of 5kΩ.

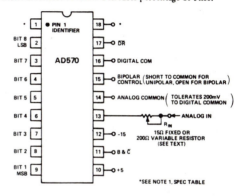

Figure 2. Standard AD570 Connections

BIPOLAR CONNECTION

To obtain the bipolar −5V to +5V range with an offset binary output code the bipolar offset control pin is left open.

A −5.00 volt signal will give a 10-bit code of 00000000 00; an input of 0.00 volts results in an output code of 10000000 00 and +4.99 volts at the input yields the 11111111 11 code. The nominal transfer curve is shown in Figure 3.

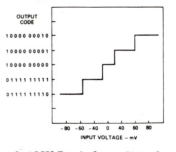

Figure 3. AD570 Transfer Curve — Bipolar Operation

NOTE: That in the bipolar mode, the code transitions are offset 1/2LSB such that an input voltage of 0 volts ±20mV yields the code representing zero (10000000 00). Each output code is then centered on its nominal input voltage.

An eight-bit analog-to-digital converter data sheet (*continued*).

Full Scale Calibration

Full Scale Calibration is accomplished in the same manner as in Unipolar operation except the full scale input voltage is +4.99 volts.

Negative Full Scale Calibration

The circuit in Figure 4a can also be used in Bipolar operation to offset the input voltage (nominally –5V) which results in the 00000000 00 code. R2 should be omitted to obtain a symmetrical range.

ZERO OFFSET

The apparent zero point of the AD570 can be adjusted by inserting an offset voltage between the Analog Common of the device and the actual signal return or signal common. Figure 4 illustrates two methods of providing this offset. Figure 4A shows how the converter zero may be offset by up to ±3 bits to correct the device initial offset and/or input signal offsets. As shown, the circuit gives approximately symmetrical adjustment in unipolar mode. In bipolar mode R2 should be omitted to obtain a symmetrical range.

Figure 4B shows how to offset the zero code by 1/2LSB to provide a code transition between the nominal bit weights.

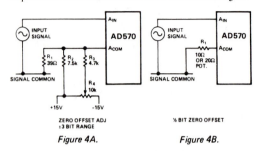

Figure 4A.   Figure 4B.

CONTROL AND TIMING OF THE AD570

There are several important timing and control features on the AD570 which must be understood precisely to allow optimal interfacing to microprocessor or other types of control systems. All of these features are shown in the timing diagram in Figure 5.

The normal stand-by situation is shown at the left end of the drawing. The BLANK and CONVERT (B & C) line is held high, the output lines will be "open", and the DATA READY (DR) line will be high. This mode is the lowest power state of the device (typically 150mW). When the (B & C) line is brought low, the conversion cycle is initiated; but the DR and Data lines do not change state. When the conversion cycle is complete (typically 25μs), the DR line goes low, and within 500ns, the Data lines become active with the new data.

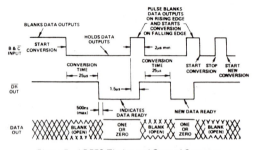

Figure 5. AD570 Timing and Control Sequence

About 1.5μs after the B & C line is again brought high, the DR line will go high and the Data lines will go open. When the B & C line is again brought low, a new conversion will begin. The minimum pulse width for the B & C line to blank previous data and start a new conversion is 2μs. If the B & C line is brought high during a conversion, the conversion will stop, and the DR and Data lines will not change. If a 2μs or longer pulse is applied to the B & C line during a conversion, the converter will clear and start a new conversion cycle.

CONTROL MODES WITH BLANK AND CONVERT

The timing sequence of the AD570 discussed above allows the device to be easily operated in a variety of systems with differing control modes. The two most common control modes, the Convert Pulse Mode, and the Multiplex Mode, are illustrated here.

Convert Pulse Mode — In this mode, data is present at the output of the converter at all times except when conversion is taking place. Figure 6 illustrates the timing of this mode. The BLANK and CONVERT line is normally low and conversions are triggered by a positive pulse.

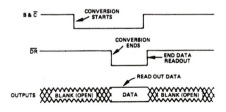

Figure 6. Convert Pulse Mode

Multiplex Mode — In this mode the outputs are blanked except when the device is selected for conversion and readout; this timing is shown in Figure 7.

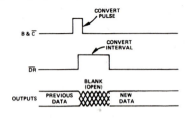

Figure 7. Multiplex Mode

This operating mode allows multiple AD570 devices to drive common data lines. All BLANK and CONVERT lines are held high to keep the outputs blanked. A single AD570 is selected, its BLANK and CONVERT line is driven low and at the end of conversion, which is indicated by DATA READY going low, the conversion result will be present at the outputs. When this data has been read from the 8-bit bus, BLANK and CONVERT is restored to the blank mode to clear the data bus for other converters. When several AD570's are multiplexed in sequence, a new conversion may be started in one AD570 while data is being read from another. As long as the data is read and the first AD570 is cleared within 15μs after the start of conversion of the second AD570, no data overlap will occur.

An eight-bit analog-to-digital converter data sheet (*continued*).

Index

507